GKSS School of Environmental Research

Series editors: H. von Storch · E. Raschke · G. Flöser

Springer

Berlin
Heidelberg
New York
Barcelona
Hong Kong
London
Milan
Paris
Singapur
Tokyo

Hans von Storch and Götz Flöser (Eds.)

Anthropogenic Climate Change

With 124 Figures and 15 Tables

 Springer

Volume editors:
Prof. Dr. Hans von Storch
Dr. Götz Flöser

GKSS Forschungszentrum
Max-Planck-Straße
D-21502 Geesthacht
E-mail: storch@gkss.de
E-mail: floeser@gkss.de

ISSN 1437-028X
ISBN 3-540-65033-4 Springer-Verlag Berlin Heidelberg New York

CIP data applied for

Die Deutsche Bibliothek - CIP-Einheitsaufnahme
Anthropogenic climate change : with 15 tables / Hans von Storch and Götz Flöser (eds.) -
Berlin ; Heidelberg ; New York ; Barcelona ; Hong Kong ; London ; Milan ; Paris ;
Singapur ; Tokyo : Springer, 1999
     (GKSS school of environmental research)
     ISBN 3-540-65033-4

Typesetting: Data conversoion: Büro Stasch, Bayreuth
Cover: Erich Kirchner, Heidelberg
SPIN:10680187   32/3020 - 5 4 3 2 1 0 - Printed on acid -free paper

Foreword

GKSS SCHOOL OF ENVIRONMENTAL RESEARCH

The National Research Laboratory GKSS (member of the Hermann von Helmholtz-Association of German Reserach Centres) located in Geesthacht, near Hamburg, is engaged in environmental research. The main interest of the research center focuses on regional climatology and climate dynamics, interdecadal variations in the state of the Baltic and North Sea and related estuaries, and the flow of heavy metals, nutrients, and other materials in river catchments to the coastal zones. This research aims at developing an understanding of changes in the environment, both as a result of internal (natural) dynamics and as a result of anthropogenic interference. In an effort to disseminate the results of these research activities, as well as to initiate a broad discussion among senior scientists in the field, and younger colleagues from all areas of the globe, the Institutes of Hydrophysics and Atmospheric Physics at GKSS have instituted the GKSS School of Environmental Research.

Applied environmental research has always contained an element of awareness of the societal implications and boundary conditions associated with environmental concerns. Consequently, the School of Environmental Research adheres to the philosophy that all discussion regarding environmental change should incorporate a social component. This necessity has been well acknowledged and is apparent by the incorporation of social scientists into the series of lectures.

Senior scientists from Europe and North America were invited to give lectures to "students" from all parts of the globe. The participation of these "students", many of whom were post docs and have already made significant contributions to their respective sciences, was made possible by the financial support of the following sponsors:

- The German Climate Computer Center, Hamburg, Germany

- The European Commission, DG XII, Climate and Environment Programme

- Verein der Freunde und Förderer des GKSS Forschungszentrums, Germany

- Fachbereich Geowissenschaften der Universität Hamburg, Germany

The lectures of the Spring School comprise the body of this book.

The topics of this Spring School series of lectures were designed to incorporate the expertise of both the Institute of Hydrophysics and the Institute of Atmospheric Physics. Future lecture series in the form of Schools are proposed to deal with more specific topics, regional climate modeling, and the concepts and roles of models in environmental research, for example.

Hans von Storch
   Series Editor

Preface

We initiated our program of Schools with the theme of "Anthropogenic Climate Change" as a result of the belief that the field of climate research and, in particular, the aspect of anthropogenic climate change, has several intriguing features which bear upon the more encompassing discourse of environmental research.

First, the climate sciences have been marked by significant progress in the recent decades, spawned by the introduction of fundamentally new and powerful techniques made possible by the age of the supercomputer, the rapid evolution of remote sensing techniques, and the availability of long (homogenized) time series of instrumentally recorded and proxy data. These developments pushed climate research to the forefront of scientific endeavors, transforming the traditional approaches of dynamical meteorology and descriptive climatology into a modern, multidisciplinary conglomerate incorporating such diverse fields as meteorology, oceanography, hydrology, geochemistry, glaciology and geology, and this list is not exhaustive, under a single blanket of research. In the course of this progress, a multitude of new techniques and concepts have evolved, some now to the state of maturity. One such example is the design of quasi-realistic models representing an (almost) infinity of degrees of freedom, acting and interacting on a wide range of spatial and temporal scales, and the coupling of such models with different response times. These models are used for simulating scenarios of possible future developments and for the reconstructions of past (historical and paleoclimatic) developments.

The role of observational data has also changed. Traditionally, such data were gathered in experimental campaigns aimed at the understanding of specific processes. The focus has now changed towards monitoring, i.e., the collection of quality-controlled data over decades of years. Such monitoring data, available from a limited number of observations at any given time, are used for the determination of the spatially continous state of the climatic environment ("analyses") with the help of the above mentioned quasi-realistic models. Such analyses provide indispensable information for the understanding of climate dynamics, for the ability to discriminate between anthropogenic impacts and natural variation, as well as for successful short-term forecasts.

The second reason the Schools were initiated with the topic of "Anthropogenic Climate Change" has to do with the extension of the discussion to well beyond academic circles. Whereas such data and knowledge was once mostly confined to the scientific discourse of the likes of the dynamics of the waxing and waning of ice ages, or the impact of solar variability or the presence of volcanic aerosols, the topic of climate change is now well versed in both public and political realms. Terms such as "global warming" and "climate change" are now a part of household vocabulary, and a constant source of topics for the public media. Thus, "anthropogenic climate change" has acquired a second identity, namely a social identity which shapes social and political consciousness.

Finally, "Climate Change" was chosen as the School's theme because climate can be interpreted as a limiting resource, or a limiting factor, acting upon many other social and natural systems, be it marine ecosystems, tourism, the stability of slopes, coastal defense or offshore structures, for example. Such an extension of the role of climate and climate change into such diverse concerns has resulted in the emergence of a broad, interdisciplinary and multidisciplinary agenda of research, addressing not only the sensitivity of a broad spectrum of systems to climate variability and climate change, but also the determination of accommodating managerial options. This link between natural and social spheres raises a specific challenge. For example, it is necessary to transform the knowledge about a larger, better understood, and better monitored, state, such as the "global" climate to a societally relevant, and typically smaller, scale such as a river catchment area or the Wadden Sea.

In conclusion, climate research could be labeled as "the mother of all environmental research" (well, maybe not all, but close), and the decision to begin with the inclusive topic "Anthropogenic Climate Change" we hope, will provide examples that will benefit many areas of environmental research in terms of concepts and methods, the degree to which interdisciplinarity is necessary, and the essential need to incorporate the social dimension into environmental analysis.

The Editors
Geesthacht
Hans von Storch and Götz Flöser

Contents

List of Contributors

Dennis Bray
Institute of Hydrophysics, GKSS Research Centre, Max-Planck-Straße, 21502 Geesthacht, Germany

William Cotton
Atmospheric Science, Colorado State University, Fort Collins, CO 80523, Colorado, USA

Klaus Hasselmann
Max-Planck-Institut für Meteorologie, Bundesstraße 55, 20146 Hamburg, Germany

Nico Stehr
Green College, University of British Columbia, 6201 Cecil Green Park Road, Vancouver, British Columbia, Canada

Hans von Storch
Institute of Hydrophysics, GKSS Research Centre, Max-Planck-Straße, 21502 Geesthacht, Germany

Jin-Song von Storch
Meteorologisches Institut der Universität Hamburg, Bundesstraße 55, 20146 Hamburg, Germany

Warren Washington
National Center for Atmospheric Research, P.O. Box 3000, Boulder 80307, Colorado, USA

Eric Wood
Princeton University, Department of Civil Engineering and Operations Research, Princeton 08544, New Jersey, USA

Francis Zwiers
Canadian Centre for Climate Modelling and Analysis, P.O. Box 1700, Victoria, BC V8W 2Y2, British Columbia, Canada

Part I

The Climate System

Chapter 1

The Global and Regional Climate System

by Hans von Storch

Abstract

The relationships between planetary and regional or local scales of the climate system are discussed.

The main features of the planetary scale climate may be understood as the response of the climate system to the planetary scale forcing (heating by the sun), the distribution of continents and largest mountain ranges. The regional climate, then, may be understood as the result of both the planetary scale climate and the regional features such as secondary mountains, marginal seas, land use, and the like.

The small scale features have little effect on the planetary scale in the sense that *details* are rather unimportant; but the statistics of the small scale features modify the planetary scale state significantly. This modification may be described by parameterizations, i.e., through statistics of small scale processes, which are *conditioned* by the large-scale states.

The concept of randomized parameterizations is introduced.

The discourse is illustrated with various examples, ranging from energy balance models, the emergence of the general circulation of the atmosphere from a state at rest to downscaling applications.

1.1 Prologue

Seen as a resource or limiting factor, climate is a local or regional object. In areas where temperatures fall below the freezing point, palm trees do not grow, and the agronomist is advised not grow citrus trees. Therefore, not surprisingly, climatology as a science began with the description of local and regional climates, and the various regional charts were combined in world maps. In that sense, the global climate was perceived as the sum of regional climates. Famous representatives of this line of research are Vladimir Köppen, Eduard Brückner or Julius von Hann. Köppen's (1923) maps are even nowadays famous; they classify the surface of the world into climatic zones, which are determined by the amount of precipitation they receive, and the temperature regime. Dynamical quantities such as the wind are considered secondary who deserve attention mostly since they determine the primary variables, namely precipitation and temperature. For demonstration, we have reprinted one of Köppen's maps in Fig. 1.1.

The weather services pursue this traditional line of research, by providing "climate normals" for planning exercises in traffic, agriculture, tourism, and other applications. Such normals do not only refer to mean values derived from, say, a 30 year interval, but also the probability for extreme events (such as 100-year storm surges).

On the other hand, a different line of reasoning has developed in the course of centuries – aimed at the understanding of why climate is as it is. Examples of research of this type are George Hadley's explanation of the trade wind system[1] (see Fig. 1.1), and Emanuel Kant's postulation of a continent south of the Indonesian archipelago – at that time unknown – based on wind observations from merchant vessels. In the same category is Arrhenius (1896) hypothesis about the impact of air-borne carbon dioxide on near-surface temperature.

The two views have long been separated, and are re-conciliated now after the emergence of "global warming", which has its societal implications on the local and regional scale, but which prospect was born in the global realm of climate dynamics.

A major progress of climate research is the understanding that the local climate is only partially determined by local features, such as geographical locations, local topography, proximity to the ocean, land use and the like. Globally averaged features are defined as the arithmetic mean of many local features - but with the help of *energy balance models* these averaged quantities, such as the global mean *near-surface temperature*[2], can be determined directly without knowledge about any local aspects (Section 1.2). Also, the major features of the general circulation of the atmosphere, such as the tropical meridional cells, the jet streams associated with baroclinic instability

---

[1] For an interesting review of the history of ideas concerning the general circulation of the atmosphere, of which the trade wind system is part of, refer to Lorenz (1967).

[2] With this expression we mean the globally averaged air temperature at 2m height.

Fig. 1.1. One of Vladimir Köppen's climate maps of the world. The figure caption is in German

Fig. 1.2. Hadley's concept of thermally driven vertical cells, which are deflected by the rotation of the earth, creating the equator-ward directed trade winds. From Lorenz (1967)

and the formation of storms, can be simulated well for a planet entirely covered by water without any mountains, coasts and vegetation (see Section 1.3). The global-scale features of the climate are modified by the continental distribution and the presence of the largest mountainous ranges, such as the Himalaya, the Andes, the Rocky Mountains, Antarctica, and Greenland. Since these factors are of planetary scale, their impact can be described by standard "low-resolution" general circulation models.

After the planetary scale state of the atmosphere is set, the regional and local climates emerge as the result of an interplay between the planetary climatic state and the local features. This point is illustrated in Section 1.4 with two examples: one about rainfall in Romania and a second one about high sea states in the North Atlantic.

The statement that *the global climate would not be the sum of all regional climates* does not imply that the regional scale dynamics is inconsequential for the global scales. Many processes, such as convection in the atmosphere and in the ocean, play a crucial role in the formation of the global scale state. These processes usually cannot be resolved by the climate models because they take place on scales much too small for such models. However, the details in space and time of these processes are not of importance but only their overall impact. The limited importance of regionally limited changes of the boundary conditions is demonstrated by a numerical experiment on the effect of the transformation of the North American prairies into crop- and farmland (Section 1.5). The concept of parameterizations of sub-grid scale processes is outlined, and the idea of a randomized design is sketched and motivated.

1.2 The Global View: Energy Balance Models

The simplest conceptual model for the earth's climate, or any other planet's climate is based on the conservation of energy: In order to have a stationary climate, the time-mean incoming radiation (energy) must be balanced by the same amount of energy radiated back to space. Or, formally:

$$R_{\text{incoming}} = R_{\text{outgoing}} \tag{1.1}$$

Models based on this ansatz[3] are called *energy balance models*, and can achieve considerable complexity (e.g. Crowley and North, 1991).

The sun, being the source of energy in this system, radiates mostly in the short wave band of the spectrum. Part of this incoming radiation is reflected back to space, but another part is absorbed by the earth system, and eventually re-radiated. Since the earth is a relatively cool body, this radiation takes place in the long wave spectral band. Then, the balance (1.1) reads:

$$R_{sw} = \alpha R_{sw} + R_{lw} \tag{1.2}$$

with the *albedo* α and the short wave and long wave radiations R_{sw} and R_{lw}. The incoming short-wave radiation is $R_{sw} = 342 \ Wm^{-2}$. The long-wave radiation is a function of the temperature of the radiating body (obeying the Stefan-Boltzmann law)

$$R_{lw} = \kappa \sigma T^4 \tag{1.3}$$

with $\sigma = 5.67 \cdot 10^{-8} Wm^{-2}K^{-4}$ representing a universal constant and κ describing the transmissivity of the earth atmosphere with respect to long-wave radiation (the amount of long-wave radiation emitted at the earth surface which escapes to space). If the system is in equilibrium, we find as a temperature for the radiating earth:

$$T_{eq} = \left[\frac{(1-\alpha)R_{sw}}{\kappa \sigma} \right]^{\frac{1}{4}} \tag{1.4}$$

If there were no atmosphere, κ would be unity, and the albedo would be about 15%, so that the resulting equilibrium temperature would be about

---

[3]This technical term of German origin is well known among physicists and mathematicians but not entirely so in the meteorological and oceanographic community. "To make an ansatz" means to assume a certain plausible relationship (in this case, equation (1.1)), and to examine if this relationship can be detailed so that it explains certain observed or theoretical "facts".

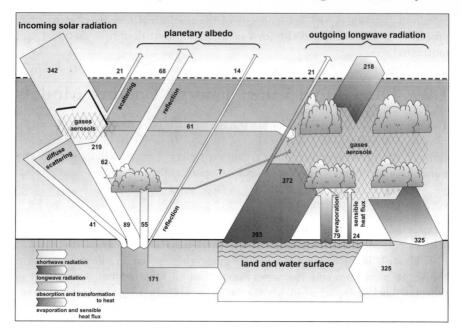

Fig. 1.3. Energy balance diagram. Units of Wm^{-2}. From von Storch (1997)

-4°C. The presence of the atmosphere leads to an increased albedo (clouds) of about $\alpha = 0.30$ but also to a reduced transmissivity of $\kappa = 0.64$. The resulting equilibrium temperature is about 15° C, being close to the observed value.[4]

Figure 1.3 summarizes the globally averaged energy balance. It shows the incoming short wave and outgoing long wave radiative terms, as well as the various complications introduced by the presence of gases and particles in the atmosphere which absorb and re-emit radiation. Also some of the energy is absorbed by the surface and released as latent and sensible heat which eventually is radiatively exported from the system at higher levels, in part after condensation. All these processes are hidden in the number κ – or, in other words, these processes are *parameterized* by adding the tuning constant κ in the Stefan-Boltzmann radiation law.

The dynamics of the system may be made more "interesting", by incorporating a nonlinear term, describing the dependency of the albedo on the presence of snow and ice. If more of the earth surface is covered by snow and ice, then the albedo is enhanced and the surface of the earth is colder, and vice versa. An ad-hoc ansatz for this dependency is displayed

---

[4]The similarity of the modeled temperature and the real global mean near-surface temperature does not provide validation of the estimated value of $\alpha = 0.3$; the parameter κ has been chosen so that meaningful numbers emerge - however the choice of κ is fully consistent with physical reasoning about the involved processes.

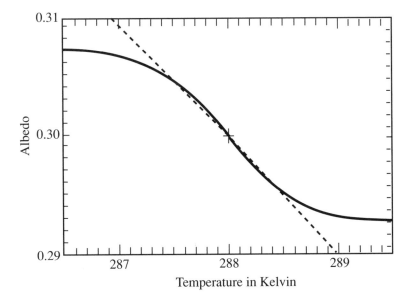

Fig. 1.4. Assumed nonlinear dependency of albedo α upon the global mean temperature. From von Storch (1997)

in Fig. 1.4. The intersections with the dashed lines are the three solutions $T_{eq} \approx 287.5K, 288.5K$ and $288K$ of equation (1.4). The first two are stable solutions, i.e, if the system is moved away from the stable solution it returns to it within a characteristic time. The middle solution is unstable, so that any miniscule disturbance will drive the temperature to the nearest equilibrium stable solution temperature. The behavior is described in Fig. 1.5.

Further complication can be introduced by noise (see Section 1.5.2) and latitudinal and vertical structures (Crowley and North, 1991). However, all these models have something in common, that is, they do not model the *circulation* of the atmosphere. Obviously the circulation is a significant climatic feature, which, among others, causes our daily experience of weather.

1.3 The Global Atmospheric Circulation: Emergence from a State at Rest

In the previous section we have dealt with the climate system as a thermodynamic system which continuously receives energy and re-emits this energy to space. In fact, the thermodynamic system "climate" may be seen as a thermodynamic machine which transforms thermal energy into mechanical energy – in the form of winds in the atmosphere and currents in the ocean. Eventually, the mechanical energy is dissipated to thermal energy which is

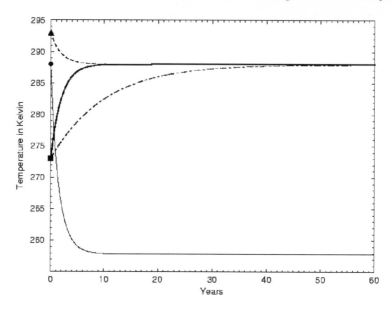

Fig. 1.5. Convergence towards stable equilibrium solution of the Energy Balance Model (1.4) with temperature dependent albedo. From von Storch (1997)

radiated to space. In this section we will demonstrate what kind of motions our machinery performs – and we will see again that the regional details are insignificant for the planetary scale patterns of the circulation.

Washington (1968) [5] and later Fischer et al. (1991) have performed interesting *numerical experiments* with a *General Circulation Model* (GCM) of the atmosphere (for the design of such models, refer to Washington's contribution to this book in Chapter 2). They both initialized their multi-month simulation of the atmospheric circulation from a state at rest. Washington used realistic topography, land-sea distribution and winter sea surface temperature (SST) distribution, whereas Fischer et al. (1991) used an *aqua planet*, i.e., a planet entirely covered by water with zonally symmetric SST distribution. In both simulations, a realistic planetary scale circulation emerged within one month, or so.

Figure 1.6 displays the evolution of the air-pressure in Washington's experiment - the pressure is falling over the warmer sea and rising over land. However, after about 10 days the system becomes unstable at mid latitudes and weather systems (storms) begin to emerge. After about 40 days, the characteristic macro-turbulent behavior, in particular on the Southern hemi-

---

[5] A summary of the experiment is given in the monograph by Washington and Parkinson, 1986, page 209-210.

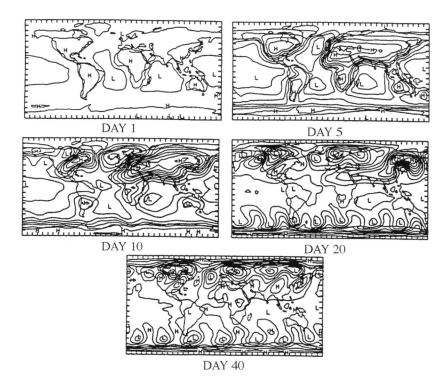

Fig. 1.6. Emergence of the general circulation of the troposphere from an isothermal atmospheric state at rest with a realistic land-sea distribution. The variable shown is the air-pressure. From Washington (1968)

sphere, has emerged in the mid latitudes while in the tropics weak pressure gradients prevail.

In Fischer's experiment the initial fingerprint of the land-sea distribution on air pressure does not appear – as there are no continents – and during the first 10 days the atmosphere stays mostly motionless (apart of small meridional winds close to the equator; Fig. 1.7b), and almost all transports of heat taken up by the atmosphere at the ocean surface (which represents the input of the short term incoming radiation) takes place by diffusive mechanisms. At about day 10 an almost explosive change takes places in the tropics and wanders within a couple of weeks across the mid and polar latitudes. First, the tropical cells and the trade wind system are built within a couple of days. Then, about 20 days after initializing the indirect Ferrell cell with westerly and northerly near-surface winds appear (Fig. 1.7a,b) and very vaguely a polar cell. Turbulent motions, as revealed by the energy of disturbances (Fig.

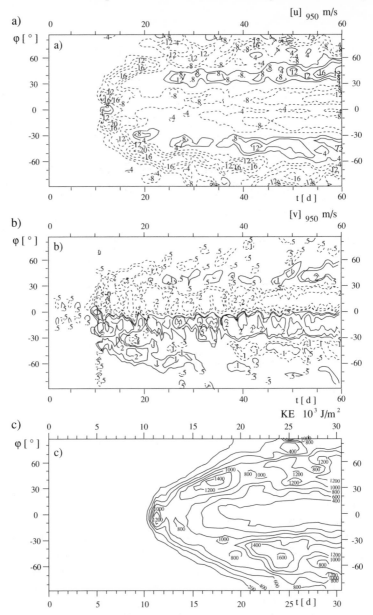

Fig. 1.7. Emergence of the general circulation of the troposphere from an isothermal state at rest on an aqua-planet. From Fischer et al. (1991). All variables are shown as zonal averages.
a) zonal wind at 950 hPa
b) meridional wind at 950 hPa
c) kinetic energy of the disturbances (i.e. deviations from the zonally averaged state)

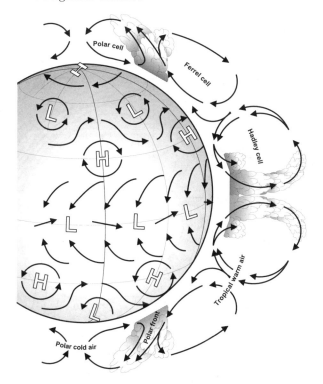

Fig. 1.8. Sketch of the general circulation of the atmosphere. Note that the mid latitudes are characterized by unsteady (but statistically quasi-stationary) states with short term baroclinic disturbances. From von Storch et al. (1999)

1.7c), first show up in the vicinity of the equator, but "drift" within about 10 days to its observed location at mid latitudes.

The eventual state reached within one to two months is in both, Washington's and Fischer's case, very similar to the observed state (as sketched in Fig. 1.8). Thus, the pattern of the "planetary scale atmospheric circulation" is unrelated to regional details such as topography or land use. The main elements, such as the lowered geopotential height in polar areas and the lifted height in tropical areas with southwesterly flows across the mid latitude oceans, with elongated mid-latitude jet streams near the tropopause and enhanced baroclinic activities at the exit of the jet-streams (Fig. 1.9) are conditioned by planetary scale forcing factors such as global radiation and the planetary scale topography.

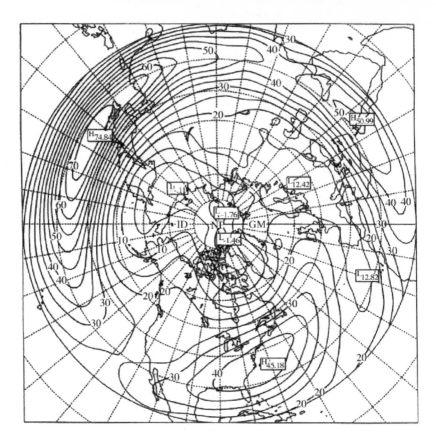

Fig. 1.9. Northern hemisphere long-term mean distributions of geopotential height at 500 hPa. From Risbey and Stone (1996)

1.4 The Regional Climate: Controlled by Planetary Scales

After the planetary scale features of the atmospheric circulation (tropical cells, jet streams, baroclinic zones) are established, these features interact with regional features, such as secondary mountain ranges, coast lines, inland waters. These interactions lead to modifications of the planetary climate on the regional scales - the regional climates. For example, secondary storm tracks may form, for instance in the lee of the Alps over the warm waters of the Mediterranean Sea (both the Alps and the Mediterranean are insignificant features for the formation of the planetary scale climate). Maritime storms dump abundant rainfall on coastal areas, and the like.

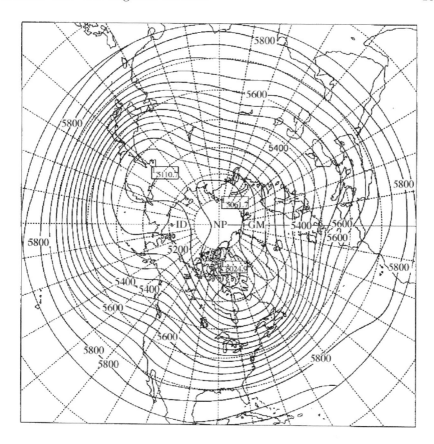

Fig. 1.10. Northern hemisphere long-term mean distributions of zonal wind. Other explanations s. Fig. 1.9

As a demonstration, we show in Fig. 1.12 a *composite analysis* of rainfall in the Sacramento Valley (California, USA). The Northern Hemisphere mean distribution of air pressure (at sea level), of 500 hPa height and zonal wind at 200 hPa averaged over all days with maximum local precipitation were calculated, as well as one day before and after this event. Not surprisingly, rainfall tends to be maximal when on the large-scale a storm travels across the area (day 0), coming from the west (where the storms are formed – day -1). After the precipitation, the storms weakens (day +1). The low pressure system is also visible at the middle of the troposphere at 500 hPa by a trough displaced by several tens of degrees to the east compared to its climatological mean state (not shown), and the jet stream, as visualized by the near-tropopause zonal wind, is elongated far to the east (compare with Fig. 1.9). Thus, a planetary scale event – the displacement of the trough at

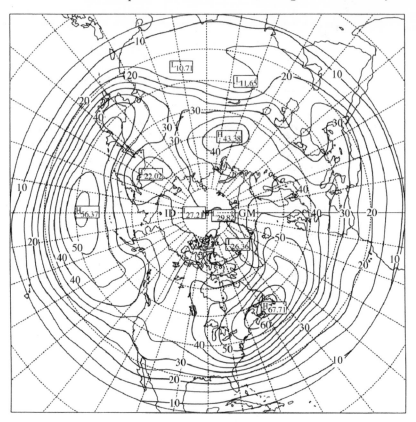

Fig. 1.11. Northern hemisphere long-term mean distributions of the band-pass filtered variance of 500 hPa geopotential height. The filter is selected to retain variability on time scales related to baroclinic processes, i.e. mostly midlatidude storms. Other explanations s. Fig. 1.9

500 hPa and the elongation of the jet stream – eventually lead to a regional event, namely a strong rainfall storm in the Sacramento Valley.

After having illustrated synoptically how the planetary scale variability influences the regional variability, we will demonstrate how the statistics of regional variability, or in another term: the regional climate, is conditioned by the statistics of the planetary scale climate. To do so, we need to formally introduce the concept of conditional statistical models in the following subsection.

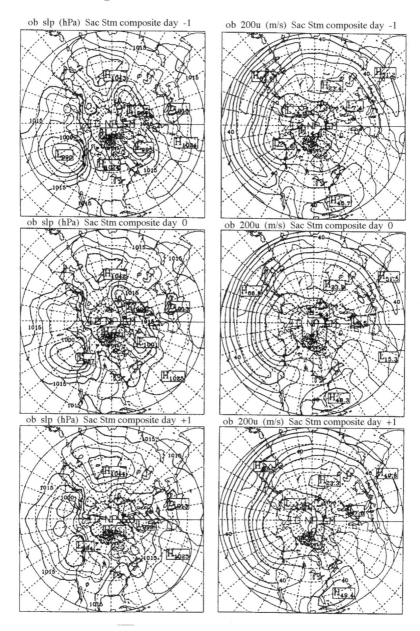

Fig. 1.12. Composite analysis of air pressure and 200 hPa zonal wind for 10 days with maximum precipitation ("day 0" at a location in the Sacramento Valley (middle) and for the 10 days one day before (top, "day -1") and after (bottom, "day +1") maximum precipitation. From Risbey and Stone (1996)

1.4.1   Conditional Statistical Models

If a random variable $\vec{\mathbf{X}}$ is *conditioned* upon another random variable $\vec{\mathbf{G}}$,[6] then the probability density function $F_{\vec{X}}(\vec{x})$ of $\vec{\mathbf{X}}$ may by partitioned such that

$$f_{\vec{X}}(\vec{\mathbf{x}}) = \int f_{\vec{X}|\vec{G}=\vec{g}}(\vec{\mathbf{x}}) f_G(\vec{\mathbf{g}}) d\vec{\mathbf{g}} \tag{1.5}$$

where $f_{\vec{X}|\vec{G}=\vec{g}}(\vec{x})$ is the *conditional* probability function of $\vec{\mathbf{X}}$ provided that the random variable $\vec{\mathbf{G}}$ takes the value $\vec{\mathbf{g}}$, and $f_{\vec{G}}$ is the probability density function of $\vec{\mathbf{G}}$ (cf. Katz and Parlange, 1996).

The expectation and the variance of $\vec{\mathbf{X}}$ may be decomposed:

$$\mathrm{E}_X\left(\vec{\mathbf{X}}\right) = \mathrm{E}_G\left(\mathrm{E}_X\left(\vec{\mathbf{X}}|\vec{\mathbf{G}}\right)\right) \tag{1.6}$$

$$\mathrm{VAR}_X\left(\vec{\mathbf{X}}\right) = \mathrm{E}_G\left(\mathrm{VAR}_X\left(\vec{\mathbf{X}}|\vec{\mathbf{G}}\right)\right) + \mathrm{VAR}_G\left(\mathrm{E}_X\left(\vec{\mathbf{X}}|\vec{\mathbf{G}}\right)\right) \tag{1.7}$$

where the subscript indicates with respect to which random variable the operation "expectation" and "variance" is to be executed.

Thus, the overall expected value of $\vec{\mathbf{X}}$ is a weighted mean of the conditional expectations; the overall variance is seen to be attributable to two different sources, namely the mean uncertainty of the conditional distributions, and the variability of the different conditional means.

For further demonstration of this effect, let us consider the regression case. To do so, we assume that the univariate variable $\mathbf{X}$ is normally distributed with mean μ and variance σ_x^2: $\mathbf{X} \sim \mathcal{N}(\mu, \sigma_x^2)$. Let us further assume that the mean state μ depends linearly on a large scale time-dependent state $\mathbf{G}_t$:

$$\mu = \mu_0 + \beta\mathbf{G} \tag{1.8}$$

and that the variability around μ is independent of $\mathbf{G}$. Then

$$\mathbf{X}_t = \mu_0 + \beta\mathbf{G}_t + \mathbf{N}_t \tag{1.9}$$

with a normally distributed variable $\mathbf{N} \sim \mathcal{N}(0, \sigma_n^2)$. If the "driving" process $\mathbf{G} \sim \mathcal{N}(0, \sigma_g^2)$, then

$$
\begin{aligned}
\mathrm{E}(\mathbf{X}) &= \mu_0 \\
\mathrm{E}(\mathbf{X}|\mathbf{G}_t) &= \mu_0 + \beta\mathbf{G}_t \\
\mathrm{VAR}(\mathbf{X}) &= \mathrm{E}\left((\mathbf{X} - \mu_0)^2\right) = \mathrm{E}\left((\beta\mathbf{G}_t + \mathbf{N}_t)^2\right) = \beta^2\sigma_g^2 + \sigma_n^2 \quad (1.10) \\
\mathrm{VAR}(\mathbf{X}|\mathbf{G}_t) &= \mathrm{E}\left((\mathbf{X} - \mu_0 - \beta\mathbf{G}_t)^2\right) = \sigma_n^2
\end{aligned}
$$

This decomposition is a special version of equation (1.7) and attributes part of the $\mathbf{X}$-variance to the internal variability (σ_n^2) unrelated to the driving

---

[6]For an overview about statistical analysis in climate research, refer to von Storch and Navarra (1995) and von Storch and Zwiers (1998).

process, and the remaining variance to the variability of the driving process itself (σ_g^2).

If we describe the regional climate $\vec{\mathbf{X}}_t$ as the outcome of a stochastic process[7], i.e., when we formally write

$$\vec{\mathbf{X}}_t \sim \mathcal{P}(\vec{\alpha}) \tag{1.11}$$

with a vector of parameters $\vec{\alpha} = (\alpha_1 \ldots \alpha_K)$. The probability distribution $\mathcal{P}$ has to be chosen from a suitable family of distribution. In many cases $\mathcal{P}$ will be Gaussian so that $\alpha_1 = \vec{\mu}$ and $\alpha_2 = \mathbf{\Sigma}$, with $\vec{\mu}$ representing the mean vector and $\mathbf{\Sigma}$ the variance-covariance matrix. In other cases, $\mathcal{P}$ may be of considerably more complex form, for instance in case of daily amounts of rainfall (cf. Lettenmaier (1995) or Katz and Parlange (1996)).

Following our previous reasoning, that the regional climate is determined by the planetary scales, we may consider the random process $\vec{\mathbf{X}}_t$ being *conditioned* by the planetary scale variable $\mathbf{G}_t$, or:

$$\vec{\alpha} = \mathcal{F}(G_t) \tag{1.12}$$

so that equation (1.11) is replaced by

$$\vec{\mathbf{X}}_t \sim \mathcal{P}(\mathcal{F}(G_t)) \tag{1.13}$$

In the special case of the regression (1.10), the variability of the regional climate variable is caused by the local uncertainty unrelated to the large-scale dynamics, and to the effect of the large-scale variability.

Different techniques may be used to actually estimate the conditioning relationship (1.13) from observational data. The classical approach of synoptic climatology is to determine *Großwetterlagen*, i.e. a finite number of typical weather situations, and to determine for each of these Großwetterlagen the regional weather, which can be done empirically (Bárdossy and Plate, 1992; Conway et al., 1996; Enke and Spekat, 1997) or with dynamical models (Frey-Buness, 1995; Fuentes and Heimann, 1996). Nonlinear techniques refer to neural networks (Hewitson and Crane, 1992, 1996), classification and regression trees or analog specifications (Zorita et al., 1995).

A widely used technique is (linear) regression, which is often combined with linear filter techniques (von Storch et al., 1993; Kaas et al., 1996). A typical representative of this class is *Canonical Correlation Analysis* (CCA; von Storch and Zwiers, 1998), which identifies in the paired vectors $(\vec{\mathbf{G}}_t, \vec{\mathbf{X}}_t)$ vectors $\vec{p}_j^G$ and $\vec{p}_j^X$ so that

$$\vec{\mathbf{G}}_t = \sum_{j=1}^{J} g^j(t) \vec{p}_j^G \tag{1.14}$$

$$\vec{\mathbf{X}}_t = \sum_{j=1}^{J} x^j(t) \vec{p}_j^X \tag{1.15}$$

---

[7]This is done explicitly by hydrologists when they build weather generators, see for instance Lettenmaier (1995).

The coefficients $g^j(t)$ and $x^j(t)$ with different indices are pairwise uncorrelated, and coefficients with identical indices share maximum correlations ρ_j. Then, the regression model is

$$\mathrm{E}\left(\vec{\mathbf{X}}_t | \vec{\mathbf{G}}_t\right) = \sum_{j=1}^{J} \rho_j g^j(t) \vec{p}_j^X \tag{1.16}$$

Usually, the number J of retained terms in (1.14,1.15) is small, so that the truncated expansions (1.14,1.15) represent efficient filtering operations. Note that Canonical Correlation Analysis is intrinsically symmetric, and puts the same weight on the "predictor" $\vec{\mathbf{G}}_t$ and on the "predictand" $\vec{\mathbf{X}}_t$.

Another technique, so far hardly used to this end, is *Redundancy Analysis* (RDA; Tyler, 1982; von Storch and Zwiers, 1998). In this case, also expansions like (1.14,1.15) are determined, but the predictor and predictand are treated differently. For any set of linearly independent patterns $\{\vec{p}_j^G; j = 1 \ldots J\}$ there is a uniquely determined regression operator $\mathcal{R}_J$ on $\vec{\mathbf{X}}_t$ so that

$$\hat{\vec{\mathbf{X}}}_{tJ} = \mathcal{R}_J \sum_{j=1}^{J} g^j(t) \vec{p}_j^X \tag{1.17}$$

represent a maximum of variance of $\vec{\mathbf{X}}_t$. The redundancy pattern set $\{\vec{p}_j^G, j = 1 \ldots J\}$ is defined to be that set with minimum $\mathrm{VAR}\left(\vec{\mathbf{X}}_t - \hat{\vec{\mathbf{X}}}_{tJ}\right)$. Then, an orthonormal set of vectors $\{\vec{p}_j^X; j = 1 \ldots J\}$ may be determined so that the regression $\mathcal{R}_J$ maps the predictor patterns $\vec{p}_j^G$ on the predictand patterns $\vec{p}_j^X$. After these manipulations, equation (1.16) may be written as (1.17).

In both cases, CCA and RDA, the resulting patterns may be interpreted such that the emergence of the predictor pattern $\vec{p}_j^G$ on the planetary scale atmospheric state coincides with the emergence of the predictand pattern $\vec{p}_j^X$ on the regional scale state. The difference between the two techniques is that CCA maximizes the correlation of the link, whereas RDA maximizes the amount of variance of the regional scale variable describable by the planetary scale variable. In the examples considered so far, the patterns identified by CCA and RDA were rather similar, with slightly larger correlation coefficients in the case of CCA and somewhat larger proportions of variances of the predictand described by the predictor in the case of the RDA.

In the following applications of CCA and RDA are demonstrated.

1.4.2   The Case of Rumanian Precipitation

In several studies, the control of regional precipitation by the large scale flow has been studied and demonstrated. In the following we describe results obtained for Romania (Busuioc and von Storch, 1996). The predictand variable is the winter mean precipitation amount at 14 stations and the predictor

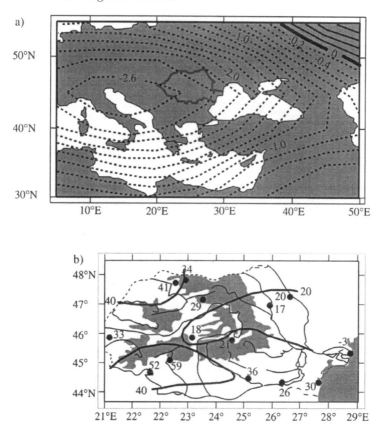

Fig. 1.13. The patterns of the first canonical pair of the winter mean SLP (top; contour 0.2 mb; the area of Romania is encircled by a heavy line) and total winter Rumanian precipitation (bottom; contour 20 mm; the Carpathian mountain range as well as the Black Sea area is marked by stippling). Continuous lines mark positive values, and dashed lines negative values. From Busuioc and von Storch (1996)

variable is chosen to be the winter mean air pressure distribution (SLP) in $30-55°N$ and $5-50°E$. The data used in this study, from 1901-88, may be considered reasonably homogeneous.[8]

The first CCA pair exhibits a correlation between the precipitation and SLP coefficient time series of 0.84. They explain 35% of the total seasonal mean SLP variance and 47% of the total precipitation variance. The patterns

---

[8]Air pressure is often used in this type of "downscaling" studies (cf. von Storch et al., 1993; Hewitson and Crane, 1996). This pressure data set (Trenberth and Paolino, 1980) does not represent "observed" data but represents an analysis of instrumental data. Apart from near surface temperature, this pressure data set is the only one for which a century long observational record with planetary scale coverage and few inhomogeneity problems exist.

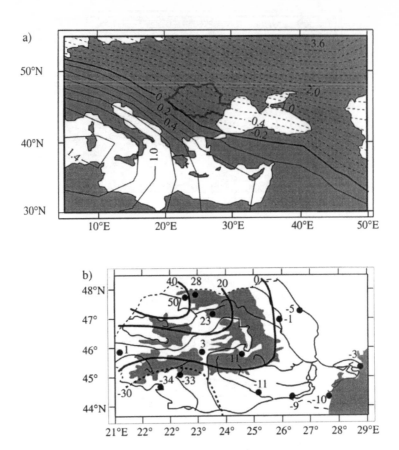

Fig. 1.14. The patterns of the second canonical pair of the winter mean SLP (contour 0.2 mb) and total winter Rumanian precipitation (contour 20 mm). Continuous lines mark positive values, and dashed lines negative values. From Busuioc and von Storch (1996)

(Fig. 1.13) represent a link that is reasonable from the physical point of view: Low pressure over Europe and the Mediterranean basin guides maritime air and precipitating weather systems into Romania, such that above normal precipitation is recorded. The maximum values of almost 59 mm are in the southwest and the minimum values of 17 mm in the northeast, that shows the orographic perturbation effect of the Carpathian mountains.[9]

The second CCA (0.65 correlation) explains 31% of the total SLP variance and 20% of the total precipitation variance. The patterns (Fig. 1.14) suggest

---

[9]Note that the sign of the two patterns is arbitrary. If both patterns are sign-reversed, the same link is described - high pressure over Europe guide continental air into Romania, and there the precipitation is reduced by 17 to 59 mm.

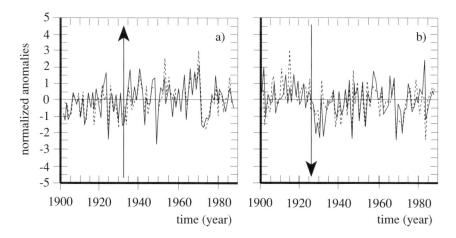

Fig. 1.15. Normalized time components of the first (a) and second (b) CCA patterns of SLP anomalies (continuous line) and Rumanian precipitation anomalies (dashed line). The vertical arrows mark change points in both coefficient time series. From Busuioc and von Storch (1996)

another physically plausible link: The SLP pattern describes a northwesterly flow that affects mostly the intra-Carpathian region where the positive precipitation anomalies are emphasized, the highest (of almost 50 mm) being in the northwest.

In Fig. 1.15a the time coefficients of the first CCA pair are shown. The year-to-year variations are coherent. An inspection with the Pettitt-test of both time coefficient time series reveals for both, independently, the presence of an upward change-point, i.e., a displacement of the time mean, at about 1933. This finding suggests that the change point detectable in the precipitation series, is not due to an inhomogeneity[10], but is a real feature; in fact *this systematic change in precipitation may be traced back to a change of the large-scale circulation.*

A second downward change-point at about 1969 is signaled by the Pettitt statistic[11] when the time interval is limited to 1935 – 1988. This second change point is again consistent with the analysis of the individual precipitation time series (not shown).

Figure 1.15b shows the time coefficients of the second CCA pair. The year-to year variations are fairly coherent. Both curves share a simultaneous downward change-point in the mid 1920's.

---

[10]For problems with observational records due to changes in the instrumentation, observational practices, displacements of the instruments and other reasons, refer to, for instance, Karl et al. (1993) or Jones (1995).

[11]An objective technique for the identification of change points was introduced by Pettitt (1979); for a critique, see Busuioc and von Storch (1996).

The results presented so far, support the notion that Rumanian winter precipitation statistics are to a large extent determined by the large-scale European SLP distribution.

1.4.3   The Case of Short Term Event Statistics

The background of this example is the following: Data about wave height (sea state) are available from reports about visual assessments from ships of opportunity and lighthouses, from wave rider buoys and ship borne wave riders at ocean weather stations; also wave heights maps have been constructed for the purpose of ship routing from wind analyses (WASA, 1998). These data are sparse, and suffer from inhomogeneities of various kinds, or their records are too short for allowing an assessment about changes in the past century.

Thus, using observational data alone is hardly sufficient for getting information about interdecadal variability of wave statistics. An option for overcoming this problem is a combined statistical/dynamical reconstruction which makes use of a "hind-cast" simulation with a dynamical wave model, forced with observed wind fields over a limited time for which the wind data are believed to be of sufficient quality and little affected by improving analysis routines.[12]

As a second step, the wave heights, derived from the hind-cast simulation, are considered as "quasi-observations" and are used for building a statistical model, linking the wave height data at one location to a variable which has been monitored for more than a century, namely mean air pressure distribution. In the last step, the constructed statistical model is fed with the observed air pressure from the beginning of the century onward, and a plausible estimate of wave height statistics for the entire century is obtained. In the following we present the statistical model.

In this case we bring together "apples" and "oranges", i.e. two vector quantities which are not directly linked together. One vector time series, $\vec{\mathbf{S}}_t$, represents the winter (DJF) monthly mean air-pressure distributions in the North Atlantic. The other vector time series, $\vec{\mathbf{X}}_t$, is formed by the 50%, 80% and 90% quantiles of the *intra-monthly distributions* of significant wave height at the oil field *Brent* (approximately $61°N, 1.5°E$; northeast of Scotland)

$$\vec{\mathbf{X}}_t = \begin{pmatrix} q_{50\%} \\ q_{80\%} \\ q_{90\%} \end{pmatrix}_t \tag{1.18}$$

---

[12]Note that the question of whether weather maps, and their surface winds, are homogeneous or introduce artificial signals, such as increasing frequencies of extreme events, into the hind-cast, is a problem difficult to assess - and a possibility which can not be ruled out in general.

Improving analyses procedures, be it because of more or better observations or because of more intelligently designed dynamical and statistical analysis tools, lead to the emergence of more details in weather maps and, therefore, larger extremes.

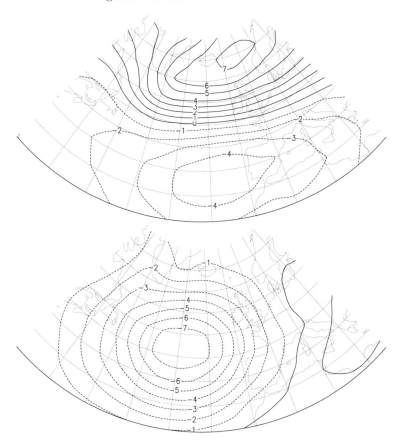

Fig. 1.16. First two monthly mean air pressure anomaly distributions identified in a Redundancy Analysis as being most strongly linked to simultaneous variations of intra-monthly quantiles of significant wave height at Brent $(61°N, 1.5°E)$. The anomalies of the quantiles at that position are listed in Table 1.1

Both vector time series are centered, so that the air pressure values and percentiles are deviations from their long-term mean.[13]

The *monthly mean* of North Atlantic SLP is only indirectly linked to the *intra-monthly percentiles*: the storms affect both, the monthly mean air pressure distribution, as well as, the distribution of wave heights within a month at a specific location. Of course, the storm activity may also be seen as being conditioned by the monthly mean state.

The daily wave height data are taken from a hind-cast simulation over 40 years (Günther et al., 1998).[14] A Redundancy Analysis (RDA) of the

---

[13]Therefore the percentiles are no longer ordered and it may happen that $q_{50\%} > q_{90\%}$.

[14]The following analysis is based on the assumption that the hind-cast is describing the

Table 1.1. Characteristic anomalies of intra-monthly percentiles of significant wave height at the oil field "Brent" $(61°N, 1.5°E)$ northeast of Scotland in winter (DJF) as obtained in a Redundancy Analysis.
The j-row is the j-th redundancy vector $\vec{p}_j^X$. This vector accounts for ϵ_j of the variance of $\vec{\mathbf{X}}$

$\kappa =$	50%	wave height 80%	90%	ϵ_j
j		[cm]		[%]
1	-81	-107	-114	94
2	-32	-2	25	5

Table 1.2. Correlation between hind-casted and reconstructed quantile time series, and proportion of described variance of wave height at "Brent" (see Fig. 1.16).

quantile	50%	wave height 80%	90%	units
correlation	83	82	77	%
described variance	70	66	60	%

two vector time series is performed in order to detect the dominant coupled anomaly patterns in the mean air pressure and in the intra monthly wave height percentiles. The SLP patterns $\vec{p}_j^G$ are shown in Fig. 1.16, and the pattern for the intra-monthly percentiles $\vec{p}_j^X$ are given in Table 1.1 together with the correlations and the proportion of $\vec{\mathbf{X}}$-variance represented by the $\vec{p}_j^X$-patterns. The RDA coefficients are normalized to unity so that the three components of $\vec{p}_j^X$ may be interpreted as anomalies which occur typically together with the "field distribution" $\vec{p}_j^G$.

The first air pressure pattern is closely related to the *North Atlantic Oscillation* (van Loon and Rogers, 1978). A weakened NAO in the monthly mean is associated with an enhancement of all intra monthly percentiles of significant wave heights. In effect, this pattern describes a shift of the intra-monthly distribution towards smaller waves.

The second pattern describes a mean easterly flow across the northern North Sea; the 50% quantile of the wind sea is reduced by 30 cm, whereas the 90% is enhanced by 30 cm. When this monthly mean air pressure pattern prevails then there is a tendency for the wave height distribution to be widened, while the reversed pattern goes with a narrowed intra monthly distribution of wave heights.

With this regression model (1.17), the observed monthly mean air pressure anomaly fields $\vec{\mathbf{G}}_t$ from 1899 until 1994 were used to estimate the time series

---

real evolution sufficiently well and that modifications in analysis of the driving wind field have no significant effect on the statistics of the hind-cast.

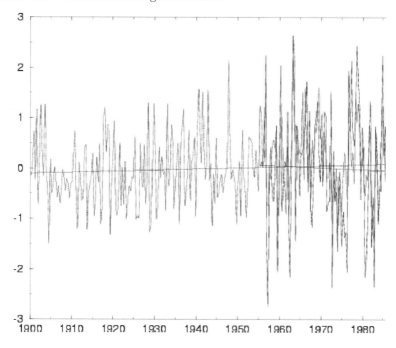

Fig. 1.17. Reconstructed (continuous line) and hind-casted (dashed line; 1955-94) anomalies of 90% quantiles of significant wave heights at "Brent" ($61°N$, $1.5°E$). Units: m

of the percentiles of significant wave height at the oil field "Brent". The last 40 years may be compared with the hind-cast data, whereas the first five decades represent our best guess and can not be verified at this time. For 90% quantiles of wind sea height, the reconstructed time series 1899-1994 and the hind-casted time series 1955-94 are displayed in Fig. 1.17.

In the past four decades, the correlation between hind-cast and statistically derived heights is good (Table 1.2) As with all regression models, the variance of the estimator is smaller than the variance of the original variable: $\text{VAR}\left(\hat{\vec{\mathbf{X}}}\right) < \text{VAR}\left(\vec{\mathbf{X}}\right)$, which makes sense, as the details of the wave action in a month are not completely determined by the monthly mean air pressure field (see Fig. 1.6). Thus, the vector of percentiles is made up of an externally determined component $\hat{\vec{\mathbf{X}}}$ and a component not or only indirectly related to the large-scale state of the atmosphere. To first order approximation, the difference $\vec{\mathbf{X}} - \hat{\vec{\mathbf{X}}}$ may be modeled as white, or maybe red, noise (cf., von Storch, 1997).[15]

---

[15]See also the concept of randomized parameterizations discussed in Section 1.5.2.

1.5 Feedback of Regional Scales on Global Scales

Earlier in this Chapter we have demonstrated the validity of Hadley's approach for understanding climate, namely that the climatic features form in a cascade of decreasing spatial (and temporal) scales. But what about the geographers' Köppen-ansatz, namely that the small scale pieces add to a global entity? Is the global climate conditioned by the regional climates?

The answer is, in principle, positive. First, since the dynamics of the atmosphere are highly non-linear, we may expect each detail to matter for the dynamics – the famous metaphor of the butterfly flapping its wings and by that changing rainfall in Paris (cf. Inaudil et al., 1995). However, we must expand the metaphor and take into account that myriads of butterflies are flapping their wings all the time, and then the question is whether an individual butterfly matters in practical terms.

To give a real-world example: Is there a significant impact if somebody ignites many oil wells in the Kuwait area? As everybody knows, this really happened in 1991 - and before the event the public was greatly concerned about this perspective. Respectable scientists voiced grave concerns that such an event would lead to a situation comparable to the disastrous nuclear winter scenario. The effect was indeed locally disastrous, but the overall effect on the planetary scale turned out to be insignificant (for a photographic account, see Cahalan, 1992; for simulations refer to Bakan et al., 1991; Browning et al., 1991).

Another case refers to a case of an anthropogenic climate change experiment, namely the conversion of the North American prairies into farm land in the last century. Even though no representative observational record from the time before the conversion is available, neither on the regional scale nor on the planetary scale, the case may be studied through a GCM experiment (Copeland et al., 1996). They ran a paired simulation under identical conditions apart of the specified land-use in North America. One simulation was done with present day land-use, and the other with *potential vegetation*. Such a potential vegetation is determined like Köppen's climate map (Fig. 1.1), namely through an empirically determined bivariate function, operating with temperature and precipitation, specifying typical vegetation. No planetary scale changes are induced (not shown) only some regional changes mostly in the vicinity of the altered landscape.

On the other hand, local processes, such as the impact of a meteorite or a major eruption of a volcano, such as the Krakatoa eruption in 1883, may have a significant effect on the planetary scale state of the climate – mainly by injecting large amounts of aerosols on a planetary scale into the stratosphere.

If we disregard such dramatic effects, though, the details of regional processes usually are unimportant. Let us discuss this problem in a somewhat more systematic manner. Let us assume that the climate state is given

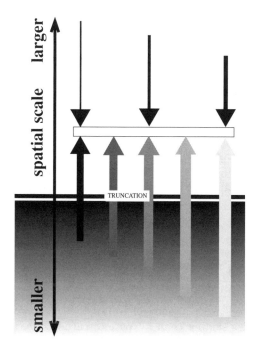

Fig. 1.18. Number of influences on a spatial scale, originating from smaller and larger scales.
(a) full system
(b) truncated system

by a variable Φ and its dynamics by

$$\frac{\partial \Phi}{\partial t} = \mathcal{R}(\Phi) \tag{1.19}$$

with a suitable operator $\mathcal{R}$. We may expand the climate variable into an infinite series

$$\Phi = \sum_k \Phi_k = \sum_k \sum_{j=1}^{k} \phi_{k,j} \tag{1.20}$$

where the $\phi_{k,j}$ may be spherical harmonics or other suitable orthogonal functions. The index k represents the spatial scale L/k with some characteristic length scale L. The index j counts the independent processes with spatial scale L/k. Note that the number of these processes increases with decreasing spatial scale. Thus, the dynamics (1.20) may be expressed for each spatial scale

$$\frac{\partial \Phi_k}{\partial t} = \sum_{j \leq k} \mathcal{R}_{k,j}(\Phi_j) + \sum_{j > k} \mathcal{R}_{k;j}(\Phi_j) \tag{1.21}$$

$$= \sum_{j \leq k} \sum_{l=1}^{j} \mathcal{R}_{k,j;l}(\phi_{j,l}) + \sum_{j=k+1}^{\infty} \sum_{l=1}^{j} \mathcal{R}_{k,j;l}(\phi_{j,l})$$

The first sum in the last expression features a finite, and often small number of terms, whereas the second features infinitely many terms (cf. Fig. 1.18a). If we assume that processes of comparable scales have similar influences in (1.21), then a specific process (j, l), as represented by $\mathcal{R}_{k,j;l}$ cannot be important for the spatial scale L/k with $k \ll j$, even if the sum over all l may be so.

In General Circulation Models of the atmosphere and of the climate system as a whole, the dynamics cannot be expressed through (1.21), requiring *truncation* of the range of j:

$$\frac{\partial \Phi_k}{\partial t} = \sum_{j \leq k} \mathcal{R}_{k,j}(\Phi_j) + \sum_{j=k+1}^{N} \mathcal{R}_{k;j}(\Phi_j) \qquad (1.22)$$

$$= \sum_{j \leq k} \sum_{l=1}^{j} \mathcal{R}_{k,j;l}(\phi_{j,l}) + \sum_{j=k+1}^{N} \sum_{l=1}^{j} \mathcal{R}_{k,j;l}(\phi_{j,l})$$

i.e., only processes with a scale of at least L/N are considered, while all processes at scales smaller than L/N are discarded - the situation is sketched in Fig. 1.18b. The effect of this truncation is worst for scales L/k close to L/N, and least for scales $k \gg N$. However, since the differential equations are non-linear and are integrated forward in time, the errors, which are limited in the beginning of the integration to the small resolved scales eventually are transported to larger and larger scales and finally ruin the entire simulation (cf, Roeckner and von Storch, 1980). Thus, the disregarded sub-grid scale processes must somehow be brought into the equation (1.22). The technique for doing so is called *parameterization*. The idea of these parameterizations will be explained in the next subsection.

1.5.1   Parameterizations

Let us write a climate variable Φ as a sum of the large-scale resolved component $\bar{\Phi}$ and an unresolved part Φ'

$$\Phi = \bar{\Phi} + \Phi' \qquad (1.23)$$

Then, our basic differential equation (1.19) is replaced by

$$\frac{\partial \bar{\Phi}}{\partial t} = \mathcal{R}_{\Delta x}(\Phi) \qquad (1.24)$$

with a modified operator $\mathcal{R}_{\Delta x}$ resulting from the full operator $\mathcal{R}$ after introducing a truncated spatial resolution Δx (this could be the global or regional average). In general, this operator may be written as

$$\mathcal{R}_{\Delta x}(\Phi) = \mathcal{R}(\bar{\Phi}) + \mathcal{R}'(\Phi') \qquad (1.25)$$

with an operator $\mathcal{R}'$ describing the net effect of the sub grid scale variations represented by Φ'. With this set-up, the system (1.24) is no longer closed and can therefore no longer be integrated. To overcome this problem, conventional approaches assume that the "nuisance" term $\mathcal{R}'(\Phi')$ is either irrelevant, i.e.,

$$\mathcal{R}'(\Phi') = 0 \qquad (1.26)$$

or may be *parameterized* by

$$\mathcal{R}'(\Phi') \approx \mathcal{Q}(\bar{\Phi}) \qquad (1.27)$$

with some empirically determined or dynamically motivated function $\mathcal{Q}$.

This is the conventional approach, which is routinely and massively used not only in all climate models, but also in many other environmental models. Processes which are parameterized in climate models are the turbulence in the planetary boundary layer, mixing of matter, momentum and energy, convection, interaction of radiation with clouds, aerosols and gases, the emission of matter and energy from the surface, the routing of rain water into rivers and many more (cf. Fig. 1.3). Papers dealing with the design and the test of the effect of different parameterizations are manifold – examples are Brinkop and Roeckner (1995), Lohmann and Roeckner (1993) or Hense et al. (1982).

In the following subsection a somewhat different concept of parameterizations is introduced, which considers the effect of the unresolved scales as only partially determined by the resolved scales (von Storch, 1997). This concept has not yet been tested and only preliminary examples can be offered.

1.5.2 Randomized Parameterization

While both specifications (1.26,1.27) return an integrable equation (1.24), they both have to assume that the local scale acts as a deterministic slave of the resolved scales. However, as we have seen, in reality there is variability at local scales *unrelated to the resolved scales*. Thus, equation (1.24) should take into account that $\mathcal{R}'(\Phi')$ can not completely be specified as a function of $\bar{\Phi}$, but that formulation (1.27) should be replaced by

$$\mathcal{R}'(\Phi') \sim \mathcal{S}(\vec{\alpha}) \qquad (1.28)$$

with a random process $\mathcal{S}$ with parameters $\vec{\alpha}$ which are conditioned upon the resolved state $\bar{\Phi}$:

$$\mathcal{R}'(\Phi') \sim \mathcal{S}(\mathcal{F}(\bar{\Phi})) \qquad (1.29)$$

When the mean value μ is the only parameter in the vector $\vec{\alpha}$ which depends on $\bar{\phi}$, then the distribution $\mathcal{S}$ may be written as

$$\mathcal{S}(\mathcal{F}(\bar{\Phi})) = \mu(\bar{\Phi}) + \mathcal{S}' \qquad (1.30)$$

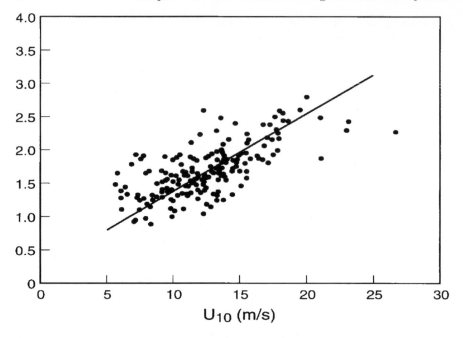

Fig. 1.19. Scatter of various simultaneous measurements of the drag coefficient c_D and of the wind speed at 10 m height. The straight line is a regression line for the scatter. After De Cosmo et al. (1996).

with a conditional mean value and a random component with zero mean value $(E(\mathcal{S}') = 0)$ and uncertainty unrelated to the resolved scales. Specification (1.27) equals specification (1.30) if $\mathcal{S}' = 0$ and $\mu(\bar{\Phi}) = \mathcal{Q}(\bar{\Phi})$.

For demonstration of the difference between the two specifications, (1.27) and (1.30), we discuss Fig. 1.19 displaying various "measurements" of the drag coefficients c_D of the sea surface sorted according to the value of the la"10 m wind" $|\vec{u}_{10}|$ during neutral conditions (from De Cosmo et al., 1996). In this case, we consider the wind measured at a height of 10 m as the "resolved scale" parameter $\bar{\Phi}$, which is representative for a certain spatial and temporal scale and is readily observable. The transfer of momentum $\vec{\tau} = \mathcal{R}(\Phi')$ through the interface of ocean and atmosphere, however, depends on the variance of short-term and smallest scale variations of the wind. The latter quantity can be determined only in expensive observational campaigns, but it has long been known that it can be approximated by the "bulk formula"

$$\vec{\tau} \approx c_D \rho |\vec{u}_{10}| \vec{u}_{10} \tag{1.31}$$

where c_D is considered to depend on the thermal stability and the wind speed. Formula (1.31) represents the classical case of a parameterization, namely the frictional effect of small-scale short-term fluctuations of wind on the atmospheric flow. Figure 1.19 displays a scatter of points, each representing one observation, and a summarizing regression line. Thus, to completely specify the parameterization (1.31), disregarding the dependency on the thermal stratification for the time being, the drag coefficient is specified as a linear function of $|\vec{u}_{10}|$, namely

$$\widehat{c_D} = a + b \cdot |\vec{u}_{10}| \tag{1.32}$$

where a and b are not relevant in the present discussion. This type of specification has been used in countless simulations with numerical ocean circulation models and ocean surface wave models forced with observed wind fields.

The question is what to do about the scatter around the regression line in Fig. 1.19. The application of the bulk formula (1.31) with the regressed $\widehat{c_D}$ implies that the scatter is considered inconsequential or artificial, reflecting observational errors. The alternative interpretation is, however, that unknown processes (such as the sea state, gustiness of the wind, secondary flows) influence the value of c_D so that it exhibits unaccounted variations almost symmetrically around the regression line with a standard deviation of $\sigma \approx 0.5$ units. Therefore, a randomized version of the bulk formula (1.31) would use

$$c_D = \widehat{c_D} + \mathbf{N} \tag{1.33}$$

with

$$\mathbf{N} \sim \mathcal{N}(0, \sigma) \tag{1.34}$$

provided that the bulk formula is used with sufficiently large time steps so that the temporal correlation may be disregarded.

The question is, of course, if the use of the randomized parameterization (1.30) has an effect on the performance of the model in which the parameterization is used. One may argue that the introduction of noise, represented by $\mathcal{S}'$, is inconsequential since the contributions will just be averaged out. This may indeed be so in many cases, in particular in diffusive systems, but the situation is similar to the case of the *stochastic climate model* (Hasselmann, 1976), where the noise acts constructively in building red noise variance. We argue that the use of the randomized design (1.30) will enhance the variability of the considered model on all time scales, and will demonstrate this prospect by means of the energy balance model introduced in Section 1.2.

Differently from the deterministic model considered in Section 1.2, the process of long-wave radiation is not set to be constant, but it is randomized by adding a random term, with zero mean and a standard deviation of 3%:

$$\kappa = 0.64 + \mathbf{N} \tag{1.35}$$

with a random variable $\mathbf{N} \sim \mathcal{N}(0, \ 3\%)$ which may represent the effect of variable clouds and vegetation, for instance. The effect of this randomization on the global mean temperature $\bar{T}$ is shown in Fig. 1.20.

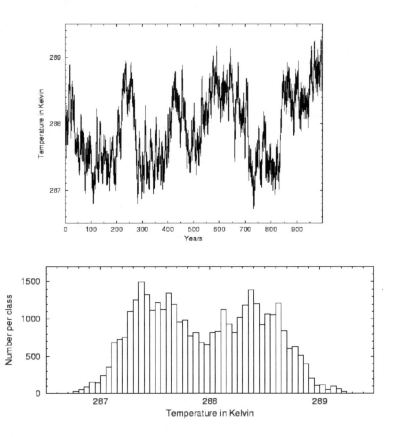

Fig. 1.20. Realization of randomized EBM
Top: time series;
Bottom: Frequency distribution of the temperature

Note that the resulting model itself has become a stochastic model, since it delivers a different path in its phase space for each simulation. Differently from the smooth convergence towards one of the stable equilibria in Fig. 1.5, the trajectory performs an irregular wandering between two regimes, which correspond to the two stable equilibria of the non-randomized system. Overlaid are short-term erratic variations around these equilibria. The existence of the two statistical equilibria is documented clearly from the bimodal histogram of the temperature shown in the lower panel of Fig. 1.20. Thus, the addition of noise transforms the simple dynamically inactive deterministic system into a much richer dynamically active system.

The noise does not act as a nuisance, or a veil blurring the dynamics of the system; instead, the noise contributes a significant component of the dynamical system. Therefore, it appears likely that the randomization of sub-grid scale parameterizations in numerical dynamical models will have an effect on the simulated space-time statistics; in particular, one may expect the overall level of variability to be enhanced and that more often transitions between different sub-regimes, if they exist, will occur. Also, the extreme values may become larger.

1.6 Epilogue

The major result of the present paper is that regional climates do not create the global climate. Instead, the regional climate should be understood as the result of an interplay between the global climate and the regional details. The local processes are important for the formation of the planetary climate, but not in terms of their details, only through their overall statistics.

This understanding has two important implications:

- It is possible to model the planetary scale climate with numerical models with limited, and even rather coarse, spatial and temporal resolutions of several hundred kilometers.

- The success of such models on the planetary scale implies by no means that such models are skillful in the simulation of regional and local features. [16]

Indeed, this dual fact has been noticed by weather forecasters already in the first half of our century, as is documented by Victor Starr's statement from 1942 :

"The general problem of forecasting weather conditions may be subdivided ... into two parts. In the first place, it is necessary to predict the state of motion of the atmosphere ... and, secondly, it is necessary to interpret this expected state of motion in terms of the actual weather which it will produce at various locations. The first of these problems is essentially of a dynamics nature ... The second problem involves a large number of details because, under exactly similar conditions of motion, different weather types may occur, depending upon the temperature of the air involved ... and a host of local influences."

In more modern terms, Starr's statement is that we can describe the large-scale climate dynamics well with our models, but that we need specific *downscaling* techniques to arrive at meaningful statements about the regional

---

[16]The failure of GCMs in this respect was usually not explicitly mentioned, and non-specialist users of information provided by GCM data were unaware of this limitation. This caused sometimes bizarre situations, as for instance an analysis of the differences of climate change on the Northern and Southern slope of the Alps in a GCM too coarse to resolve the Alps.

and local states. Indeed, the mathematics of modern downscaling techniques have their roots in the synoptic climatology and statistical techniques such as *Perfect Prog* which were developed already in the 1950s and 1960s.

However, since climate modeling is not merely an extension of weather forecasting, climatic evolutions on the small scale can not be completely described (or optimally guessed) by evolutions on the large-scale; instead, such local evolutions must be understood as being partially random in character. Therefore, parameterizations as well as weather generators, as opposed to local weather forecasts, should be randomized.

1.7   Acknowledgements

Stefan Güß supplied me with the EBM example. Enjoyable discussions with Peter Müller, Klaus Hasselmann and Kristina Katsaros helped clarify the concept of randomized parameterizations. Warren Washington helped with a review of the manuscript.

Chapter 2

Three Dimensional Numerical Simulation of Climate: The Fundamentals

by Warren Washington

Abstract

Three dimensional computer climate models have been developed over the past thirty years. These models use the same basic physical equations, however, the numerical methods for solving the equations have evolved. This chapter provides a derivation of the fundamental equations, discussion of the basic physical processes such a radiation, precipitation and cloud processes, and discussion of the ocean and sea ice model components of the climate system. At the end of the chapter there are suggestions for interesting internet web sites, where more information can be found and suggestions for future developments.

2.1   Introduction

The purpose of this chapter is to discuss the development and status of three-dimensional computer models of the global climate system. Present day climate models are much better than earlier generation models and are increasingly more complex. They simulate virtually all major features of the atmosphere, land, oceans and sea ice, but there is still room for substantial improvement. We will try to provide in this chapter the fundamentals and references to many internet sources for additional information. Two examples of new generation climate models will be presented. One is the National Center for Atmospheric Research (NCAR) Climate System Model (CSM), which is mostly a National Science Foundation effort involving universities and NCAR scientists. This model has a complex structure of community working groups to guide its development and usage. The other effort that will be discussed is a distributed United States Department of Energy effort involving the Los Alamos National Laboratory, the Naval Postgraduate School and NCAR. The emphasis is on developing a Parallel Climate Model (PCM) that has a massively parallel computer architecture and high ocean and sea ice resolution in the polar regions. In addition, this model has high latitudinal resolution in the ocean tropics in order to better simulate the tropical sea surface temperature anomalies.

The chapter discusses climate models from a point of view of fundamentals so that the reader can develop an understanding of what the models are attempting to simulate, how they are put together, what the models have succeeded in simulating, and a few examples of how they are used for climate change experiments.

Parts of this overview have been adapted from Washington and Parkinson (1986) in which they give an introduction to Three Dimensional Climate Models. More recently Trenberth (1992) edited "Climate System Modeling", which gives a more detailed description of climate modeling.

2.2   Model Equations

The fundamental equations that govern the motions of the atmosphere, oceans, and sea ice are derived from the basic laws of physics, particularly the conservation laws for momentum, mass, and energy. They are Newton's second law of motion, the conservation of momentum, the law of conservation of mass and the law of conservation of energy. The equations for the conservation of energy require incorporation of both internal and external energy sources, which vary with the climate component being considered.

Climate models also require an equation of state relating several of the parameters in the other equations, plus a moisture equation. And in some cases equations for salinity and chemical constituents. For the atmosphere, the equation of state relates the pressure, density, and temperature, and for the ocean the equation of state relates the pressure, temperature, density, and salinity. This equation, together with the moisture equation and the equations for the conservation of momentum, the conservation of energy, and the conservation of mass, constitute the basic equations used in modeling the climate system.

In order to use Newton's second law, for atmospheric, oceanic, and sea ice motions, we need expressions for the forces acting on a volume or mass of the atmosphere, ocean, or sea ice. Such major forces are the frictional forces, the gravitational force due to the attraction of the earth's mass, the pressure gradient force due to the pressure differences on the separate sides of the air, water, or sea ice volume, and the forces resulting from the earth's motion. For the time being, the frictional forces per unit mass in the three primary directions will be labeled F_λ, F_ϕ, and F_z. The gravitational force is essentially unidirectional toward the center of the earth, so its components in the λ and ϕ direction are 0 and in the w direction $-g$, the gravitational acceleration. The remaining forces will require more extensive discussion.

Using a spherical coordinate system with three space dimensions (λ, ϕ, z) at right angles to each other, Newton's second law of motion in each of the three directions can be written

$$a_\lambda = \frac{du}{dt} = \sum \text{forces}_\lambda \qquad (2.1)$$

$$a_\phi = \frac{dv}{dt} = \sum \text{forces}_\phi \qquad (2.2)$$

$$a_z = \frac{dw}{dt} = \sum \text{forces}_z \qquad (2.3)$$

where u, v, and w represent velocity components in the λ, ϕ, and z directions, t is time, and the forces in each of the three directions are forces per unit mass. The total derivatives can be expressed as

$$\frac{du}{dt} = \frac{\partial u}{\partial t} + \frac{u}{r \cos \phi} \frac{\partial u}{\partial \lambda} + \frac{v}{r} \frac{\partial u}{\partial \phi} + w \frac{\partial u}{\partial z} \qquad (2.4)$$

Corresponding expressions for dv/dt and dw/dt can be derived similarly.

The major forces affecting the motion are the centripetal and Coriolis forces due to the rotation of the earth, gravitation, pressure forces, and frictional forces.

The pressure forces can be expressed as

$$-\frac{1}{\rho} \frac{\partial p}{r \, \partial \phi} \qquad \text{and} \qquad -\frac{1}{\rho} \frac{\partial p}{\partial z} \qquad (2.5)$$

The apparent force caused by rotation of earth in λ direction is

$$(2\Omega \sin \phi)v - (2\Omega \cos \phi)w \qquad (2.6)$$

in the ϕ direction is

$$-(2\Omega \sin \phi)u \tag{2.7}$$

and in the z direction is

$$(2\Omega \cos \phi)u \tag{2.8}$$

where Ω is the earth's angular velocity. For convenience, definitions are made as follows: $f = 2\Omega \sin \phi$ and $\hat{f} = 2\Omega \cos \phi$.

Using the above expressions for the force terms, (2.1)–(2.8) can be expressed as follows for the u, v, and w components of the wind:

$$\frac{du}{dt} - \frac{uv \tan \phi}{r} + \frac{uw}{r} = -\frac{1}{\rho r \cos \phi}\frac{\partial p}{\partial \lambda} + fv - \hat{f}w + F_\lambda \tag{2.9}$$

$$\frac{dv}{dt} + \frac{u^2 \tan \phi}{r} + \frac{vw}{r} = -\frac{1}{\rho r}\frac{\partial p}{\partial \phi} - fu + F_\phi \tag{2.10}$$

$$\frac{dw}{dt} - \frac{u^2 + v^2}{r} = -\frac{1}{\rho}\frac{\partial p}{\partial z} - g + \hat{f}u + F_z. \tag{2.11}$$

Equations (2.9)–(2.11) can be simplified further by energy consistency and scale considerations, and that many of the terms are generally small since the atmosphere and oceans are shallow envelopes on the earth.

The resulting set of equations is

$$\frac{du}{dt} - \left(f + u\frac{\tan \phi}{a}\right)v = -\frac{1}{a \cos \phi}\frac{1}{\rho}\frac{\partial p}{\partial \lambda} + F_\lambda \tag{2.12}$$

$$\frac{dv}{dt} + \left(f + u\frac{\tan \phi}{a}\right)u = -\frac{1}{\rho a}\frac{\partial p}{\partial \phi} + F_\phi \tag{2.13}$$

$$\frac{dw}{dt} = -\frac{1}{\rho}\frac{\partial p}{\partial z} - g + F_z \tag{2.14}$$

with

$$\frac{d}{dt} = \frac{\partial}{\partial t} + \frac{u}{a \cos \phi}\frac{\partial}{\partial \lambda} + \frac{v}{a}\frac{\partial}{\partial \phi} + w\frac{\partial}{\partial z} \tag{2.15}$$

and

$$u = a \cos \phi \frac{d\lambda}{dt} \tag{2.16}$$

$$v = a\frac{d\phi}{dt} \tag{2.17}$$

$$w = \frac{dz}{dt}. \tag{2.18}$$

The forces that drive the motion are the local gradients (or differences per unit distance) of pressure, gravity, the Coriolis terms, and the frictional terms F_λ, F_ϕ, and F_z.

The principal terms in (2.12)–(2.18) for large-scale atmospheric motions are generally the pressure gradient, Coriolis, and gravity terms.

The second conservation law explicitly modeled in climate models is the conservation of mass, sometimes referred to as the equation of continuity is

$$\frac{d\rho}{dt} + \frac{\rho}{a \cos\phi}\left[\frac{\partial u}{\partial \lambda} + \frac{\partial}{\partial \phi}(v \cos\phi)\right] + \rho\frac{\partial w}{\partial z} = 0. \tag{2.19}$$

Equation (2.19) is the fundamental equation of the conservation of mass, often termed the equation of mass continuity. In the atmosphere the time derivative term is very important since the air is compressible for large-scale motions; however, water and sea ice are nearly incompressible, so that the total derivative term in (2.19) can generally be ignored for ocean and ice modeling.

The first law of thermodynamics for a gas is a statement of how the thermal energy of a system is related to the work done by compression or expansion (that is, by changing its volume). In mathematical form it can be written as

$$C_v\frac{dT}{dt} = -p\frac{d}{dt}\left(\frac{1}{\rho}\right) + Q. \tag{2.20}$$

The term on the left represents the time change of internal energy, where C_v is the specific heat for air at constant volume and T is temperature in Kelvins. The first term on the right of (2.20) is work done upon a unit mass of air by compression or expansion of the volume, reflecting the elementary physics rule that if a gas is compressed without the addition or subtraction of external heat then its temperature will increase, and if a gas is expanded it will cool. In the ocean the $(d/dt)(1/\rho)$ term is essentially negligible since seawater is almost incompressible; and in sea ice studies the term is usually neglected relative to the term Q. In (2.20), Q is the net heat gain or loss to the system from external sources, such as heating due to solar insolation, heating due to longwave radiation, latent heating due to condensation of water vapor into liquid water, and sensible heating due to conduction and convection. The system being considered in the case of atmospheric models is a small particle or volume of gas. When there is no external gain or loss of heat, so that $Q = 0$, processes are termed *adiabatic*. For many purposes atmospheric and oceanic motions can be considered essentially adiabatic, especially over short time periods. However, for climate studies, the assumption of exclusively adiabatic processes is not appropriate since the amount of heat added or lost to a unit volume of air or water over a long period of time can be substantial. The term Q is called the nonadiabatic (or diabatic) term.

The equation of state relates the pressure, temperature and density to each other, such that as temperature and density increase, for example, then the pressure increases.

$$p = \rho RT \tag{2.21}$$

where R is the so-called gas constant for dry air. A further refinement to (2.21) can be made by modifying either R or T to take into account moisture.

2.3   Vertical Coordinate Systems

There are several alternative ways of treating the vertical coordinate in atmospheric models and the treatment of mountains (orography).

Phillips (1957) devised a transformed coordinate system that avoided the orographical difficulties of the pressure coordinate system. In Phillips' system, termed the sigma-coordinate system, the vertical coordinate, σ, is defined as

$$\sigma = \frac{p}{p_s} \tag{2.22}$$

instead of z or p. In (2.22) p signifies the atmospheric pressure at the point in question and p_s, which is a function of λ, ϕ, and t, signifies the atmospheric pressure at the earth's surface vertically below the point in question. Note that $\sigma = 0$ at the top of the atmosphere, where $p = 0$, and $\sigma = 1$ at the earth's surface, where $p = p_s$. Hence, there is no longer the problem of having a vertical coordinate level intersect mountains, because the $\sigma = 1$ level by definition follows the model's orography precisely.

The basic predictive equations are converted to a σ-coordinate system, which is used in some form or other as a variant of (2.22).

Another way of writing the equations of motion is particularly convenient for understanding the dynamics of the atmosphere and for some forms of numerical solution. As mentioned earlier, the equations of motion yield predictive equations for u and v, the two perpendicular components of horizontal velocity. These components together make up the horizontal velocity vector, which gives the magnitude and direction of flow.

The horizontal wind vector, $\mathbf{V}$, can be separated into two scalar terms by the Helmholtz theorem:

$$\mathbf{V} = \mathbf{k} \times \nabla \psi + \nabla \chi \tag{2.23}$$

where ψ is the streamfunction (a parameter of two-dimensional nondivergent flow whose value is constant along a streamline, which is a line following the flow of the fluid) and χ is the velocity potential (a scalar function whose gradient equals the velocity vector of an irrotational flow). The streamfunction is similar to the pressure field in that wind speed associated with it is proportional to the gradient. It turns out that, except in the tropics, the first term on the right-hand side of (2.23), involving the streamfunction, is generally the largest contributor to the velocity, while the velocity potential, or divergent component term $\nabla\chi$, is usually much smaller for large-scale motions.

By performing the $\mathbf{k}\cdot\nabla\times(\quad)$ and $\nabla\cdot(\quad)$ vector operations on the horizontal equations of motion we obtain the vertical component vorticity equation

$$
\underbrace{\frac{\partial\zeta}{\partial t}}_{\text{large}} = -\underbrace{\mathbf{V}\cdot\nabla(\zeta+f)}_{\text{large}} - \underbrace{w\frac{\partial\zeta}{\partial z}}_{\text{small}} - \underbrace{(\zeta+f)D}_{\text{medium}} + \underbrace{\mathbf{k}\cdot\nabla w\times\frac{\partial\mathbf{V}}{\partial z}}_{\text{small}}
$$

$$
+ \underbrace{\mathbf{k}\cdot\nabla p\times\nabla\left(\frac{1}{\rho}\right)}_{\text{small}} + \underbrace{\mathbf{k}\cdot\nabla\times\mathbf{F}}_{\text{small}} \tag{2.24}
$$

where $\zeta\equiv\mathbf{k}\cdot\nabla\times\mathbf{V}$ is the vertical component of vorticity, and the divergence equation

$$
\underbrace{\frac{\partial D}{\partial t}}_{\text{small}} = -\underbrace{\nabla\cdot[(\mathbf{V}\cdot\nabla)\mathbf{V}]}_{\text{small}} - \underbrace{\nabla\cdot(f\mathbf{k}\times\mathbf{V})}_{\text{large}} - \underbrace{\nabla w\cdot\frac{\partial\mathbf{V}}{\partial z}}_{\text{small}} - \underbrace{w\frac{\partial D}{\partial z}}_{\text{small}}
$$

$$
- \underbrace{\nabla\cdot\left(\frac{1}{\rho}\nabla p\right)}_{\text{large}} + \underbrace{\nabla\cdot\mathbf{F}}_{\text{small}} \tag{2.25}
$$

where D is the horizontal divergence $\nabla\cdot\mathbf{V}$. The size designations above the equations indicate typical relative magnitudes of the terms when considering large-scale weather and ocean circulation patterns.

By performing the same curl and dot operations on (23), the vorticity and divergence can be expressed in terms of streamfunction and velocity potential

$$
\zeta = \mathbf{k}\cdot\nabla\times\mathbf{V} = \frac{1}{a\cos\phi}\frac{\partial v}{\partial\lambda} - \frac{1}{a}\frac{\partial u}{\partial\phi} + \frac{u\tan\phi}{a} = \nabla^2\psi \tag{2.26}
$$

$$
D = \nabla\cdot\mathbf{V} = \frac{1}{a\cos\phi}\frac{\partial u}{\partial\lambda} + \frac{1}{a}\frac{\partial v}{\partial\phi} - \frac{v\tan\phi}{a} = \nabla^2\chi \tag{2.27}
$$

where the ∇^2 on the right-hand side is called the Laplacian operator and is defined as

$$
\nabla^2(\quad) \equiv \frac{1}{a^2\cos\phi}\frac{\partial}{\partial\phi}\left(\cos\phi\frac{\partial(\quad)}{\partial\phi}\right) + \frac{1}{a^2\cos^2\phi}\frac{\partial^2(\quad)}{\partial\lambda^2}. \tag{2.28}
$$

For most nontropical large-scale motions, $\zeta\gg D$.

2.3.1   Solar Radiation

Before deriving the radiation equations for solar heating and determining the effect of solar radiation on the atmosphere, land, oceans, and sea ice, it is necessary to compute the solar intensity at the top of the atmosphere. The distribution of the radiative flux arriving at the top of the atmosphere can be determined from spherical trigonometry, taking into account the date, time of day, and latitude. The zenith angle, Z, at any location is the angle between the vector perpendicular to the earth's surface at that location and the incoming solar rays. This angle can be seen to equal the angle between the plane tangent to the surface of the earth and the plane normal to the sun's rays. By the use of standard trigonometric equations, the cosine of this zenith angle can be calculated from

$$\cos Z = \sin \phi \sin \delta + \cos \phi \cos \delta \cos H \qquad (2.29)$$

where ϕ is latitude, δ is solar declination, which is the angular distance of the sun north of the equator and varies from about 23.5° on June 22 (summer solstice) to -23.5° on December 22 (winter solstice), and H is the hour angle, which is the longitudinal distance from the point in question to the meridian of solar noon and, therefore is 0 at any point experiencing solar noon.

The solar flux, S, arriving at the top of the atmosphere is a function of $\cos Z$ and the distance from sun to earth, d, such that

$$S = S_0 f(d) \cos Z \qquad (2.30)$$

where S_0 is the *solar constant* and the factor $f(d)$ is 1.0344 in early January and 0.9674 in July for the present astronomical conditions. In paleoclimate studies, where the orbital parameters are quite different from those of the present, both δ and $f(d)$ must be changed. This is an important point since there is increasing evidence that orbital parameters are a major cause of climate change over long time scales such as the ice ages and longer.

The parameterization of the absorption of solar radiation in the climate system is critical for a proper simulation of the climate and its sensitivity. One of the principal absorbers in the atmosphere is stratospheric ozone, which absorbs very effectively in the ultraviolet ($\lambda < 0.4$ μm) and in the visible (0.4 μm$< \lambda < 0.7$ μm) portions of the electromagnetic spectrum. Water vapor is the primary absorber in the troposphere in the near-infrared (0.7 μm $< \lambda < 4$ μm). Figure 2.1 shows the spectral energy distributions of a blackbody with $T = 6000$ K, the solar energy at the top of the earth's atmosphere, and, for clear-sky conditions, the solar energy reaching the earth's surface. As shown, the most effective atmospheric absorber of solar energy in the shorter wavelengths is ozone, whereas water vapor and carbon dioxide are important absorbers for the longer wavelengths.

In climate models, the exact absorption, scatter, and transmission of solar energy are never used because of the enormous numerical calculations

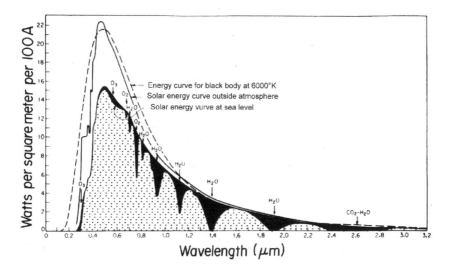

Fig. 2.1. Spectral energy distribution as a function of wavelength, at the top of the atmosphere and at sea level. The figure includes absorption by various atmospheric gases. Also shown is the shape of the blackbody spectral energy distribution for a temperature of 6000 K, although the vertical scale in the figure is not appropriate for this latter (dashed) curve. From Stephens (1984) who modified an earlier version from Lacis and Hansen (1974)

involved. The amount of absorption of solar energy in each model layer can be computed based on clouds, ozone, water vapor, zenith angle of the sun, and the local albedo of the earth's surface. This simplified computation must account for multiple-beam reflections from other layers, as well as, the earth's surface. For example, Rayleigh scattering of solar radiation by atmospheric molecules is an important part of the shortwave treatment in climate models.

The absorption of solar radiation by water vapor is a very strong factor in the heating of the lower troposphere, especially in the tropics. There are many formulations for this absorption based upon the optical path lengths of water vapor, ozone (O_3), carbon dioxide (CO_2), and oxygen (O_2).

The computation of diffuse versus direct solar radiation must be taken into account, as well as, reflections and absorption by clouds and the earth's surface. There are many uncertainties about the accuracy of the parameterizations used. Climate models do not accurately take into account varying amounts of aerosols such as dust (volcanic and otherwise), one of the reasons being that little is known about the geographical aerosol size distribution and composition.

2.3.2   Infrared Radiation

Terrestrial radiation is mostly in the infrared region of the electromagnetic spectrum with wavelength $\geq 4\mu m$ (4 micrometers). It is known that all gases emit and absorb radiation. In the atmosphere, most of the infrared radiation is affected by molecules rather than individual atoms. The triatomic molecules such as ozone, carbon dioxide and water vapor are very strong absorbers and emitters of infrared energy for the atmosphere. Much of this is caused by rotational and vibrational modes of interaction of the molecules. The absorption and emission from more complex molecules such as chlorofluorocarbon molecules is even larger, however, the quantity of such molecules is lower so in an absolute sense they have less radiative impact than ozone, carbon dioxide, and water vapor.

For a complete derivation of the infrared equations used typically in climate atmospheric model components, it is suggested the readers consult various radiative texts. It is also found in Washington and Parkinson (1986). The basic infrared equation in emissivity form is as follows for upward and downward fluxes

$$F^{\uparrow}(z) = \pi B(0) + \int_0^z \tilde{\epsilon}(z,z')\,d[\pi\,B(z')] \tag{2.31}$$

and

$$F^{\downarrow}(z) = \pi B(z_0)\epsilon(z,\infty) + \int_{z_0}^z \tilde{\epsilon}(z,z')\,d[\pi\,B(z')] \tag{2.32}$$

where the emissivities are

$$\tilde{\epsilon}(z,z') = \sum_i A_i(z,z')\frac{dB_i(z')}{dB(z')} \tag{2.33}$$

and

$$\epsilon(z,z') = \sum_i A_i(z,z')\frac{B_i(z')}{B(z')} \tag{2.34}$$

and i sum overall wavelength intervals, B is the Planck function; note that for a perfect blackbody $B(T) \simeq T^4$. Z is the vertical coordinate and Z_o is the earth surface (land, sea ice, or ocean).

2.3.3   Net Heating/Cooling Rates

The net heating or cooling of an atmospheric layer due to solar and infrared radiation is given by

$$\rho C_p \frac{dT}{dt} = \frac{d(F_N)}{dz} \tag{2.35}$$

where ρ is air density, C_p is the specific heat of air at constant pressure, and F_N is the total net solar and infrared radiative flux.

The largest contributors to the heating/cooling rates in K/day are water vapor, CO_2, and O_3.

The latitudinal distribution of the annually-averaged emitted infrared flux, planetary albedo, and net flux at the top of the atmosphere are shown in Fig. 2.2. The planetary albedos are far greater in high latitudes than elsewhere, in large part due to more ice and snow, along with moderate-to-heavy cloud covers. The albedo also undergoes large temporal changes because of high sensitivity to the nature of the surface, especially snow and ice distribution, and clouds. The emitted longwave flux $F^\uparrow$ is determined largely from temperature and moisture distributions, both of which vary greatly between tropics and poles. The net flux, which is the difference between the absorbed solar flux and the emitted infrared flux, is the energy source for atmospheric and oceanic circulations that carry heat from the tropics to the poles.

2.3.4 Physical Processes

Clouds are a major aspect of any climate model. In the simplest terms, if the air is saturated but not convective then a stratiform (layer) cloud or fog exists. Usually it is assumed that stratiform clouds occupy all or nearly all of any grid cell where they appear. If on the other hand, a cloud is of convective origin, then usually it is assumed to occur over some small fraction of a grid cell. The reason for the difference is that convective clouds normally occur over horizontal areas that are smaller than the typical horizontal grid areas used in present climate models, whereas stratiform clouds normally cover large horizontal regions and are fairly uniform. Often climate models make use of either relative humidity (defined in the next section) or some other moisture parameter in conjunction with empirical factors to relate cloud amount to moisture. Cloud albedo is generally determined as a function of zenith angle and optical depth.

Often the radiation treatment in a model assumes that clouds at different levels are overlapped randomly so that the cloud cover C_i at each layer i can be combined as follows to give the total cloud cover C_T:

$$C_T = 1 - \prod_{i=1}^{N}(1 - C_i). \tag{2.36}$$

For example, if there is a two cloud-layer system ($N = 2$) with 0.5 cloud cover in each layer, then the total cloud cover C_T is 0.75.

In the infrared part of the spectrum, clouds usually are treated as blackbodies, meaning that their emissivity is assumed to be unity, although there is observational evidence that thin ice and water clouds such as cirrus clouds have emissivities considerably less than 1. Most climate models have some difficulty in properly treating small amounts of moisture, and thus optically

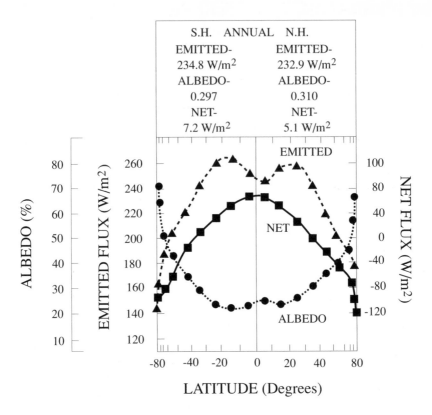

Fig. 2.2. Latitudinal distribution of the annually averaged emitted (outgoing) infrared flux at the top of the atmosphere, the net flux (absorbed solar minus emitted) at the top of the atmosphere, and the net albedo of the earth/atmosphere system. From Stephens et al. (1981)

thin cirrus-type clouds often are treated somewhat arbitrarily. There remains a great deal of uncertainty in parameterizing clouds properly in numerical models. In order to improve this important aspect of climate modeling, continuing research needs to be conducted on cloud/radiation interactions. It should be pointed out that even a very small percentage change in clouds will lead to a big climate change. For example, a typical solar flux is several hundred W/m^2, which is multiplied by the albedo of clouds. A 1 percent change in cloud albedo results in several W/m^2 change in climate forcing. Such a change can be equivalent to a doubling of greenhouse gases.

2.3.5 Precipitation and Cloud Processes

The treatment of moisture processes is one of the more difficult aspects of constructing climate models. There are several reasons for this: First, the precipitation-cloud physics is not fully understood, i.e., how condensation, sublimation, and freezing of water drops and ice particles occur; second, the several hundred kilometers horizontal and the kilometer or so vertical resolution of climate models are much larger than the scales in which most clouds are formed; and third, there are some unique problems in approximating the precipitation-cloud aspects numerically, such as the wide range of moisture values, which can extend over several orders of magnitude. Small amounts of moisture tend to be difficult to handle in climate models.

In order to discuss the precipitation and cloud physics used in climate models, several basic concepts need to be introduced. The ratio of the density of water vapor, ρ_w, to the density of dry air, ρ, is defined as the mixing ratio, q:

$$q = \frac{\rho_w}{\rho} \tag{2.37}$$

The prediction equation for moisture is written in flux form by combining with the equation of continuity to obtain

$$\frac{\partial(\rho q)}{\partial t} + \nabla \cdot (\rho q \mathbf{V}) + \frac{\partial}{\partial z}(\rho q w) = M + \rho E \tag{2.38}$$

where M is the time change of water vapor per unit volume due to condensation and E is evaporation and subgrid scale diffusion. The precipitation rates are another major aspect of climate models that needs marked improvement. The regional precipitation amounts are in error by a factor of two in most models and because much of the precipitation is linked to geographical features it is often displaced from its true location due to the crude mountain/valley representation in climate models. At this point the details of precipitation cannot be relied upon for use in detailed watershed or agriculture prediction models.

2.4 Surface Processes

2.4.1 Boundary Fluxes at the Earth's Surface

Many atmospheric modeling studies make use of simple bulk formulae for the transfer of momentum, moisture, and sensible heat within the constant flux or Prandtl layer, which is 100 to 200 m above the surface. All assume that the transfer process is proportional to the local gradient between the surface and atmosphere, multiplied by the wind speed.

The boundary flux formulae for momentum, sensible heat and moisture are

$$\tau_\lambda = \rho C_D V_a u_a \tag{2.39}$$

$$\tau_\phi = \rho C_D V_a v_a \tag{2.40}$$

$$H = \rho C_p C_H V_a (\theta_s - \theta_a) \tag{2.41}$$

$$LE = \rho L C_E V_a (q_s - q_a) \tag{2.42}$$

where ρ is air density, C_D is the drag coefficient or momentum transfer coefficient, C_H is the transfer coefficient for sensible heat, C_E is the transfer coefficient for moisture, L is the latent heat of evaporation, $\mathbf{V}_a = (u_a, v_a)$ is air velocity, V_a is velocity magnitude,

$$V_a = |\mathbf{V}_a| = \sqrt{u_a^2 + v_a^2}\,, \tag{2.43}$$

and θ_a and q_a are boundary layer potential temperature and mixing ratio. θ_s and q_s are potential temperature and mixing ratio at the earth's surface. Over the ocean θ_s is computed from the temperature of the ocean surface, and the surface is assumed saturated, so that q_s is a function of temperature alone and is computed by knowing the saturation mixing ratio and pressure.

Approximate transfer coefficients have been determined from both theoretical considerations and observational studies. For ocean surfaces and smooth surfaces

$$C_D \simeq C_h \simeq C_E \simeq 10^{-3} \tag{2.44}$$

with larger values of about 3×10^{-3} over rough terrain or regions where there is considerable boundary convection. More detailed drag coefficient formulations have been devised that take into account whether the boundary layer is stable or unstable. Another important factor is the effect of mountains in forming additional drag on the atmosphere. This can be direct in terms of additional roughness of the surface layers and also the generation of vertically moving gravity waves that can cause turbulence at higher altitudes.

2.4.2   Examples of Computations of Surface Hydrology[1]

One of the simplest equations for predicting surface temperature for a simple surface layer without a snow cover is

$$\rho c \frac{\partial T_*}{\partial t} = \frac{[S_* + F^\downarrow - F^\uparrow - H - LE]}{\Delta z} \tag{2.45}$$

where ρ is the density of the surface, c the specific heat of the surface type, T_* the temperature of the surface layer, S_* the absorbed solar flux at the

---

[1] See also Section 3.3.

earth's surface, $F^\downarrow$ the downward infrared flux, $F^\uparrow$ the upward infrared flux, H and LE the sensible and latent heats, respectively, and Δz the thickness of the surface layer.

As mentioned earlier the prediction of surface physics and its interaction with vegetation usually is treated crudely in present climate models. One aspect is the treatment of surface hydrology. The surface albedo and transfers of latent and sensible heat will be quite different over a snow-covered region than over dry land. The difference in albedo often leads to large differences in temperature and transfers of heat and moisture to the atmosphere. Likewise, the fact that one surface is saturated with moisture, while another (such as a desert) is dry could have a large effect on the regional climate, again due to the differences in the transfer of moisture to the atmosphere. Dickinson (1992) reviews the evapotranspiration processes that should be included in climate models. See Fig. 2.3, which shows in simple schematic form the various processes.

One of the earliest used methods of incorporating snow cover and soil moisture in numerical climate models follows that first introduced by Manabe (1969a,b) resulting from Soviet observational studies by Romanova (1954). Romanova found that for plains and forest regions most of the moisture change takes place in the top 1 m of soil, which usually encompasses the root zone of most vegetation. Obviously, the surface in reality is much more complex than the treatment by Manabe, but the method at least has the virtue of incorporating the fundamental processes of precipitation, evaporation, and water storage in the soil.

The basic equation for soil moisture prediction is

$$\frac{\partial W}{\partial t} = P - E + S_m \qquad (2.46)$$

where W is total soil moisture in meters stored in a surface layer of soil, P is the rain precipitation rate, E is the surface evaporation rate, and S_m is the snow melt rate, which is computed from the surface energy balance. The evaporation rate can be related to the soil moisture in the following manner

$$\text{if} \quad W \geq W_c \qquad \text{then} \quad E = E_{ap} \qquad (2.47)$$

and

$$\text{if} \quad W < W_c \qquad \text{then} \quad E = E_{ap}\frac{W}{W_c} \qquad (2.48)$$

where W_c is a critical value of soil moisture and E_{ap} is the potential evaporation rate for a saturated surface. The above equations state that if the soil moisture is greater than W_c, then the evaporation rate is a maximum, E_{ap}, and if the soil moisture is less than W_c, then the evaporation rate is reduced linearly as a function of W. Manabe assumes that field capacity for soil moisture, W_{Fc}, is 0.15 m and W_c is 75% of that value.

In a similar simplified approach Manabe uses the following snow cover prediction equation

$$\frac{\partial S}{\partial t} = P - E - S_m \tag{2.49}$$

where S is the snow cover amount in liquid water equivalent, P is snow precipitation rate, and E is rate of sublimation. It is possible for the snow cover to melt, thereby increasing soil moisture (2.46) and decreasing S (2.47). Also, if rain falls on snow, it is assumed that the snow does not hold the water, but that the water instead seeps through to contribute to soil moisture. If snow covers the soil, no evaporation of soil moisture is allowed to take place, although sublimation is allowed. The determination of whether snow or rain is falling, usually is based upon whether the lower tropospheric model layers have temperatures below or above freezing.

The existence of snow melt is determined by the sign of S_m, calculated as

$$S_m = \frac{1}{L_f}[S_* + F^\downarrow - F^\uparrow - H - LE] \tag{2.50}$$

where L_f is the latent heat of fusion and the other terms are those of (2.45). If, in the presence of a snow cover, setting the surface temperature T_* at $0°\mathrm{C}$ yields $S_m > 0$.

Then melting is inserted in (2.46) and (2.49). Otherwise, S_m is set equal to zero. The runoff, R_f, into the ocean is computed whenever $W > W_{Fc}$ as

$$R_f = P - E. \tag{2.51}$$

Usually the runoff can be translated into a river flow because that assumes a detailed description of the terrain, which most climate models do not yet incorporate.

Figure 2.3 shows a schematic diagram of land surface processes used in many climate models. This scheme uses a simplified land vegetation treatment. Dickinson (1992) shows an idealized treatment where water is taken up from the roots through the stems into the leaves and through transpiration gives the moisture back to the atmosphere. Also shown on the diagram is the sun's radiation, which is absorbed on the vegetation and the land surface. The precipitation is intercepted by the leaves and is evaporated from the vegetation. In addition, there is leaf drip, which is essentially water that collects on leaves and drips to the ground. On the right hand side of the figure the wind is shown which is very important for the rate of evaporation, sensible heat, and momentum exchanges between the vegetation and the atmosphere. The biological and ecological aspects of the climate system are becoming an integral part of the climate models albeit with very large uncertainties about how they should be incorporated and to what degree. There are several ongoing field activities which should lead to better parameterizations of this component of the climate system.

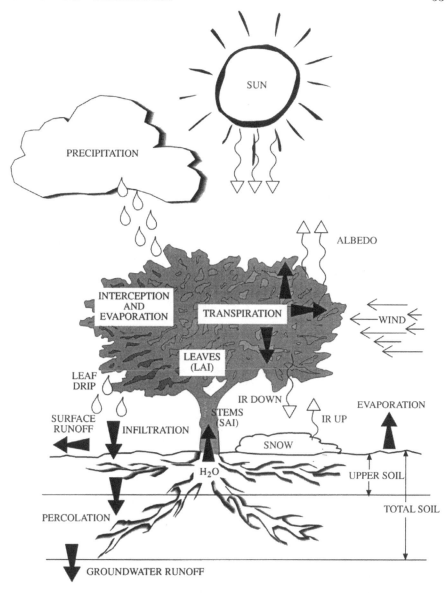

Fig. 2.3.

2.5 Ocean Models

The formulation of ocean models used by most climate groups was developed by Bryan (1969). The hydrostatic approximation is used, as well as, the incompressible assumption. Thus, the equation of continuity can be simplified to

$$\frac{\partial w}{\partial z} = -\frac{1}{a \cos \phi} \frac{\partial u}{\partial \lambda} - \frac{1}{a \cos \phi} \frac{\partial}{\partial \phi} (v \cos \phi) \tag{2.52}$$

Note that (2.52) is diagnostic in that, if u and v are known, then w can be determined.

The equation of state for the ocean relates seawater density, ρ_w, to ocean temperature, T, salinity, S, and pressure, p

$$\rho_w = f(T, S, p). \tag{2.53}$$

The ocean equation of state has a more complex form because of the salinity effects on the density. Most ocean models have an empirical form of the equation of state.

Following Bryan (1969) and Semtner (1974), the remaining basic equations for ocean modeling will be presented in the spherical coordinate system, with depth, z, defined as negative downward from $z = 0$ at the surface. In this system the horizontal equations of motion are

$$\frac{\partial u}{\partial t} + L(u) \quad -\frac{uv \tan \phi}{a} - fv = -\frac{1}{\rho_0 a \cos \phi} \frac{\partial p}{\partial \lambda} + \mu \frac{\partial^2 u}{\partial z^2}$$

$$+ A_m \left\{ \nabla^2 u + \frac{(1 - \tan^2 \phi)u}{a^2} - \frac{2 \sin \phi}{a^2 \cos^2 \phi} \frac{\partial v}{\partial \lambda} \right\} \tag{2.54}$$

$$\frac{\partial v}{\partial t} + L(v) \quad +\frac{u^2 \tan \phi}{a} + fu = -\frac{1}{\rho_0 a} \frac{\partial p}{\partial \phi} + \mu \frac{\partial^2 v}{\partial z^2}$$

$$+ A_m \left\{ \nabla^2 v + \frac{(1 - \tan^2 \phi)v}{a^2} + \frac{2 \sin \phi}{a^2 \cos^2 \phi} \frac{\partial u}{\partial \lambda} \right\} \tag{2.55}$$

where

$$L(\alpha) = \frac{1}{a \cos \phi} \left[\frac{\partial}{\partial \lambda}(u\alpha) + \frac{\partial}{\partial \phi}(\cos \phi \, v\alpha) \right] + \frac{\partial}{\partial z}(w\alpha) \tag{2.56}$$

and α is a dummy variable, ρ_0 is a constant approximation to the density of seawater, μ is the vertical eddy viscosity coefficient, and A_m is the horizontal eddy viscosity coefficient.

The hydrostatic equation for the ocean is

$$\frac{\partial p_w}{\partial z} = -\rho_w g. \tag{2.57}$$

The first law of thermodynamics for the oceans is

$$\frac{\partial T}{\partial t} + L(T) = \kappa \frac{\partial^2 T}{\partial z^2} + A_H \nabla^2 T \tag{2.58}$$

where κ and A_H are the vertical and horizontal eddy diffusivity coefficients. The prediction equation for salinity follows a similar for heat and is given by

$$\frac{\partial S}{\partial t} + L(S) = \kappa \frac{\partial^2 S}{\partial z^2} + A_H \nabla^2 S, \tag{2.59}$$

so that vertical motions at the bottom are induced by horizontal motions (u, v) interacting with the bottom topography, H. Horizontal flow impinging on an upslope will generate upward motion, and horizontal flow over a downslope will generate downward motion.

At the top of the ocean, fluxes of momentum (wind stress), heat, and moisture are either specified from observed fields or an atmospheric model such that

$$\rho_0 \mu \frac{\partial}{\partial z}(u, v) = (\tau_\lambda, \tau_\phi) \tag{2.60}$$

and

$$\rho_0 \kappa \frac{\partial}{\partial z}(T, S) = \left[\frac{1}{c_{pw}} H_{\mathrm{ocn}}, \nu_s(E - P)S \right] \tag{2.61}$$

where H_{ocn} is the net heat flux into the ocean (positive for heating and negative for cooling), P is precipitation rate, E is evaporation rate, S is surface salinity, c_{pw} is specific heat of water, and ν_s is an empirical conversion factor. H_{ocn} can be obtained from a straightforward surface energy balance:

$$H_{\mathrm{ocn}} = S_* + F^\downarrow - F^\uparrow - H - LE \tag{2.62}$$

where the terms on the right-hand side were defined after (45), in the discussion of surface temperature computation.

In the early 1980's, Semtner and Chervin (1988, 1992) developed a version of the Bryan type of model that was effectively used on multi-processor parallel-vector computers, it is termed the Parallel Ocean Climate Model (POCM). Another effort started at the Los Alamos National Laboratory was the Parallel Ocean Program (POP). POP had several innovations including a free surface and a transformed horizontal grid that avoided the pole problem found in latitude-longitude grid structure (see Smith et al. (1992, 1996)). There are several survey and review articles that the reader can refer to, such as Semtner (1995, 1997) and McWilliams (1996) for more details on recent developments. One important aspect to keep in mind is, that the ocean has a great thermal heat capacity compared to the atmosphere. This means that in order to obtain equilibrium in the climate system, that the ocean must be brought into equilibrium. The real ocean is still changing in response to earlier climates. Thus, in the beginning of a climate simulation it is necessary to "spin up" the ocean so that it is in agreement with the atmospheric component of the climate system. This is usually done by assuming some atmospheric state an then running the ocean or ocean/sea ice model to a quasi-equilibrium with the atmosphere. This is a very important aspect of climate modeling in order to avoid large "drifts" of the climate system.

2.6   Sea Ice Models

There are two major aspects of climate oriented sea ice models: (1) the computation of the thermodynamics of the ice cover and (2) the computation of ice dynamics. The first determines the thickness and temperature structure of the ice and uses the principle of the conservation of energy. The second aspect, the dynamic part, which determines sea ice motion, is based on the principle of the conservation of momentum. Both sets of calculations contribute to determining the open water or lead area within the ice cover. Most climate models only use thermodynamics, however, some are using some form of dynamics.

The one-dimensional (vertical) sea ice calculations stem from Maykut and Untersteiner (1971) for the Central Arctic, and they have been simplified by Semtner (1976). There has been further development by Flato and Hibler (1992) and Lemke et al. (1997) on arriving at a consistent formulation with leads and conservation of energy. This involves balancing incoming and outgoing energy fluxes at the air/snow, snow/ice, and ice/water interfaces. The fluxes included are solar radiation, incoming and outgoing longwave radiation, sensible and latent heat, conduction through the ice and snow layers, an ocean heat flux, and the absorption and emission of energy due to the change of state between ice and water.

The reader is referred to Semtner (1976) for a comparison of the results of ice calculations at one point in the Central Arctic from (1) a very highly resolved model (that of Maykut and Untersteiner, 1971), (2) a model with two levels in the ice and one in the snow, and (3) a model with one level in the ice and one in the snow. Each of the latter two models is computationally efficient enough to be used in large-scale simulations.

In some models, there is a simplification of having sea ice and snow unresolved in the vertical, including the zero-layer model of Semtner (1976) and the models of Bryan et al. (1975) and Parkinson and Washington (1979), the temperature at the bottom of the ice generally is set precisely at the freezing point of seawater. This is realistic both conceptually and from observations.

As the sea ice calculations proceed, at every time step the snow melt term is initially set equal to zero and are solved for the temperatures at the upper snow surface and at the snow/ice interface. An adjustment is made at each grid square where the calculated temperature at the upper surface exceeds the melting point for snow, 273.15 K and snow melt is computed. A similar calculation is done for the melting of sea ice.

2.6.1   Ice Dynamics

The five major dynamic forces that determine sea ice motion are: the air stress from above the ice, τ_a, steering the ice in the direction of the surface wind; the water stress from below the ice, τ_w, steering the ice in the direction of the water motion; the gravitational stress from the tilt of the sea surface,

D, termed the dynamic *topography*, and steering the ice downward from higher to lower surface levels; the stress induced by the earth's rotational motion, **G**, termed the *Coriolis force* and steering the ice to the right of its otherwise-induced motion in the Northern Hemisphere and to the left in the Southern Hemisphere; and the pressure stresses, **I**, from within the ice cover. The equation for the conservation of momentum becomes

$$m\frac{d\mathbf{V}_i}{dt} = \tau_a + \tau_w + \mathbf{D} + \mathbf{G} + \mathbf{I} \tag{2.63}$$

where m is the ice mass per unit area and $\mathbf{V}_i$ is the ice velocity.

One of several formulations of water stress following McPhee (1975) is

$$\tau_w = \rho_w c_w |\mathbf{V}_d| (\mathbf{V}_d \cos\theta + \mathbf{k} \times \mathbf{V}_d \sin\theta) \tag{2.64}$$

where c_w is the ocean drag coefficient, and θ is the turning angle in the ocean boundary layer.

The wind stress term in (63) is specified from observed data or is given by the atmospheric component of a coupled climate model.

The Coriolis force is given as:

$$\mathbf{D} = \rho_i h_i f \mathbf{V}_i \times \mathbf{k} \tag{2.65}$$

where ρ_i is ice density, and the Coriolis parameter.

The stress from the dynamic topography, **G**, depends on the gradient ∇ of sea surface height:

$$\mathbf{G} = -\rho_i h_i g \nabla H \tag{2.66}$$

where g is the acceleration due to gravity, H is the sea surface height field, and ∇H is the vector gradient of that field.

The last term in (63), the stress, **I**, due to internal ice resistance. This term is very complex, because in a concentrated ice pack such as occurs in the central Arctic Ocean the internal stress can be large and it is difficult to know in detail how to handle them. This term has been modeled as a viscous, plastic, or elastic material, sometimes being considered incompressible and sometimes partly compressible.

Hibler (1979) devised a constitutive law for sea ice that treats it as a viscous/plastic medium such that it behaves as a linear viscous fluid for very small deformation rates and as a rigid plastic for larger deformation rates.

Later Zhang and Hibler (1997) devised an efficient numerical scheme based on line relaxation methods for solving the complex equation. Zhang et al. (1997) have shown some very high resolution sea ice/oceans results for the Arctic regions. They found remarkable agreement with observations. There is a new approach to sea ice, which takes into account an anelastic approximation by Hunke and Dukowicz (1997). This novel approach is likely to change the methodology for solving for the dynamic aspects of sea ice models and is computationally efficient, as well as, it is capable of simulating the rapid motions of sea ice from synoptic weather events.

2.7 Suggestions for Interesting Climate Model Websites

Note: some of this information is not updated, however, there is usually a useful description of ongoing climate modeling research and results from various modeling groups. There also are some interesting animations.

http://hydra.tamu.edu/~baum/climate_modeling.html

http://www.scd.ucar.edu/info/CCM.html

http://ww-pcmdi.llnl.gov/phillips/PCMDIoverview.html

http://vislab-www.nps.navy.mil/~braccio

http://everest.ee.umn.edu/~sawdey.micom.html

http://www.cgd.ucar.edu/ccr/pcm

http://www.cgd.ucar.edu/cms/csm

2.8 Future Development

It can be expected that in the future, climate models will continue to improve. Part of the improvement will come from having faster computers, which will allow for improved resolution when really addressed. The other part will be from better understanding of the components of the climate system. Historically, the primary components of three dimensional climate models were atmospheric models coupled to simple ocean, land and sea ice models. Now the component models are much more faithful simulators of the various components of the overall system. In fact, the present day models have land-vegetation-ecological systems coupled to the atmospheric component. Some climate models have chemical cycles in the ocean and the atmosphere such as the carbon and sulfur cycles. By adding these features and others, scientists can explore climate change caused by changes in solar radiation, atmospheric composition such as carbon dioxide, ozone and other greenhouse gases, atmospheric aerosols, land use, vegetation, and volcanism. Finally, it must be realized that climate models are tools that can be used to better understand the complex interactions in the climate system. These will always be non-perfect tools. These tools must be compared to extensive observational databases and when differences are identified, the discrepancies must be explained. In order to make advances, the use of observed data in the improvement of models is essential. The various international climate organizations have discussed the weaknesses of present generation climate models and they have developed scientific programs to address the uncertainties. A lowering of the model uncertainties and increased model spatial resolution will improve the climate simulations of not only the mean climate but the variability inherent in the climate system.

2.9 The Climate System Model and Parallel Climate Model

The Climate System Modeling (CSM) project was started in 1994 with the purpose of building and maintaining a comprehensive model of the climate system with respect to the physical system and, eventually, biochemical aspects of the climate system. The CSM is a set of component models made up of atmosphere, ocean, land surface and sea ice. These components are coupled through a flux coupler using message passing as a means of transferring fluxes in the computer. All of the fluxes such as sensible heat, latent heat, momentum and other fluxes pass through the flux coupler. This model is mostly supported by the United States National Science Foundation (NSF) and it will be widely used by the academic community. The above list of internet web sites has the address for the CSM.

Another climate model development that is briefly discussed is the Parallel Climate Model (PCM). This effort is associated with the CSM except that it is a distributed effort involving Los Alamos National Laboratory (LANL), the Naval Postgraduate School (NPS), and NCAR. The atmospheric component is the parallel version of the Community Climate Model (CCM) Version 3 and the Parallel Ocean Program (POP) model that was developed at LANL and a sea ice model component from NPS. These components are designed to run efficiently on massively parallel computers such as the CRAY T3D/E, parallel Hewlett Packard, and Silicon Graphics computers. Also, some of the other differences with the CSM are that it has a much higher resolution ocean and sea ice treatment. Again the update of the progress on this model can be found in the web site list mentioned above.

2.10 Acknowledgements

This work is supported by the National Science Foundation through a contract to the National Center for Atmospheric Research (NCAR) and by the National Aeronautics and Space Administration's (NASA) Goddard Space Flight Center and supported by NASA's Oceanic Processes Branch. A portion of the research involved is supported by interagency and cooperative agreements to NCAR from the Department of Energy (DOE), Office of Energy Research. The Parallel Climate Model Model Effort (PCM) is a joint effort to develop a DOE-sponsored PCM between Los Alamos National Laboratory, the Naval Postgraduate School and NCAR.

Chapter 3

Hydrological Modeling from Local to Global Scales

by Eric F. Wood

Abstract

There is a need for greater understanding of the seasonal, annual and interannual variability of water and energy cycles at continental-to-global scales, and in doing so, a greater understanding of the role of the terrestrial hydrosphere-biosphere in Earth's climate system. It is well known that land surface processes control local climate through the partitioning of precipitation and incoming radiation into their budget components. What is not well understood are the mean values and variability of the water and energy budget terms over a range of temporal and continental-to-global spatial scales. This chapter presents a multi-scale observational-modeling strategy and its testing for the development and validation of process-based, terrestrial water and energy balance models that can be applied from local point scales to global scales. This strategy includes (i) the development, calibration and validation of process-based hydrologic models using data from small scale land surface experiments; (ii) the use of the developed models to scale up to larger spatial scales on the order of 10^5 km^2, with foci on model parameterizations at regional scales and on testing remote sensing products in the hydrological models; and (iii) the simulation of continental-to-global fields of water and energy fluxes through modeling and remote sensing, and their validation using large scale hydrologic data.

3.1   Introduction

Understanding the role of the terrestrial hydrosphere-biosphere in Earth's climate system, especially in the coupling of land surface hydrologic processes to atmospheric processes over a range of spatial and temporal scales, the role of the land surface in climate variability and climatic extremes, and its role in climate change and terrestrial productivity are integral components of the World Climate Research Program's (WCRP) research plan under GEWEX, and in particular the GEWEX continental scale experiments (GCIP, BALTEX, MAGS, LBA, and GAME).

The inherent research strategy for WCRP/GEWEX for investigating these questions is through process-based, terrestrial water and energy balance models. There is the recognition that questions regarding terrestrial hydrology within the climate system can not be answered through ground-based observations alone - the historical record is too short and the required spatial and temporal observations are too extensive, costly and logistically impractical. Large scale applications of energy and water balance models are greatly complicated by the scarcity of land surface observations and difficulties in representing hydrological processes at large scales. This recognition in the early 1980's resulted in the establishment of climate experimental programs like HAPEX (Hydrology Atmospheric Parameterization EXperiment) and ISLSCP (International Satellite Land Surface Climatology Programme) as initiatives of the WMO/ISCU Joint Scientific Committee for WCRP. In the late 1980s, ICSU established a core project on the Biospheric Aspects of the Hydrological Cycle (BAHC) with a research focus on the spatial and temporal integration of biosphere-hydrosphere interactions.

Through these programs, the land surface climate modeling community has focused primarily on improved process models at small scales utilizing data collected through experiments like HAPEX-MOBILY, HAPEX-Sahel, and FIFE'87. Subsequent intercomparisons of these parameterizations, like the Project for the Intercomparison of Land Surface Parameterization Schemes (PILPS) (Henderson-Sellers et al. 1993, 1995), have been forced, by the lack of large scale data, to resort primarily the use of data collected at point or small area experiments.

The Project for Intercomparison of Land-surface Parameterization Schemes (PILPS), a joint research activity sponsored by the Global Energy and Water Cycle Experiment (GEWEX) and the Working Group on Numerical Experimentation (WGNE), has as its goal the improvement in land surface schemes used in climate and numerical weather prediction models. PILPS is especially concerned with the parameterizations of hydrological, energy, and momentum exchanges that are represented in these models. Its approach is to facilitate comparisons between models, and between models and observations, so as to diagnose shortcomings for model improvements. In general, there tends to be a mismatch between the measurement scale for data used to develop these models and the application scale within climate and weather prediction models. With the desire of having hydrologic process-based parameterizations, there is a need for systematic testing of models across a range of spatial and temporal scales.

At the regional to global scales, which are of concern to the WCRP/ GEWEX goals, there is the belief that remote sensing observations offer the potential to provide the required inputs for hydrological modeling. This belief supported remote sensing initiatives that has resulted in the recent availability of consistent, long-term remote sensing records, such as AVHRR (Agbu, 1993), GOES (Young, 1995) and SSM/I (Hollinger et al., 1992) pathfinder data sets; the compilation of remote sensing data as part of ISLSCP initiatives (Sellers et al., 1994, Meesen et al., 1995); recent advances in remote sensing algorithms for deriving forcing variables such as radiation, humidity and surface air temperature; and the development of new remote sensing instruments such as the Tropical Rainfall Measurement Mission (TRMM) for precipitation.

In this chapter, we review recent activities carried out by the author, and his research collaborators and students, in trying to develop models that provide the necessary water and energy flux data for in-depth analyses of the seasonal, annual and inter-annual variability of the terrestrial water and energy balances at regional to global scales. In the next section, we review the modeling strategy that integrates data collected from field experiments on hydrological processes, with catchment-to-regional hydrological modeling and finally, through remote sensing, to continental scales.

Following the explanation of the modeling strategy over these scales, section 3.3 presents results that demonstrate that process-based hydrological models can be accurately applied across a range of spatial scales-from field to regional scales. To date, there has been only limited progress in applying process-based hydrological models at continental-to-global scales. In this context, we mean utilizing either ground observations or remote sensing to estimate terrestrial water and energy fluxes and storage terms. The reasons for the lack of progress at these very large scales are threefold:

1. Insufficient data. To date there have been insufficient data, either ground based or remote sensing, for water and energy balance modeling at regional to continental scales. Specifically, operational data collection has focused primarily on precipitation, temperature, and streamflow, with less intensive data collection of other surface hydrometeorological variables. Until recently, there have been no consistent operational data collection programs for radiation, and even today these programs are limited to a few locations. Only recently has remote sensing-based solar incoming radiation become available through the GOES Pathfinder initiative.

2. Insufficient models. Process-based, coupled water and energy balance models, and the resulting insights they provide to climate variability and climate change, is a rather new interest for both hydrologists

and atmospheric scientists interested in land-atmospheric exchanges. Recent HAPEX and ISLSCP experiments have focused on improving process parameterization at small scales (scales from a "tower" up to about 100 km - small compared to continental scales.)

3. Insufficient validation data. Validation data at continental-to-global scales are limited, and the use of a water and energy balance climatology is insufficient for the needs of process-based hydrological modeling envisioned in the WCRP.

3.2   The Modeling Strategy Across a Range of Scales

The goal of estimating the terrestrial water and energy fluxes and storage terms at continental-to-global scales results in an observation-modeling strategy that consists of three elements:

- development, calibration and validation of a process-based hydrologic model using data at small scales,

- application of the model to scale-up to large scales, and

- validation of the large-scale modeling results using large scale hydrologic data.

Table 3.1 summarizes the modeling strategy as it was implemented over a range of modeling and observational scales.

The strategy takes advantage of data from a range of World Climate Research Programme (WCRP) experiments carried out with the objective of improved representation of key hydrological processes. These experiments include the First ISLSCP Field Experiment (FIFE), the Boreal Ecosystem-Atmospheric Study (BOREAS), Intercomparison of Land Parameterization Schemes (PILPS), as well as, NASA's Shuttle Radar Laboratory (SRL) as part of the SIR-C mission. The strategy utilizes data from the Red-Arkansas River basins available under the WRCP/GEWEX Continental-scale International Project (GCIP) to investigate the scaling of the parameterization to basin scales, and data reported by Schnur and Lettenmaier (1997) and Messen (1995) to develop an initial 15-year global water balance simulation.

Finally, initial testing of remote sensing data within the distributed water and energy balance is carried out over the 566,000 sq. km Red-Arkansas River basins (Dubayah et al., 1997). The above strategy has resulted in significant progress towards producing well-validated, distributed water and energy flux and storage fields at continental-to-global scales.

Table 3.1. Implementation of observation-modeling strategy

Modeling Observations	Scale (km^2)	Research focus	Research objective	Data set/ experiment
Small Scale	10^2 – 10^3	Process modeling	Land-atmosphere interactions	FIFE, BOREAS, SIR-C, PILPS
Medium Scale	10^4 – 10^5	Coupled models	Scaling	GCIP-SW
		Remote sensing	Test algorithms	Red Arkansas basins
Large Scale	10^5 – 10^6	Budget analysis	continental scale variability	PILPS-2c, ISLSCP (Schnur and Lettenmaier 1997)

3.3 Modeling Water and Energy Fluxes Across a Range of Scales [1]

To model consistently across a range of spatial scales, a fundamental modeling element is defined at the 'local' or grid-scale that consists of a local-scale process-based water and energy balance model, and these grids make up the spatially variable large scale domain. The spatially distributed model can be structured either as an explicitly distributed model, suitable for large scale modeling at fine resolution of distributed inputs such as GOES-based solar radiation and WSR 88D-radar estimated rainfall (Peters-Lidard and Wood, 1997; O'Neill et al., 1996), or as a statistically distributed model where spatial processes such as infiltration capacity and soil moisture are represented by their statistical distribution through a functional form (Liang et al., 1994; Peters-Lidard et al., 1997a). The development and testing of this modeling framework, and the parameterization of the underlying hydrological processes, is reported in Liang et al. (1996a, b); Peters Lidard et al. (1997a, b).

The primitive water balance equation is:

$$\Delta S = P - E - Q \tag{3.1}$$

where ΔS is the change in soil moisture over a specified time interval; P is precipitation; E represents evapotranspiration, which is the sum of evaporation from bare soil, E_S, and transpiration from vegetation, E_V; and Q is runoff, which is the sum of surface, or direct storm, runoff, and subsurface

---

[1]See also Section 2.4.

or base flow. The corresponding energy balance equation is

$$R_n = \lambda E + H + G \tag{3.2}$$

where R_n is net surface radiation (solar and longwave), λE is the latent heat, H the sensible heat, and G the ground heat flux.

Equations (3.1) and (3.2) are not directly usable for determining the terrestrial water and energy fluxes and states; the flux terms ($\Delta S, P, E, Q, H$ and G) must be parameterized in terms of hydrometeorological state variables, including the soil moisture profile, the surface, ground and near-surface air temperatures, and near-surface humidity. The purpose of Soil-Vegetation-Atmosphere-Transfer (SVAT) models is to provide the necessary parameterizations. Figure 3.1 shows the data required for such land surface models. Basically the inputs to the model can be divided into three categories:

- *Forcing variables needed to drive the model:* These include precipitation (both liquid and solid), incoming solar (shortwave and near-infrared), and down-dwelling longwave (thermal) radiation from the atmosphere.

- *Surface meteorological variables needed to parameterize evaporation, transpiration and sensible heat variables:* The required surface meteorology includes surface air temperature, surface humidity and surface wind.

- *Land surface, soil and vegetation variables needed to parameterize water balance processes:* These processes include interception of precipitation by vegetation and net throughfall; infiltration and bare soil evaporation; surface runoff; and base flow recession. For the energy balance, these variables are needed for canopy-scale transpiration and ground heat flux, including the energy processes associated with snow melting and frozen soils.

The output from the water and energy balance models include runoff, soil moisture, surface temperature, and latent, sensible and ground heat fluxes.

The parameterization of equations (3.1) and (3.2) is the focus of land surface SVAT model research, including the intercomparisons of such schemes under the World Climate Research Programme Project for the Intercomparison of Land Parameterization Schemes (PILPS). One such scheme is the Variable Infiltration Capacity 2-layer (VIC-2L) model (Liang et al., 1994), which was developed as an appropriate model for water and energy balance studies and for inclusion into general circulation models (GCMs). As compared to other SVATs, its distinguishing features are that it represents the subgrid variability in soil storage capacity as a spatial probability distribution, to which surface runoff is related (Zhao et al., 1977), and that base flow occurs from a lower soil moisture zone as a nonlinear recession (Dumenil and Todini, 1992). Transpiration from vegetation utilizes a Penman-Monteith

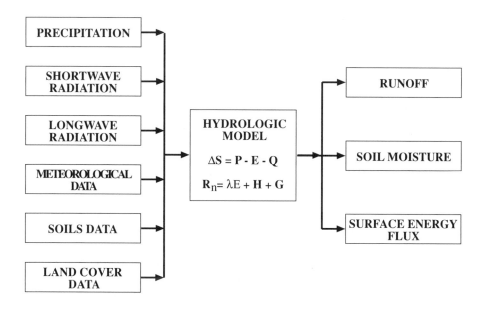

Fig. 3.1.

formulation with adjustments to canopy conductance to account for environmental factors following Jarvis (1976). This formulation is similar to that found in other schemes (see for example, Noilhan and Planton, 1989; Viterbo and Beljaars, 1995).

The VIC model has participated in the SVAT model intercomparisons under PILPS, where it has performed well relative to other schemes and to available observations (see Pitman et. al., 1993; Chen et al., 1997; Liang et al., 1996b; Lettenmaier et al., 1997; Pitman et al., 1997). When the VIC model is used to simulate only the water balance, for example, in large scale river basin studies (Nijssen et al., 1997; Wood et al, 1997), the model is driven by precipitation and air temperature (usually at the daily time step) and predicts surface and subsurface runoff, evaporation, and soil moisture. In cases where the coupled water and energy balance equations are required, but when direct observations of radiation forcing data are unavailable, estimates of the forcings are made based on air temperature and humidity as described in Nijssen et al. (1997). The model is well described by Liang et al. (1994, 1996b) and the application of the model to large basins is described in Abdulla et al (1996); Nijssen et al. (1997), and Wood et al. (1997). The reader is referred to these sources for a detailed description of the model.

Model developments have evolved through participation in PILPS and the application of the model to climate data collected as part of ISLSCP and HAPEX experiments. It is important for the reader to appreciate how data from these experiments have resulted in changes and improvements to the local model, and how these improvements affect the modeling at larger scales. As examples, the following three recent modifications are especially relevant to remote sensing applications at large scales:

- *Thin surface layer:* Recent results in PILPS simulations of the HAPEX-MOBILY data have shown that a thin surface soil layer can improve the model results during the summer period when dry conditions result in a soil control on evapotranspiration. The formulation and improvements have been reported in Liang et al., 1996b and Peters-Lidard et al., 1997a. This PILPS/HAPEX analysis resulted in the current VIC-3L version of the model.

- *Improved ground heat flux parameterization:* Accurate ground heat flux and surface temperature calculations are critical for accurate estimates of the other fluxes in the water and energy balance. It has been shown in Peters-Lidard et al. (1997b) that the often used McCumber and Pielke (1981) estimates of soil heat conductance can be better represented by using a formulation offered by Johansen (1975). This parameterization is now used in the model.

- *Sub-grid precipitation:* Including a sub-grid precipitation representation within land surface models having model resolutions greater than about 0.5° has been shown to be important (Shuttleworth, 1996). An efficient parameterization has been tested and included in the VIC-3L SVAT (Liang et al., 1996a).

3.3.1   Model Testing at Small Scales

The parameterizations of the VIC-3L model processes have been tested over the last few years through participating in the PILPS model intercomparison study where its performance has been extremely good (see Chen et al., 1997; Lettenmaier et al., 1997). In addition, the local-scale model processes have also been tested using data from FIFE (Peters-Lidard et al, 1997a), BOREAS (Pauwels and Wood, 1997) and Shuttle Radar Laboratory (SRL) on SIR-C over the Little Washita (Oklahoma) basin. Sample results are shown in Fig. 3.2-3.6.

Figure 3.2 and 3.3, using FIFE data, show the improvement in surface energy fluxes due to the improvements in the ground heat flux parameterization (Peters-Lidard et al., 1997b). Note that during the FIFE/IFC-3 period, the ground heat flux errors tend to result in offsetting sensible heat flux errors. Simulations of the IFC-4 period show that the ground heat flux errors have offsetting errors in the latent heat flux, showing the complexity of model interactions.

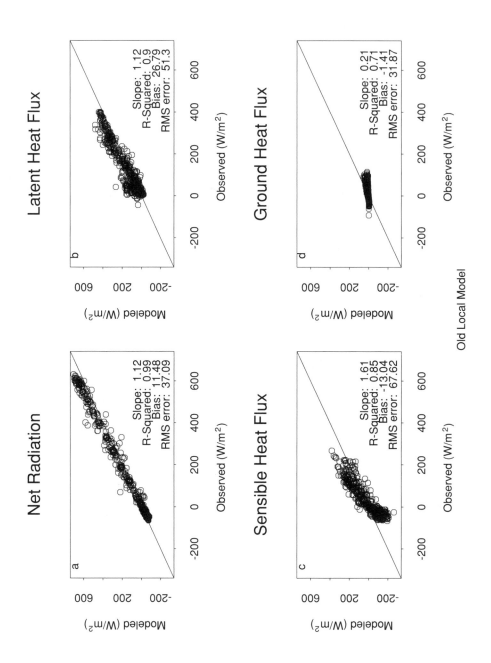

Fig. 3.2.

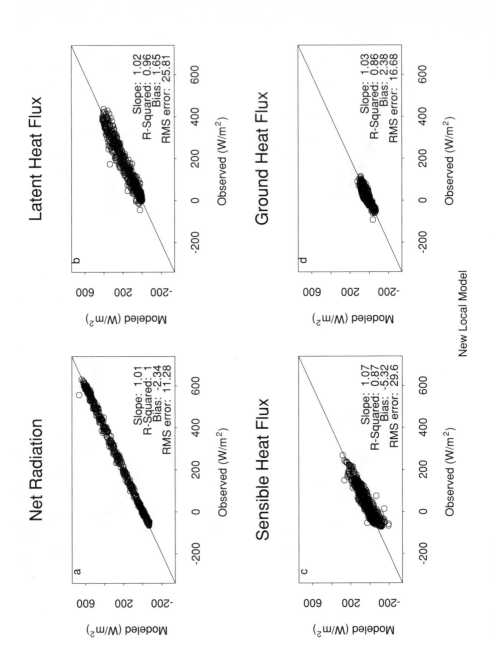

Fig. 3.3.

Figures 3.4 and 3.5 present local-scale modeling results for the ISLSCP/ BOREAS site in northern Canada (Sellers et al., 1995). Figure 3.4 presents a scatter plot that compares the model-derived fluxes with measured values for one particular 'tower' site: the Old Black Spruce (OBS) tower site the Southern Study Area (SSA) during IFC-1 of 1994. Figure 3.5 presents summary results for the Bowen ratio ($H/\lambda E$) and evaporative fraction ($\lambda E/P$) aggregated over the intensive field campaign (IFC) periods and for a number of different tower sites. As these figures demonstrate, the model provides good agreement in simulating water and energy fluxes at the local scale. High latitude sites like BOREAS are challenging because of high Bowen ratios in a non-water limited environment - compare them to those suggested in Fig. 3.2.

Simulating the dynamics of a surface soil layer, is a good test for the parameterization of the water fluxes (infiltration and evaporation) in equation (3.1). Figure 3.6 shows the model soil moisture dynamics during the 10-day April 1994 SIC-C mission for a range of fields in the Little Washita basin where detailed ground sampling took place. These results show that the model can predict soil moisture changes accurately at scales from 1 to 500 km^2. Currently the model uses a soil resistance formulation for soil evaporation. Similar performance has been reported in Peters-Lidard et al. (1997) for determining soil moisture over FIFE.

3.3.2   Recent Model Testing at Medium Scales

Application and testing of the model at medium scales has been reported in Abdulla et al. (1996); Nijssen et al. (1997); Wood et al. (1997); Lettenmaier et al. (1996) and Dubayah et al. (1997). Lettenmaier et al. (1996) report the results of a PILPS-2(c) intercomparison for the 566,000 km^2 Red-Arkansas River basins using 10-year, 1-hourly model forcing data (precipitation, solar and longwave radiation, and surface meteorology) interpolated to 1° x 1° grids. Figure 3.7 provides a basin-averaged, monthly summary of the performance of the VIC-3L model over the 1980-86 period. Model-estimated runoff was compared to gaged discharge, adjusted to account for reservoir effects, and model-estimated evapotranspiration to basin-wide evaporation estimated from a radiosonde-based atmospheric budget (for further details see Abdulla et al., 1996, and Lettenmaier et al., 1966). For this basin, the runoff ratio is estimated to be about 14.8 %. The model has a slight summertime bias of overpredicting runoff in the arid region of the basin. Based on the PILPS-2(c) intercomparison, there are plans for modifying the runoff parameterization for arid environments.

Initial testing of remote sensing products appropriate for utilization within a SVAT model was carried out using the VIC-3L model. The remote sensing products were derived from June 1987 GOES and AVHRR Pathfinder data for the Red-Arkansas River basins. GOES data were used to estimate incoming solar radiation. The AVHRR-derived variables were air temperature,

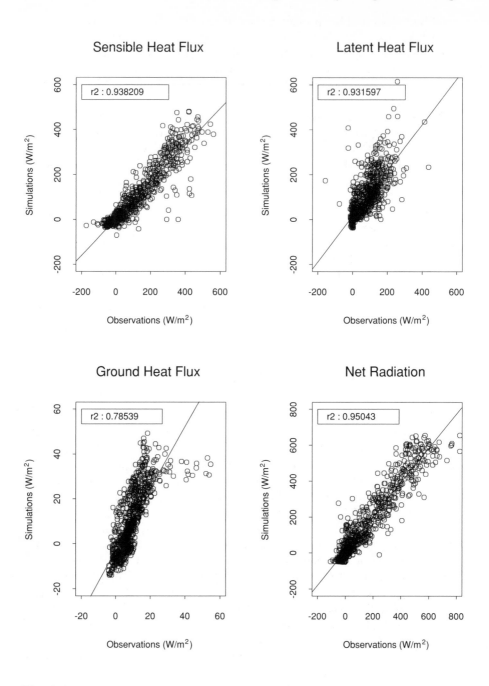

Fig. 3.4.

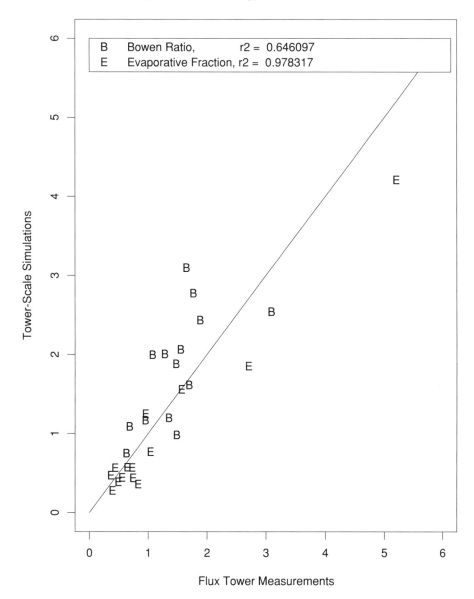

Fig. 3.5.

humidity and downwelling longwave radiation, which were time-interpolated to produce diurnal curves. Using these data, the VIC-3L model was run at a 3-hourly time-step and estimated the water and energy fluxes, which were then compared to fluxes estimated using ground-based data alone. (See

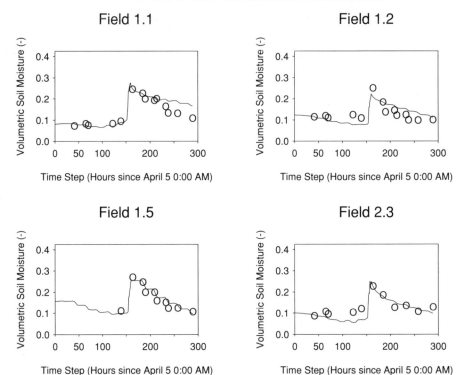

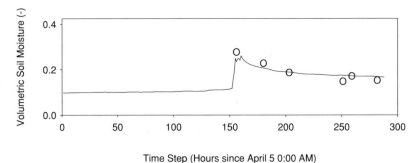

Fig. 3.6.

Dubayah et al., 1997 for details). A summary of the June, 1987 basin-average energy balance for the two model runs is presented in Table 3.2. From this

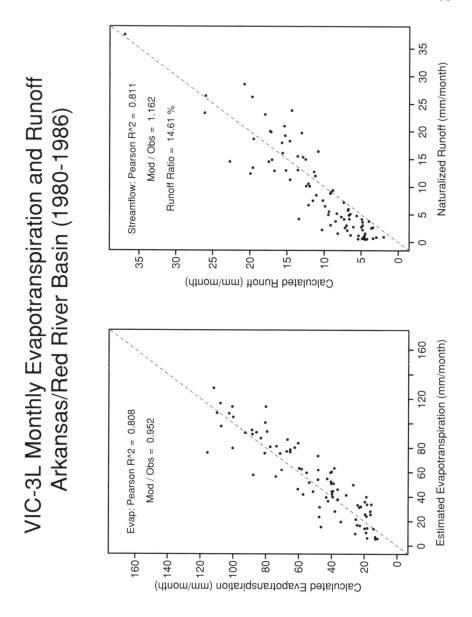

Fig. 3.7.

initial test, it appears that remote sensing data can be used to estimate radiation and important surface variables.

Table 3.2. Modeled basin average energy fluxes for June, 1987, using ground based meteorological forcings and remotely sensed forcings

	Net Radiation	Latent Heat	Sensible Heat
Ground-Based Forcings	190	106	81
Remotely-Sensed Forcings	242	112	125

3.3.3 Recent Model Testing at Global Scales

The VIC-2L water balance model was used to estimate the water and energy fluxes on a daily basis for 15 years, from 1979-1993. Ground data needed to derive the required forcing and parameterization fields were reported by Schnur and Lettenmaier (1997) and consisted of daily precipitation and minimum and maximum daily temperature. The ground data were interpolated and gridded to 2° x 2° resolution. Vegetation and soil data were based on ISLSCP data (Meesen et al., 1995). Figure 3.8 shows the model-derived climatology for the water balance components – runoff, evapotranspiration, soil moisture –, as well as, snow extent. Figure 3.9 shows examples of model-derived seasonal water balance climatology for six major river basins. Notice the variations in the fraction of precipitation that results in runoff, and the role of stored water (as either frozen precipitation or in soil moisture) and its seasonal cycle. While the figures report annual average values, the simulations were carried out on a daily basis and can therefore be used to investigate issues regarding seasonal and inter-annual variability of the components of the water balance on a spatial global scale analyses that address issues that are central to the World Climate Research Program's (WCRP) research plans, such as the role of the land surface in climate variability and climatic extremes. World Climate Research Programme (WCRP)

3.4 Remote Sensing of Environmental Forcing Variables

To truly move up to global scales, the hydrological community needs consistent time series of model inputs. The assumption underlying many international programs is that, this data will come through remote sensing. There has been great progress in the development of remote sensing techniques that provide data suitable for hydrological modeling. Approaches that predominantly rely on remote sensing are discussed by Dubayah et al. (1997), which reviews and summarizes current remote sensing capabilities and applies these techniques to a large basin in the Mississippi. The development of remote sensing algorithms and the integration of remote sensing data into macroscale

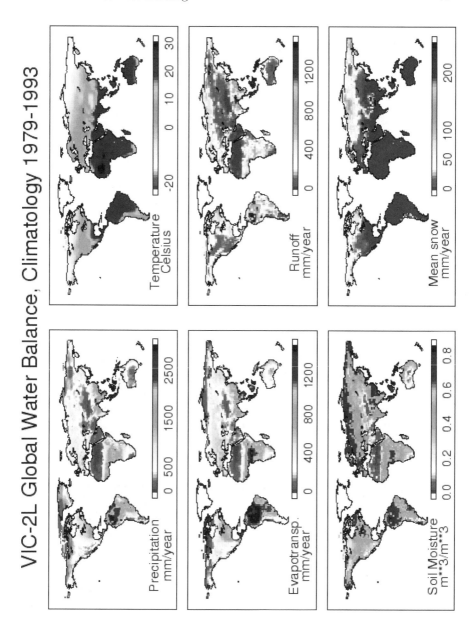

Fig. 3.8.

hydrological modeling is an important research area. A rigorous validation
of the data produced by these remote sensing techniques is of paramount

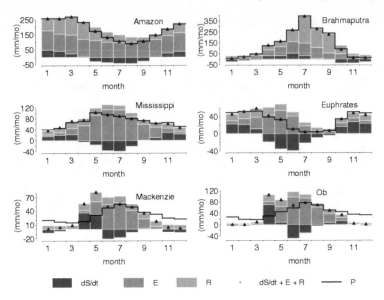

Fig. 3.9.

importance, if they are to be used with confidence in hydrological modeling. In the approach given in Dubayah et al. (1997), the following variables are derived using remote sensing and then subsequently used to drive the VIC-3L SVAT model: solar radiation; downwelling longwave radiation; air temperature; and surface humidity (vapor pressure deficit). Below we briefly describe the derivation of each.

- *Solar Radiation:* Incoming and net solar radiation are derived using the GOES-VISSR sensor, which combines high temporal resolution (30 minutes) with coarse (> 1 km) spatial resolution. Several physically-based algorithms exist for both clear and cloudy sky conditions, such as the benchmark code of Pinker and Laszlo (1992). Following Dubayah and Loechel (1997), the issue of the most appropriate aggregation size of GOES-derived insolation fields relative to the spatial modeling scale and temporal time step was resolved by using aggregation windows of 50 km to 100 km. (Subsequent analyses for the GIST/GEWEX area suggest slightly smaller window sizes for fluxes at shorter integration times. However, the effect of spatial and temporal resolution on hydrologic model outputs is unknown and of interest given the relatively coarse resolutions of existing solar archives.)

- *Downwelling Longwave:* Downwelling longwave radiation from the atmosphere to the land surface is difficult to estimate from remotely sensed data. It is dependent on several factors including clouds, air temperature, surface temperature, surface and atmospheric emissivity, near surface humidity, and the water vapor and temperature lapse rates. Statistical and physical methods may be used to estimate LW that include both cloudy and clear sky conditions. For example, radiances obtained from infrared sounders can be correlated with downwelling longwave (Wu and Cheng, 1989). Physically based methods use the vertical profile of water vapor, temperature, and cloud information, as obtained from sounders or radiosondes, to estimate the downwelling flux with radiative transfer models. In our work, we have used estimates of near-surface air temperature (from AVHRR and TOVS below), humidity (from AVHRR, GOES and TOVS below), and cloudiness (from GOES) in traditionally ground based methodologies that relate these variables to downwelling longwave (e.g. see Brutsaert, 1975).

- *Air Temperature:* Spatially-varying air temperature may be estimated using AVHRR data as shown in Goward et al. (1994), Prince and Goward (1995), Prihodko and Goward (1997), Prince et al. (1997), and Dubayah et al. (1997). Dubayah et al. (1997) have also shown how TOVS-derived air temperature may be used for hydrological modeling. Fields of air temperature derived from these sensors and compared to a variety of field campaign and surface airways data have shown good correspondence, but with somewhat large positive biases. The source of these biases and the spatial and temporal patterns of errors with respect to ground data are unknown and must be examined, if such data are to replace or supplement ground observations.

- *Surface Humidity:* Total column precipitable water may be derived in a variety of ways, including the split-window differential absorption techniques used in AVHRR and GOES methods, as well as, the sounding and data assimilation strategy using TOVS data (Dubayah et al., 1997). We use all data sources to obtain precipitable water, and then use empirical relationships that relate dew point temperature as a function of precipitable water and latitude to get near surface humidity. This humidity is then combined with air temperature (as derived above) to yield vapor pressure deficit (VPD). This technique has been applied over the entire Mississippi basin (Dubayah et al., 1997). However, because the derivation of VPD requires both air temperature and humidity, both of which in turn are dependent on the accurate retrieval of other variables such as surface temperature, which have their own errors, overall errors can be large and their spatial and temporal error propagation complex.

- *Precipitation:* Estimation of precipitation by satellite remote sensing remains problematic at time scales required by hydrologic models. While missions such as TRMM have great promise for improving climatological estimates of precipitation, and perhaps for estimating areal precipitation at monthly to seasonal time steps, the current (and planned) platforms suffer from an inability to observe the diurnal cycle directly. In the U.S., the new NOAA WSR-88D weather radars are now producing archived precipitation products at 4 km resolution. The most widely available (Level 3) products effectively incorporate some observing station data as well. We have used these products with some success over part of the Arkansas-Red basin (Arola et al., 1994). We are currently investigating the potential of satellite-based precipitation over land, where monthly estimates have been made based on SSM/I data (Adler et al., 1994; Huffman, 1996). Understanding the accuracy and usefulness of these products for land surface hydrological modeling is still unresolved.

3.5   Discussion and Summary

Over the last ten years there has been significant interest in furthering the understanding of the role of the land surface in the Earth's climate system, especially the coupling between the land surface hydrologic processes and atmospheric processes, and the role of the land surface in climate variability and climatic extremes. Hydrological processes are closely linked to terrestrial productivity. These questions have driven the research agenda of the World Climate Research Program (WCRP), and there is general agreement that these questions can be addressed through improved process-based, terrestrial water and energy balance models and remote sensing.

Nonetheless, this approach has not been widely tested and validated. This chapter presents a observation-modeling strategy that has lead to the development of coupled water and energy models that are consistent at spatial scales ranging from the local 'field' scale to continental and global scales. The strategy utilizes point scale field data collected as part of WCRP activities to develop process parameterizations at the 'local' scale. A grid-based representation of spatial complexity at larger scales is used where regional and continental scale catchments can be represented. At these larger scales, remote sensing algorithms are developed so that global-scale water and energy fluxes can be predicted at high spatial and temporal resolutions. Within five years it is expected that the NASA Earth Observing System (EOS) instruments will provide the necessary data - heralding a new era for terrestrial hydrology.

3.6   Acknowledgements

This material represents the work of students and research staff at Princeton, in particular Christa Peter-Lidard, Mark Zion and Val Pauwels, and collaborators at other universities (Dennis Lettenmaier, University of Washington, and Ralph Dubayah, U M D). The research was supported by NASA through contract NAS5-31719 and NOAA grant NA56GP0249. This support is greatly appreciated.

Chapter 4

Natural Climate Variability and Concepts of Natural Climate Variability

by Jin-Song von Storch

Abstract

Concepts of climate variability are formulated based on two frequently used definitions. One considers climate variability as the evolution of the mean climate conditions over time, the other considers second moments or frequency-decomposed covariances (i.e. cross-spectrum matrices) around the mean to be characteristic of climate variability. Variability related to the mean becomes appreciable when considering paleo time scales, whereas covariance or frequency-decomposed covariances describe the key aspect of the variability of the present-day climate, for which the mean conditions can be assumed to be stationary. Different concepts of natural climate variability can be classified according to these definitions.

Before discussing the concepts of climate variability, two estimates of frequency-decomposed covariances of the present-day climate derived from observed and simulated data are presented. The dominant features of cross-spectrum matrices of atmospheric variables can be described by standing patterns with red spectra. The most anisotropic pattern has the reddest spectrum. The features related to the El Niño Southern Oscillation phenomenon are characterized by propagating patterns with a peak spectrum. The multivariate spectra are smooth and continuous, indicating that the randomness is a dominant feature of the considered climate variables.

Typical concepts dealing with both changes in the mean and frequency decomposed covariances are discussed. For the former, Croll's Earth orbit theory (also known as the Milankovich theory) and the multiple-equilibria of the deep ocean circulation, are reviewed. For the latter, a mechanism of the generation of covariances of slow climate variables is considered.

The spatial variability is normally studied using simplified equations of motions. As an example, the processes responsible for the dominant spatial features of the two reddest modes identified in the ECHAM1/LSG atmosphere are studied.

4.1 Introduction: Definitions of Climate Variability

The term "climate variability" is frequently used by meteorologists, oceanographers and paleoclimatologists, without being explicitly defined. Even though they all refer to "climate variability" as something within the climate system, which fluctuates with time, a more careful consideration of their arguments suggests that distinctly different definitions of "climate variability" are assumed by different groups of climate researchers to formulate various concepts of climate variability. It is worthwhile to first consider the most frequently used definitions.

In order to define climate variability, one must first describe the climate system using state variables, such as temperature, velocity, and pressure. These variables are generally vectors whose dimensions correspond to the numbers of grid points needed to represent the state variables in space. A climate state variable may be denoted by

$$\mathbf{v}(t) = \begin{pmatrix} v_1(t) \\ v_2(t) \\ \dots \\ v_M(t) \end{pmatrix}$$

where the subscript i indicates the i-th grid-point, M is the total grid points, and t is time.

The most intuitive definition of climate variability is to consider it as the evolution of the state variables over time.

> Definition 1: $\mathbf{v}(t)$, for $t = -\infty...\infty$

If the state variables are one-dimensional, climate variability is defined by uni-variate time series. Figure 4.1 shows some time series of observed and simulated variables. Figure 4.1a shows monthly anomalies of sea surface temperature (SST) averaged over a small area in the equatorial eastern Pacific derived from the GISST (Global Ice and Sea Surface Temperature) data set. This quantity characterizes the El Niño Southern Oscillation (ENSO) phenomenon in the tropical Pacific. Figure 4.1b shows monthly anomalies of 30-hPa zonal velocity at the equator obtained from station data (Naujokat, 1986) describing the so-called quasi-biennial oscillation (QBO) in the equatorial stratosphere. Figure 4.1c shows monthly anomalies of the difference between North Atlantic subtropical high surface pressure (centered near the Azores) and the subpolar low surface pressure (extending south and east of Greenland), which is also referred to as the North Atlantic Oscillation (NAO) Index. Figure 4.1d represents yearly anomalies of zonally averaged stream function in the Atlantic near 30°S at about 2000 and 3000 meter depth, the region of the deep outflow of Atlantic overturning circulation. The time series is derived from a 1260-year integration with the coupled ECHAM1/LSG

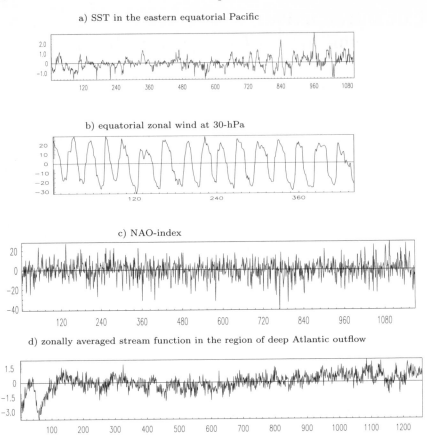

a) SST in the eastern equatorial Pacific

b) equatorial zonal wind at 30-hPa

c) NAO-index

d) zonally averaged stream function in the region of deep Atlantic outflow

Fig. 4.1. Time series of anomalies of monthly sea surface temperature (SST) averaged over a small area in the eastern equatorial Pacific for the period 1903 - 1994 obtained from GISST data set in a), composed monthly 30-hPa zonal wind for the period 1953 - 1989 in b), monthly difference between North Atlantic subtropical high surface pressure (centered near the Azores) and the subpolar low surface pressure (extending south and east of Greenland) for the period 1899 - 1996 in c), and zonally averaged stream function in the Atlantic near 30°S at 2000 to 3000 meter depth obtained from the 1260-year integration with the coupled ECHAM1/LSG general circulation model in d). SST is in degrees C, zonal wind in meters per second, pressure in hPa, and Atlantic stream function in Sverdrups (1 Sverdrup=10^6 m^3/s)

general circulation model (GCM) (von Storch et al., 1997). Anomaly time series are obtained by subtracting averaged annual cycles for monthly data, and the 1260-year mean for the yearly time series.

ENSO and the stratospheric QBO are two oscillatory phenomena, which can be clearly seen in the observational data. The oscillatory behavior is much more pronounced in Fig. 4.1b than in Fig. 4.1a. The time series of NAO-Index on the other hand represents the typical temporal behavior of many atmospheric time series. Without a time filter, no preferred oscillation can be identified. Figure 4.1d describes properties of the deep ocean for which no observations are available. Whether or not a pronounced oscillation is present, all time series shown in Fig. 4.1 have in common that they are all irregular in time; a given segment of a time series in Fig. 4.1 never exactly repeats itself during the considered time period.

Strictly speaking, if we stick to Definition 1, climate variability should be described by vector time series rather than point time series. Figures 4.2 - 4.4 show sequences of four consecutive anomaly fields of SST, 300-hPa geopotential, and zonally averaged stream function as functions of latitude and depth in the Atlantic. The latter two are obtained from the 1260-year integration with the coupled ECHAM1/LSG model. Even though SST anomalies in Fig. 4.3 are more persistent than anomalies of 300-hPa geopotential height in Fig. 4.2, month-to-month changes in the geographical distribution of anomalies are apparent for both variables. The stream function of Atlantic in Fig. 4.4 reveals also notable year-to-year changes in spatial structures. Generally, in addition to temporal irregularity, spatial irregularity evolving over time also influence climate variability. A video of these maps would document these irregularities more vividly.

Because of these irregularities, Definition 1 is, although it is the most complete definition, not helpful in making any conclusive statement about climate variability. The common feature of climate time series, namely that a given sequence of $\mathbf{v}(t)$ never repeats itself, indicates that climate state variables behave like random variables. For such variables, it is more appropriate to consider the averaged properties of $\mathbf{v}(t)$ rather than the individual time evolutions of $\mathbf{v}(t)$. For this purpose, the following mathematical construct is adapted.

> A climate state variable $\mathbf{v}$ is a multivariate random variable, and a time series $\mathbf{v}(t)$ of a climate variable is generated by a stochastic process.

This construct is consistent with the random character of climate state variables. Consider an ensemble of earth climate systems, which evolve over time under exactly the same conditions. $<\ >$ denotes the average over this ensemble. Climate variability can then be defined as the ensemble-averaged behaviors of $\mathbf{v}(t)$. One aspect of these averaged behaviors is defined by

> Definition 2: $< \mathbf{v}(t) >$

In statistics, $< \mathbf{v}(t) >$ is referred to as the first moment of the random variable $\mathbf{v}(t)$.

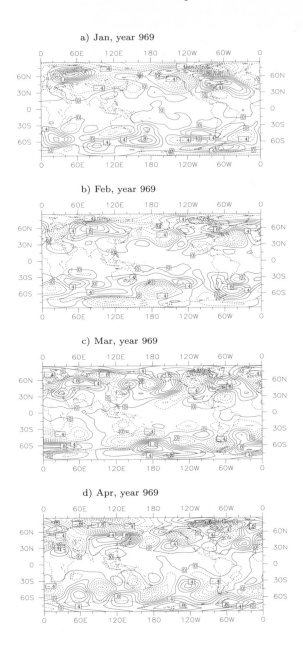

Fig. 4.2. A four-month sequence of fields of 300-hPa geopotential height anomalies obtained from the ECHAM1/LSG integration (Contour interval: 100 gpm)

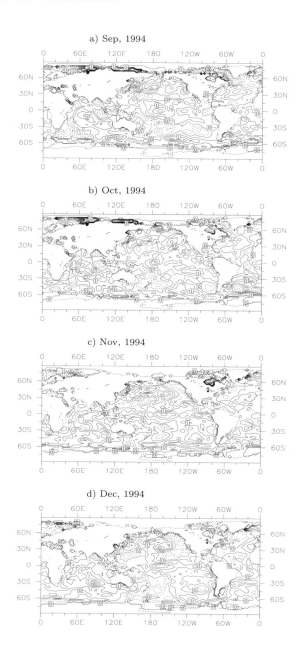

Fig. 4.3. A four-month sequence of fields of SST anomalies obtained from GISST data set (Contour interval: 0.5 degrees C)

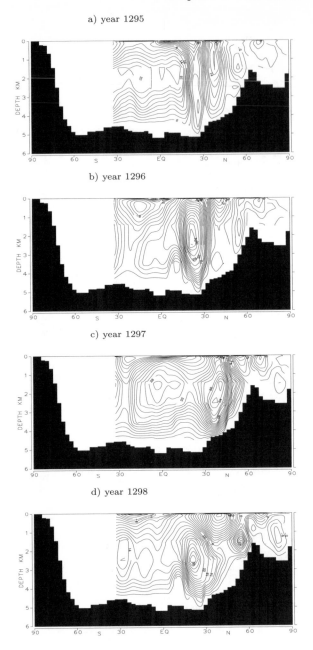

Fig. 4.4. A four-year sequence of meridional-vertical sections of zonally averaged stream function anomalies in Atlantic obtained from the ECHAM1/LSG integration (The anomalies are plotted in streamfunction x 20 Sverdrups)

Definition 2 is often used, although usually not explicitly outspoken, by paleoclimatologists and oceanographers when time evolution of *mean* climatic conditions, such as changes of mean summer temperature or changes of mean deep ocean circulation from a glacial period to the following interglacial period, are concerned. Normally, the observed realization of $\mathbf{v}(t)$ is considered as an estimator of $< \mathbf{v}(t) >$. Sometimes, if $< \mathbf{v}(t) >$ does not change much over time period $(t - T/2, t + T/2)$, the time average $< \quad >_T$ over period T is used to estimate $< \mathbf{v}(t) >$ for $t \in (t - T/2, \; t + T/2)$.

Definition 2 is not a definition of climate variability that is typically used by atmospheric scientists. They are more interested in fluctuations around a mean climate state. Denoting anomaly by

$$\mathbf{v}'(t) = \begin{pmatrix} v_1'(t) \\ \vdots \\ v_M'(t) \end{pmatrix}$$

with $v_i'(t) = v_i(t) - < v_i(t) >$, variability referred by atmospheric scientist is defined by

$$\boxed{\text{Definition 3: } \mathcal{V} = < \mathbf{v}'(t)\mathbf{v}'(t)^\dagger >}$$

where $\dagger$ denotes the vector transposition. Sometimes normalized anomalies

$$\mathbf{v}''(t) = \begin{pmatrix} v_1''(t) \\ \vdots \\ v_M''(t) \end{pmatrix}$$

with $v_i''(t) = \frac{v_i(t) - < v_i(t) >}{\sigma_i}$ and $\sigma_i = < (v_i(t) - < v_i(t) >)^2 >^{1/2}$ are considered. In this case, variability is defined by

$$\boxed{\text{Definition 3a: } \mathcal{C} = < \mathbf{v}''(t)\mathbf{v}''(t)^\dagger >}$$

$\mathcal{V}$ is referred to as the second moment of the random variable $\mathbf{v}(t)$. $\mathcal{V}$ is the covariance matrix and $\mathcal{C}$ is the correlation matrix of $\mathbf{v}(t)$. They are given by

$$\mathcal{V} = \begin{pmatrix} < v_1'^2(t) > & < v_2'(t)v_1'(t) > & \cdots \\ < v_1'(t)v_2'(t) > & < v_2'^2(t) > & \cdots \\ \vdots & \vdots & \ddots \end{pmatrix}$$

$$\mathcal{C} = \begin{pmatrix} < v_1''^2(t) > & < v_2''(t)v_1''(t) > & \cdots \\ < v_1''(t)v_2''(t) > & < v_2''^2(t) > & \cdots \\ \vdots & \vdots & \ddots \end{pmatrix}$$

The elements of $\mathcal{V}$ and $\mathcal{C}$ are familiar. The i-th diagonal element of $\mathcal{V}$ is the variance of $v_i(t)$. A map of diagonal elements of $\mathcal{V}$ describes the geographical distribution of the variance of $\mathbf{v}(t)$. Sometimes, the square root of variance, i.e. standard deviation, is considered. Figure 4.5 shows standard deviations of SST from GISST data set, 300-hPa geopotential height and zonally averaged meridional stream function in the Atlantic in the coupled ECHAM1/LSG model. The regions with the largest standard deviations in Fig. 4.5 coincide with regions with large anomalies seen in the time series in Fig. 4.2 - 4.3.

The off-diagonal elements of $\mathcal{V}$ and $\mathcal{C}$ represent, respectively, covariances and correlations of $\mathbf{v}(t)$ at two different locations. A map of the i-th column of $\mathcal{C}$, referred to as a teleconnection map, describes the correlations of a state variable $\mathbf{v}(t)$ at all grid points with the same variable at the i-th grid point. Teleconnection maps for the variables considered in Fig. 4.5 are shown in Fig. 4.6. The reference points in each map can be easily recognized by the maximum positive correlation of about 1. In Fig. 4.6c, $correlation \times 20$ is plotted. The wave train over Greenland and the North Atlantic in Fig. 4.6a resembles the observed West North Atlantic pattern (Wallace and Gutzler, 1981). Figure 4.6b characterizes the averaged spatial scales of El Niño events and the out-of-phase relationship between anomalies in the central east Pacific and anomalies to the west, north and south of these anomalies. Figure 4.6c describes to what extent variations in the region of deep Atlantic outflow in the ECHAM1/LSG model is related to variations in other regions, such estimates as the sinking region in the northern North Atlantic. The terms "deep Atlantic outflow" and "sinking in the northern North Atlantic" characterize the mean present-day deep ocean circulation. The first two hundred years, during which some initial oscillations occurred due to the coupling shock (see Fig. 4.1d), are excluded from the calculation. The correlations shown in Fig. 4.6c decrease from about 0.9 near 30^oS (which is represented by the isoline 1b in Fig. 4.6c) to about zero near 60^oN, indicating that variations in the outflow region are not coupled to those in the sinking region of the mean circulation.

The most effective way to decompose $\mathcal{V}$ is to consider the eigenvectors of $\mathcal{V}$, referred to as Empirical Orthogonal Functions (EOFs). It is well known that ordering the eigenvectors of $\mathcal{V}$ according to the amplitudes of the corresponding eigenvalues results in leading EOFs that are most efficient in representing the total covariance of $\mathbf{v}(t)$. Examples of EOFs are given in section 4.2.2.

Even though both Definitions 2 and 3 describe average behaviors of $\mathbf{v}(t)$, they provide two different kinds of information. Definition 2 represents (the evolution over time of) the mean climate and Definition 3 represents fluctuations around this mean climate. Paleoclimatic proxy data, such as $\delta^{18}O$ records from deep-sea cores, indicate that mean climate conditions have been changed significantly with time. Thus, for the past paleoclimate, both first

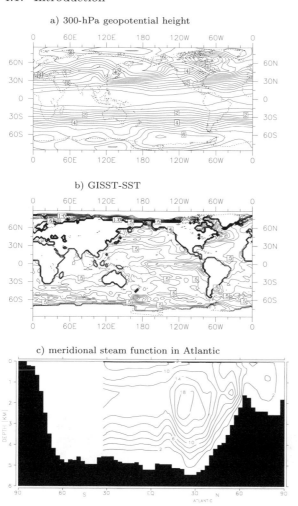

a) 300-hPa geopotential height

b) GISST-SST

c) meridional steam function in Atlantic

Fig. 4.5. Standard deviations of 300-hPa geopotential height, SST and zonally averaged stream function, which are considered in Fig. 4.1-4.4. Geopotential height is in 10 meters, SST in degrees C. For Atlantic stream function, standard deviations x 20 is plotted in c)

and second moments of $\mathbf{v}(t)$ are needed to describe climate variability. In general, the available paleo proxy data are so limited temporally and spatially, that they provide only a few hints about mean $< \mathbf{v}(t) >$, and no reasonable information about covariance $< \mathbf{v}'(t)\mathbf{v}'(t)^{\dagger} >$. For the present-day climate, which is essentially stationary, $< \mathbf{v}(t) >$ is assumed to be time independent. In this case, *covariance matrix $\mathcal{V}$ is the key quantity that characterizes climate*

a) 300-hPa geopotential height

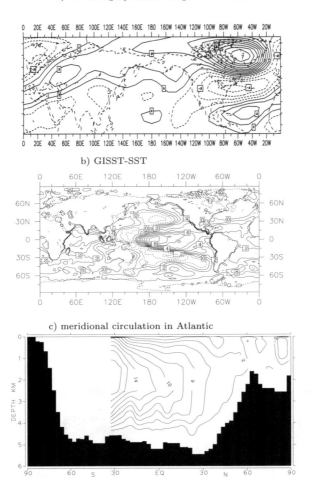

b) GISST-SST

c) meridional circulation in Atlantic

Fig. 4.6. Teleconnection patterns for the three variables considered in Fig. 4.5. As indicates by correlations close to one, the reference point is in the North Atlantic in a), in central equatorial Pacific in b) and near 30°S at 2000-3000 meter depth in c). The interval of isolines is 0.1 in a) and b). In c), correlation x 20 is plotted

variability. The ergodicity assumption used for the consideration of present-day climate variability allows the replacement of the ensemble average $< >$ by the time averaged $< >_T$.

Section 4.2 considers variability around the present climate state, as defined by the covariances $\mathcal{V} =< \mathbf{v}'(t)\mathbf{v}'(t)^{\dagger} >$ of the state variables of the present climate. Additional second moment statistics, i.e. lagged covariance matrices and cross-spectrum matrices, are introduced to deal with the frequency decomposition of the total covariance $\mathcal{V}$. Section 4.3 reviews some well-established concepts dealing with temporal variability of climate. These concepts are based on different definitions of climate variability: some concepts are constructed to explain the temporal changes of $< v(t) >$; others to explain the full second moment statistics, i.e. spectra; still others are concentrated on the individual evolution, rather than on the averaged behavior of $v(t)$. The concepts discussed in section 4.3 do not deal with spatial variability. Since different phenomena (or modes) reveal different spatial characteristics, the processes responsible for different spatial characteristics may be distinctly different. The basic idea and an example of how to deduce processes, which are responsible for observed spatial features, is discussed in section 4.4. A discussion and summary is given in the final section.

4.2 Estimation of Frequency-Decomposed Covariances of the Present-Day Climate

In this section, variability of the present-day climate is considered. Since $< \mathbf{v}(t) >$ is assumed to be essentially time independent under present-day climate condition, we concentrate on the second moments, i.e. covariance matrix $\mathcal{V} =< \mathbf{v}'(t)\mathbf{v}'(t)^\dagger >_T$. $\mathcal{V}$ describes the total temporal covariance. In order to decompose it into covariances on different time scales, we need to consider the lagged covariance matrix $\Phi(\tau)$, a function of time lag τ or its Fourier transform, i.e. cross-spectrum matrix $\Gamma(\omega)$, which is a function of frequency ω. Before we study the frequency decomposition of the total covariance $\mathcal{V}$, it is useful to first consider its analog for univariate state variables in section 4.2.1. Moreover, by considering the properties of the Fourier estimator of spectrum of a univariate time series, section 4.2.1 studies the basic feature of climate spectra. It will be shown that similar to the time-domain consideration given in the introduction, the frequency-domain consideration also suggests that randomness is the most dominant feature of climate variability.

Section 4.2.2 deals with the main purpose of this paper, namely the estimation of frequency-decomposed covariances of multivariate climate variables, which describe the key aspects of the variability of the present-day climate. A few techniques used to decompose total covariances are discussed. The dominant features of frequency-decomposed covariances as derived from a 20-year observed record and from a 500-year numerical model integration are described. The biases of these estimates, which limit our knowledge about second moment statistics of the present-day climate, are discussed.

Sections 4.2.1 and 4.2.2 consider only stationary and ergodic processes. The ensemble average $< \quad >$ will therefore be replaced by the time average $< \quad >_T$.

4.2.1   Univariate Case

For a univariate time series v'_t (t in the subscript counts numbers of time increment Δ of a discrete time series), covariance matrix $\mathcal{V}$ reduces to a number $< v'^2_t >_T$. The decomposition of variance $< v'^2_t >_T$ into variances on different time scales is done using spectrum $\gamma(\omega)$, which is the Fourier transform of lagged covariance function $\phi(\tau)$

$$\gamma(\omega) = \sum_{\tau=-\infty}^{\infty} \phi(\tau) e^{-i2\pi\tau\omega} \tag{4.1}$$

with

$$\phi(\tau) = < v'_t v'_{t+\tau} >_T \tag{4.2}$$

Since $\phi(\tau)$ is the inverse Fourier transform of $\gamma(\omega)$, one has

$$\phi(\tau) = \int_{-0.5}^{0.5} \gamma(\omega) e^{i2\pi\omega\tau} \, d\omega \tag{4.3}$$

Setting $\tau = 0$ into (4.3) yields

$$\phi(0) \equiv < v'^2_t >_T = 2 \int_0^{0.5} \gamma(\omega) \, d\omega \tag{4.4}$$

Thus, the spectrum $\gamma(\omega)$ of a discrete time series decomposes the total variance into variances across the frequency interval $[0, 0.5]$. Like the variance $< v^2_t >_T$, the spectrum $\gamma(\omega)$ is an averaged (or expected) quantity. Sometimes $\gamma(\omega)$ is also referred to as the second moment of v_t.

In practice, one deals with a finite realization of a time series v'_t. A finite time series v'_t of length N [1] can be expanded into a finite series of trigonometric functions

$$v'_t = \sum_{k=1}^{(N-1)/2} (a_k \cos(2\pi\omega_k t) + b_k \sin(2\pi\omega_k t)) \tag{4.5}$$

where $\omega_k = k/N$. Defining the periodogram I_k by

$$I_k = \frac{N}{4}(a_k^2 + b_k^2) \tag{4.6}$$

the sample variance $\frac{1}{N}\sum_t v'^2_t$ can then be written as

$$\frac{1}{N}\sum_{t=1}^{N} v'^2_t = \frac{2}{N} \sum_{k=1}^{(N-1)/2} (I_k) \tag{4.7}$$

Equation (4.7) suggests that $I_k = I_k(\omega_k)$, which decomposes the sample variance, can be considered as an estimator of spectrum $\gamma(\omega)$. However,

---

[1] For mathematical convenience, N is assumed to be odd.

Jenkins and Watts (1968) pointed out that as the length of time series increases, the behavior of I_k of a random time series differs distinctly from the behavior of I_k of a deterministic time series. For a random time series, the variance of $I_k(\omega_k)$ does not converge in a statistical sense to any limiting value as N tends to infinity. In other words, the periodogram $I_k(\omega_k)$ is not a consistent estimator of the spectrum $\gamma(\omega)$ defined in (4.1). In contrast, $I_k(\omega_k)$ of a deterministic time series converges to the spectrum $\gamma(\omega)$.

Intuitively, the lack of consistency results from the following. The information contained in $I_k(\omega_k)$ is spread over a band of frequencies with an effective width $\pm 1/N$ about ω_k. As N increases, the total information contained in the periodogram is distributed over an increasing number of bands with decreasing widths. Consequently, as N increases, the number of parameters (i.e. the number of I_k for all k) to be estimated increases, so that the efficiency of the estimate of each parameter (i.e. I_k at one particular frequency ω_k) does not improve. This situation cannot occur for a deterministic time series. Consider the time series produced by a sin function with period ω_o. For this time series, the number of parameters to be estimated does not increase with increasing length of time series N, $I_k(\omega_k)$ converges to a line spectrum.

The above discussion suggests that the true spectrum of a random time series is distinctly different from the true spectrum of a deterministic time series. The former must be smooth and continuous to ensure the frequency spreading of variance, whereas the latter has non-zero values at only a few discrete frequencies. As a consequence, the characteristic of a random spectrum is given by *the form* of the spectrum, rather than by spectral values at particular frequencies.

A climate time series, which is generated by processes with many degrees of freedom and reveals variations at many different frequencies, behaves just like a random time series. For such a time series, the second moment is more adequately described by the continuous shape of the spectrum, rather than by discrete spectral values.

We now consider the spectra derived from the time series shown in Fig. 4.1. The "chunk" spectral estimator, that is $I_k(\omega_k)$ averaged over m consecutive chunks, is used to obtain a consistent estimate. The number of chunks chosen is either two or three, depending on the length of time series. The vertical bars indicate the 95% and 99% confidence intervals. They characterize the variations of the "chunk" spectral estimator around the true spectrum.

The spectra are shown in Fig. 4.7 and 4.8. They represent three types of climate spectra. The first, shown by the spectra of eastern equatorial SST and 30-hPa zonal wind in Fig. 4.7a and b, reveals a $1/\omega^2$ high-frequency slope, which ends with a peak at a lower frequency. The peak is more significant for the QBO time series. The second, shown by the spectrum of the NAO-index in Fig. 4.8a, corresponds to a white noise spectrum. The third, the spectrum of meridional stream function in the region of deep Atlantic outflow (Fig.

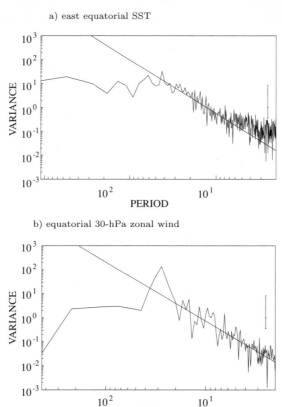

Fig. 4.7. Spectra of the first two time series shown in Fig. 4.1. The vertical bars indicate the 95% and 99% confidence intervals. The two straight lines are proportional to $1/\omega^2$

4.8b), reveals a $1/\omega$ slope at most of the resolved frequencies. As in Fig. 4.6c, the first two hundred years are removed from the analysis.

All these spectra have in common, that they are smooth and continuous. For instance, if randomness were absent, the spectrum of the QBO time series would be a delta function with a peak at the period of the QBO, which becomes infinite as the length of the time series tends to infinity. Variance would concentrate at one frequency, rather than be spread over other frequencies. Figures 4.7 and 4.8 can be taken as additional evidence supporting the idea that randomness is a dominant feature of climate variability.

Before we start to discuss the multivariate problem, we consider the spectrum of a time series generated by a first-order auto-regressive (AR(1)) process. Many climate time series can be approximately described by an AR(1) process. In this case, the spectrum of the corresponding time series can be estimated by the spectrum of the AR(1) process fitted to the time

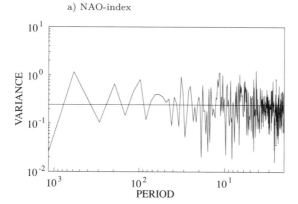

a) NAO-index

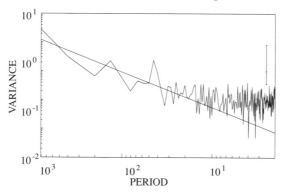

b) stream function in the Atlantic outflow region

Fig. 4.8. Spectra of the last two time series shown in Fig. 4.1. The vertical bars indicate the 95% and 99% confidence intervals. The two straight lines are proportional to $1/\omega^0$ and $1/\omega^1$, respectively

series. The advantage is that the full second moment is then described by only a few parameters of the process.

If $v'(t)$ is generated by an AR(1) process, then

$$v'_{t+\Delta} = \alpha v'_t + r_t \tag{4.8}$$

where r_t is a white noise with variance σ_r^2. The lagged covariance function $\phi(\tau)$ of such a process is

$$\phi(\tau) = \phi(0)\alpha^\tau \tag{4.9}$$

substituting (4.9) into (4.1) yields

$$\gamma(\omega) = \frac{\sigma_r^2}{1 - 2\alpha\cos(2\pi\omega\Delta) + \alpha^2} \tag{4.10}$$

where ω is in the unit of cycle per time increment Δ. For $\omega \ll 1/(2\pi\Delta)$, $\cos(2\pi\omega\Delta) \simeq 1 - (2\pi\omega\Delta)^2/2$, (4.10) becomes

$$\gamma(\omega) = \frac{\sigma_r^2}{(1-\alpha)^2 + \alpha(2\pi\omega\Delta)^2} \tag{4.11}$$

Equation (4.11) shows two characteristic spectral behaviors

$$\gamma(\omega)/\sigma_r = \begin{cases} \frac{1}{\alpha(2\pi\Delta)^2\omega^2} & \text{for } \omega \gg \omega^* \\ \frac{1}{(1-\alpha)^2} & \text{for } \omega \ll \omega^* \end{cases} \tag{4.12}$$

where

$$\omega^* = \left(\frac{1-\alpha}{2\pi\alpha^{1/2}}\right)\frac{1}{\Delta} \tag{4.13}$$

At high frequencies with $\omega \gg \omega^*$, $\gamma(\omega)$ increases with decreasing frequency at the rate of $1/\omega^2$. At low frequencies with $\omega \ll \omega^*$, $\gamma(\omega)$ reaches a constant level. The smaller ω^* is, the larger is the frequency range over which $\gamma(\omega)$ is able to increase with decreasing ω, and the higher is the final low-frequency spectral level. Thus ω^*, a function of α, determines the entire shape of the spectrum.

4.2.2   Multivariate Case

For a multivariate variable $\mathbf{v}_t$, the frequency decomposition of the total covariance into covariances on different time scales requires the consideration of the cross-spectrum matrix $\Gamma(\omega)$, which is the Fourier transform of the lagged covariance matrix $\Phi(\tau)$

$$\Gamma(\omega) = \sum_{\tau=-\infty}^{\infty} \Phi(\tau)e^{-i2\pi\tau\Delta\omega} \tag{4.14}$$

with

$$\Phi(\tau) = < \mathbf{v}'_t \mathbf{v}'^{*\dagger}_{t+\tau\Delta} >_T \tag{4.15}$$

where $*$ denotes complex conjugation. There is an inverse transformation

$$\Phi(\tau) = \int_{-0.5}^{0.5} \Gamma(\omega)e^{i2\pi\tau\Delta\omega} \, d\omega \tag{4.16}$$

Setting $\tau = 0$ in (4.16) yields

$$\Phi(0) \equiv \mathcal{V} = \int_{-0.5}^{0.5} \Gamma(\omega)d\omega \tag{4.17}$$

Equation (4.17) states that the frequency decomposition of covariances is achieved via the cross-spectrum matrix $\Gamma(\omega)$.

In the simplest case with $M = 2$ and $\mathbf{v}'(t) = \begin{pmatrix} v_1'(t) \\ v_2'(t) \end{pmatrix}$, $\Gamma(\omega)$ becomes

$$\Gamma(\omega) = \begin{pmatrix} \gamma_{v_1'}(\omega) & \gamma_{v_1'v_2'}(\omega) \\ \gamma_{v_2'v_1'}(\omega) & \gamma_{v_2'}(\omega) \end{pmatrix} \qquad (4.18)$$

where $\gamma_{v_1'}(\omega)$ and $\gamma_{v_2'}(\omega)$ are spectra of v_1' and v_2' respectively, and $\gamma_{v_1'v_2'}(\omega)$ and $\gamma_{v_2'v_1'}(\omega)$ are cross-spectra between v_1' and v_2', which are related to the lagged covariance function $\phi_{v_1',v_2'}(\tau)$ between v_1' and v_2' with $\phi_{v_1',v_2'}(\tau) = \phi_{v_2',v_1'}(-\tau) = < v_{1,t}' v_{2,t+\Delta\tau}' >_T$.

In practice only finite time series are available, and periodograms can be used to estimate the elements of $\Gamma(\omega)$. However, the number of matrices to be estimated increases with the length of the time series. One has to consider N_o $M \times M$ matrices where $N_o = (N-1)/2$ for an odd N and $N_o = N/2$ for an even N. Each contains information for one frequency $\omega_k = k/N_o$. As in the univariate case, one has be aware that periodograms are not consistent estimators of the true spectra.

Besides the estimation problem, the problem of interpreting the information contained in cross-spectrum matrices arises as the dimension M increases. With $M = 2$, the estimates of $\Gamma(\omega)$ contain only four elements and can be easily interpreted. As dimension M increases, although the structure of the estimates of $\Gamma(\omega)$ remains the same and one can still understand the meaning of each element of the estimated $\Gamma(\omega)$, it becomes extremely difficult to decompose the N_o $M \times M$ cross-spectrum matrices. One way to approach the problem is to assume that the information contained in these matrices can be expressed in terms of patterns and spectra of the coefficient time series of the patterns. But how can one derive these pairs of pattern and spectrum?

The following section discusses three techniques used to derive patterns and the corresponding spectra, which describe multivariate spectral features of the considered variable. There are also many other techniques, which deal with estimation of second moment statistics. Most of them do not consider the overall features of second moments, but concentrate only on particular phenomena of the climate system. For instance, the wavenumber-frequency analysis (Fraedrich and Böttger, 1978; Hayashi and Golber, 1993) is designed to described wave motions in the climate system. Such techniques will be not considered here.

Methods of deriving frequency decomposed covariances

a) Method I: EOF-decomposition

It is assumed that the multivariate spectral features can be described by eigenvectors of the covariance matrix $\Phi(0)$, i.e. EOFs and the spectra of the coefficient time series of these EOF patterns, i.e. principal components (PCs).

However, the relationship between such derived multivariate spectral features and the information contained in the cross-spectrum matrix is not known. Without directly considering the complicated cross-spectrum matrices, EOF-decomposition can be easily applied.

b) Method II: Consideration of $< \Gamma(\omega) >_{\delta\omega}$ - complex EOFs

Complex EOFs deal directly with estimates of cross-spectrum matrices. Generally, an average of the estimated cross-spectrum matrices over a frequency interval $\delta\omega$, denoted by $< \Gamma(\omega) >_{\delta\omega}$, is considered. It can be shown (von Storch, 1995) that if $u'_t = (v'_t + iv'^h_t)/2$ with v'^h_t being the Hilbert transform of v'_t is filtered in such a way that it contains only spectral components at frequency $\omega \in \delta\omega$, the lag-0 covariance matrix of the filtered u'_t equals the cross-spectrum matrix of v'_t averaged over $\delta\omega$. The eigenvectors, which are generally complex, are referred to as complex EOFs. The complex EOFs and the associated coefficient time series characterize the temporal evolution of the covariance structures on time scales $1/(\delta\omega)$. Thus, complex EOFs provide only information about multivariate spectral features in a small frequency range, rather than at all resolved frequencies of the considered time series.

There are two ways to estimate $< \Gamma(\omega) >_{\delta\omega}$. For the first, proposed by Wallace and Dickson (1972), the cross-spectrum matrix $< \Gamma(\omega) >_{\delta\omega}$ is estimated in frequency domain, whereas for the second, proposed by Barnett (1983, 1985), the estimation is carried out in the domain.

c) Method III: Consideration of multivariate AR(1) process - POP analysis

In the POP analysis proposed by Hasselmann (1988) and H. von Storch et al. (1988), the frequency-limited consideration of the complex EOF analysis is extended to a consideration of the entire multivariate spectra. This is achieved by approximating $\mathbf{v}'_t$ by a multivariate AR(1) process

$$\mathbf{v}'_{t+\Delta} = \mathcal{A}\mathbf{v}'_t + \mathbf{r}_t \tag{4.19}$$

where the matrix $\mathcal{A}$ is the multivariate analogy of the parameter α in (4.8). $\mathbf{r}_t$ is a multivariate white noise. For $\mathbf{v}'_t$ generated by (4.19), the lagged-covariance matrix $\Phi(\tau)$ is given by

$$\Phi(\tau) = \Phi^\dagger(-\tau) = \mathcal{A}^\tau \Phi(0) \tag{4.20}$$

which leads to $\mathcal{A} = \Phi(1)\Phi(0)^{-1}$. The matrix multiplication $\mathcal{A}^\tau$ in (4.20) can be carried out by considering the eigenvectors of $\mathcal{A}$ and their adjoint vectors. Denote the eigenvector matrix with its columns being the eigenvectors $\mathbf{p}_i$ of $\mathcal{A}$ by $\mathcal{P}$, the adjoint matrix with its columns being the adjoint vectors $\mathbf{p}_{a,i}$ of $\mathbf{p}_i$ (i.e. $\mathbf{p}^{\dagger*}_{a,i}\mathbf{p}_j = \delta_{ij}$) by $\mathcal{P}_a$, and the diagonal eigenvalue matrix with its i-th diagonal element being the eigenvalue of $\mathbf{p}_i$ by Λ. It holds

$$\mathcal{A} = \mathcal{P}\Lambda\mathcal{P}^{\dagger*}_a \quad \text{and} \quad \mathcal{A}^\tau = \mathcal{P}\Lambda^\tau\mathcal{P}^{\dagger*}_a \tag{4.21}$$

(4.20) then becomes

$$\Phi(\tau) = \mathcal{P}\Lambda^\tau \left(\mathcal{P}_a^{\dagger*} \Phi(0) \right) \tag{4.22}$$

Substituting (4.22) into (4.14), one has

$$\begin{aligned}
\Gamma(\omega) &= \sum_{\tau=-\infty}^{\infty} \mathcal{P}\Lambda^\tau \left(\mathcal{P}_a^{\dagger*} \Phi(0) \right) e^{-i2\pi\tau\Delta\omega} \tag{4.23} \\
&= \mathcal{P}\mathcal{D} \left(\mathcal{P}_a^{\dagger*} \Phi(0) \right)
\end{aligned}$$

with

$$\begin{aligned}
\mathcal{D} &= \sum_{\tau=-\infty}^{\tau=\infty} \Lambda^\tau e^{-i2\pi\tau\omega} \tag{4.24} \\
&= \begin{pmatrix} \sum_{\tau=-\infty}^{\tau=\infty} \lambda_1^\tau e^{-i2\pi\tau\omega} & 0 & 0 \\ 0 & \sum_{\tau=-\infty}^{\tau=\infty} \lambda_2^\tau e^{-i2\pi\tau\omega} & 0 \\ 0 & 0 & \ddots \end{pmatrix}
\end{aligned}$$

Equation (4.23) shows that the cross-covariance matrix $\Gamma(\omega)$ can be expressed as a product of three matrices. The left and the right matrices, i.e. the eigenvector matrix $\mathcal{P}$ and the transposed complex conjugate of the adjoint matrix $\mathcal{P}_a$ multiplied by covariance matrix $\Phi(0)$), are independent of time lags τ, whereas all information concerning the temporal characteristics are contained in the middle diagonal matrix $\mathcal{D}$.

The triple product structure of $\Gamma(\omega)$ means that the frequency decomposition of the total covariance $\mathcal{V} = \Phi(0)$ can be done by projecting $\mathbf{v}'_t$ onto the basis spanned by the eigenvectors of $\mathcal{A}$, also referred to as POPs (principal oscillation patterns). This corresponds to a coordinate transformation $\mathbf{v}'_t = \mathcal{P}\mathbf{z}_t$ with the eigenvector matrix $\mathcal{P}$ being the transformation matrix and $\mathbf{z}_t$ being the coefficients in the POP-basis and satisfying

$$\mathbf{z}_{t+\Delta} = \Lambda\mathbf{z}_t + \mathbf{s}_t \tag{4.25}$$

with $\mathbf{s}_t = \mathcal{P}^{*\dagger}\mathbf{r}_t$. von Storch (1995) carried out this coordinate transformation, and showed that the cross-spectrum matrix $\Gamma(\omega)$ of $\mathbf{v}'_t$ is related to the cross-spectrum matrix $\Gamma_z(\omega)$ of $\mathbf{z}_t$ via

$$\Gamma(\omega) = \mathcal{P}\Gamma_z(\omega)\mathcal{P}_a^{-1} \tag{4.26}$$

which leads to, under the consideration of (4.23),

$$\Gamma_z(\omega) = \mathcal{D} \left(\mathcal{P}_a^{*\dagger} \Phi(0)\mathcal{P}_a \right) \tag{4.27}$$

Using the condition derived from (4.25)

$$\mathbf{p}_{a,j}^{*\dagger} \Phi(0)\mathbf{p}_{a,j} = \sigma_{z_j}^2 = \frac{\sigma_{s_j}^2}{1 - \lambda_j\lambda_j^*} \tag{4.28}$$

where $\sigma_{s_j}^2$ is the variance of the time series $s_{j,t} = \mathbf{p}_{a,j}^{*\dagger}\mathbf{r}_t$, the j-th diagonal element of $\Gamma_z(\omega)$ in (4.27) can be written as

$$\gamma_{z_j}(\omega) = \frac{\sigma_{s_j}^2}{(e^{i2\pi\omega} - \lambda_j)(e^{i2\pi\omega} - \lambda_j)^*} \tag{4.29}$$

The shape of the spectrum is determined entirely by the corresponding eigenvalue λ_j. Since $\mathcal{A}$ is not symmetric, the POPs and the corresponding eigenvalues can be real or complex. If λ_j is real, $\gamma_{z_j}(\omega)$ is a spectrum of a univariate AR(1) process. For a complex eigenvalue $\lambda_j = |\lambda_j|e^{i\phi_j}$, $\gamma_{z_j}(\omega)$ describes a spectrum with a spectral peak at frequency $\omega_j = \phi_j/(2\pi)$. For $|\lambda_j| = 1$, a resonance peak appears with zero width. $|\lambda_j|$ of a stationary time series is smaller than unity; the smaller the $|\lambda|$, the broader the peak.

Thus, in contrast to a complex EOF analysis, a POP analysis can be used to estimate the entire multivariate spectra. By approximating a multivariate time series with a multivariate AR(1) process, the information contained in cross-spectrum matrices is expressed in terms of patterns, which are eigenvectors of the system matrix $\mathcal{A}$, and the associated spectra, whose shapes are determined by the corresponding eigenvalues of $\mathcal{A}$.

Two estimates of the second moment statistics of the present-day climate

We turn now to estimations of the variability of the present-day climate, as defined by cross-spectrum matrices of climate variables. Two examples are considered. One is derived from instrumental and analysis records, and the other from a long integration of the coupled ECHAM1/LSG atmosphere-ocean GCM, which can be considered to be one of our best guesses of the present-day climate variability. Different from many phenomenological studies, the emphasis in this paper is on the derivation of the full multivariate spectra, rather than on the consideration of covariances on a particular time scale.

a) An estimate of second moment statistics of the atmosphere-ocean system from a 20-year observational record

In order to obtain joint covariances of the atmosphere-ocean system, a combined observational data set, consisting of observations of both oceanic and atmospheric state variables, is formulated and analyzed using the POP technique (Xu, 1993). The oceanic variables are SST from the Comprehensive Ocean-Atmospheric Data Set (COADS) (Woodruff et al. 1987) (which covers the world oceans from 40°S to 60°N), subtropical Pacific ocean temperature at various depth from surface down to 200-m as archived at Far Seas Fisheries Research Laboratory (Japan) and at the Scripps Institution of Oceanography (La, Jolla, USA), and finally sea level station data in the Pacific collected

Table 4.1. Eigenvalue of each POP

	Mode 1	Mode 2	Mode 3	Mode 4	Mode 5	Mode 6
λ_j	0.15	0.4	0.66	0.86	.92	$0.93e^{i2\pi(51months)^{-1}}$

by Wyrtki et al. (1988). The atmospheric state variables are sea level pressure (SLP) and zonal winds at 200 and 700 hPa. The SLP data include the Northern (20°N-50°N) and Southern (10°S-40°S) Hemisphere SLP stored at the National Center for Atmospheric Research (Boulder, USA) and at the Bureau of Meteorology Research Center (Melbourne, Australia), and the tropical (20°N-20°S) COADS SLP. The combined data set is on a monthly basis, and covers the period from 1967 to 1986.

The POP analysis is applied to the combined data set. That is, the variables of the data set is put into one big vector $\mathbf{v}'_t$ in (4.19). The data are compressed into a smaller space spanned by the few leading EOFs of each variable, and the obtained POPs are then transformed back to the physical space.

Six dominant POPs are identified, five real and one complex. The corresponding eigenvalues are listed in Table 4.1. The time series of the POP coefficients are shown in Fig. 4.9. Following (4.29), the shapes of the spectra of the POPs are determined by the corresponding eigenvalues. Modes 1-3, which have the smallest eigenvalues, should be whiter than Mode 4 and 5, which have larger eigenvalues. The complex POP should have a spectrum which peaks around four years. A spectral analysis of the POP time series (Xu, 1993) showed that the spectra of time series shown in Fig. 4.9 are consistent with the spectra defined in (4.29). However, Fig. 4.9 suggests that the redness of Mode 4 probably comes from a downward trend until the mid seventies and a subsequent upward trend, whereas the redness of Mode 5 is likely caused by the jump in the mid seventies. The time series is certainly too short to provide reliable estimates of covariances whose dominant power are on decadal time scales.

The first three modes with flat spectra are large-scale modes in the Northern Hemisphere. Mode 1 is confined in the North Atlantic, Mode 2 to the North Pacific, and Mode 3 appears in the region from the central North Pacific to the North Atlantic. They bear strong resemblance to the "Pacific/North American" (PNA), "East Atlantic" (EA) and "West Atlantic" (WA) teleconnection patterns identified by Wallace and Gutzler (1981). They are most pronounced during the Northern winter season. For all three modes, the patterns in SST and station sea level indicate that oceanic changes are essentially passive. SST anomalies are generated by the winter-time warm marine and cold continental air advection brought about by anomalous atmospheric circulation, whereas sea level anomalies are produced by the in-

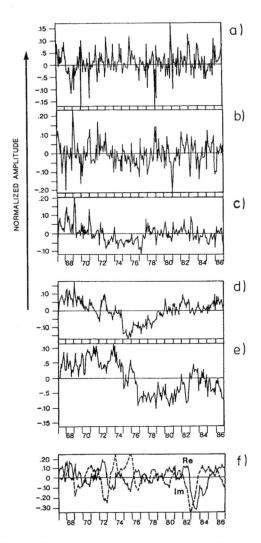

Fig. 4.9. POP coefficient time series of mode 1 to 6 ordered according to the amplitudes of the corresponding eigenvalues. From Xu (1993)

verse barometric effect. A detailed discussion is given in Xu (1993) and Zorita et al. (1992).

Mode 4 reveals positive SST anomalies over all tropical oceans with amplitudes of about a few tenth of $^{\circ}$C. The ocean temperature anomalies in the upper 50 meters are out of phase with those in the layer below. The associated atmospheric anomalies indicate changes of tropical convection over the west Pacific. Besides the tropical anomalies, changes in the subtropical and subpolar Pacific are found. The sea level anomalies along the coast of Japan are characterized by anomalies of one sign in the south and anomalies of the opposite sign in the north, indicating changes in the wind-driven circulation. These sea level anomalies are accompanied by the changes in winds over the North Pacific. Some of these decadal changes were also reported by other authors (Qiu and Joyce, 1992; Trenberth, 1990).

The most striking feature of Mode 5 is the large anomalies of winds at both 700 and 200 hPa and SLP in the Southern Hemisphere mid-latitudes.

Mode 6 describes atmospheric and oceanic changes related to ENSO. Since it is a complex mode, two patterns are obtained, the real and the imaginary parts of the POP, $\mathbf{p}^{re}$ and $\mathbf{p}^{im}$. The two corresponding time series are temporally 90° out of phase, so that $\mathbf{p}^{re}$ and $\mathbf{p}^{im}$ describe spatial propagation at the dominant frequency indicated by the eigenvalue. Figure 4.10 shows the real and imaginary patterns for sea level. The imaginary pattern (Fig. 4.10a), which describes the conditions in cold events, is followed by the real pattern (Fig. 4.10b), describing the condition prior to a warm event. Along the equator, an eastward propagation of enhanced sea level anomalies from the west Pacific in Fig. 4.10a to the central Pacific in Fig. 4.10b are visible. The propagation is related to large explained variances (shaded areas in Fig. 4.10). At least for the period 1967-1986, the eastward propagation of equatorial anomalies of sea level, zonal wind, and sea level pressure appears as the most dominant feature of this mode. A more detailed description of the eastward propagation of zonal wind is given in Xu (1992). In terms of a series of hindcast experiments, Xu and H. von Storch (1990) showed that a part of the ENSO predictability originates from the eastward propagation of tropical anomalies.

The estimates derived from the observational data have two problems. First, the shortness of the records makes the estimates of covariances on decadal time scales, such as Mode 4 and 5, questionable. Second, the result may be biased toward variability in a particular part of the atmosphere-ocean system, namely the tropical Pacific near the interface between the atmosphere and ocean, since this part of climate system is represented by more data than the other parts (sea level and near surface ocean temperature are only available for the Pacific). In general, we can only study the part of the system that is well observed.

b) An estimate of the second moment statistics of the atmosphere from a long integration with the ECHAM1/LSG coupled GCM

500-year data of a 1260-year integration with the ECHAM1/LSG model are used to study the second moment statistics of the model atmosphere (von Storch, 1997a). Different from the observational data, such long records allow better estimates of the cross-spectrum matrix at frequencies as low as one cycle per several decades. The disadvantage is that the atmosphere of the coupled model may not behave in exactly the same manner as the real

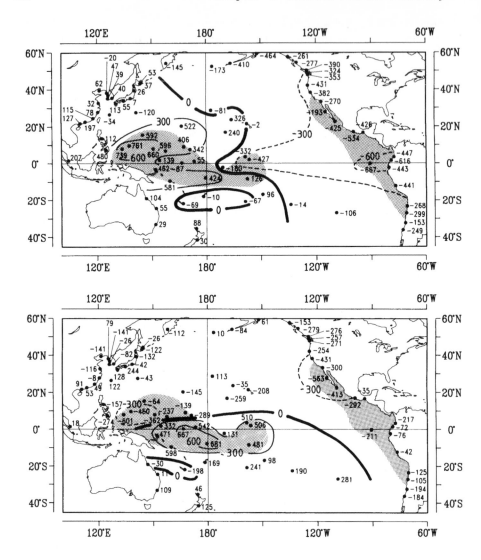

Fig. 4.10. POP patterns of the imaginary and real parts of mode 6 for sea level.
Shaded areas indicate explained variance larger than 30%. From Xu (1993)

atmosphere. In other words, features of cross-spectrum matrices of simulated
data may differ from those of the real world.

The technique used is the EOF decomposition discussed in section 4.2.2a.
To be systematic, EOF analysis is carried out for 500-year monthly anomalies of two-dimensional state variables at various levels. Subsequently, the
spectral features of the coefficient time series of the EOFs, also referred to

as principal components (PCs), are studied. It turns out that the PCs can be approximated, to different extents, by AR(1) processes. This greatly simplifies the investigation of spectral features of the PCs. As shown in section 4.2.1, the shape of the spectrum of an AR(1) process is controlled by the parameter α, or the bending time scale $1/\omega^*$, of the AR(1) process. Thus, the spectral features of all PCs can be studied by estimating α and through (4.13) the bending time scale $1/\omega^*$ from the time series.

Figure 4.11 shows the bending time scales of the first 10 PCs of velocity potential and steam function. These two variables are chosen because they completely describe horizontal motions. To capture possible vertical variations, the two variables are considered at 850, 500 and 300 hPa.

Except for the first two PCs of stream function, the bending time scales of all PCs are about two to three months. The large bending time scale of PC6 of velocity potential at 850 hPa is related to the drift of model sea ice. The largest bending time scales are obtained for PC1 and PC2 of stream function. However, the corresponding EOFs vary differently with height. EOF1 of 200-hPa stream function describes a mode in the upper troposphere (see also section 4.4.2) and looses its dominance at lower levels. In contrast, EOF2 of 200-hPa stream function is a mode of the entire troposphere and appears as the leading EOF at lower levels. In fact, EOF1 of 500 hPa and 850-hPa stream function bear some resemblances to EOF2, rather than EOF1 of 200-hPa stream function. One can conclude that the longest bending time scales of about a half year to a year are related to the first two PCs of 200-hPa stream function. Figure 4.12 shows the spectra of the first two PCs of 200-hPa stream function estimated using "chunk" spectral estimators. In order to obtain high spectral resolution, half-daily data obtained by projecting 200-year half daily data into the EOFs of monthly data are considered. Consistent with the estimates obtained from the approximated AR(1) processes, the power increases with decreasing frequency at the rate of $1/\omega^2$ and becomes essentially flat at frequencies beyond one cycle per year.

Some of the EOF patterns are shown in Fig. 4.13. A comparison of all EOFs suggests that the difference between patterns with the longest bending time scales and the patterns with shorter bending time scales lies in the degree of isotropy, i.e. the ratio of meridional scale to zonal scale (a precise definition is given in section 4.4.2). The two reddest pattern, i.e. the first two EOFs of 200-hPa stream function (Fig. 4.13a and b), are much more anisotropic than the other patterns Fig. 4.13c and d.

One might ask whether the modes found in the ECHAM1/LSG model appear in the observation, and to what extent the features of the second moment in the model are realistic. The teleconnection modes (Mode 1 to 3) of section 4.2.2a can be identified as the high-order EOFs of the stream function in the ECHAM1/LSG model. For instance, EOF3 shown in Fig. 4.13c bears some resemblances to the PNA pattern. The mode 5 discussed in section 4.2.2a might be related to the second EOF of the 200-hPa stream

(a) velocity potential

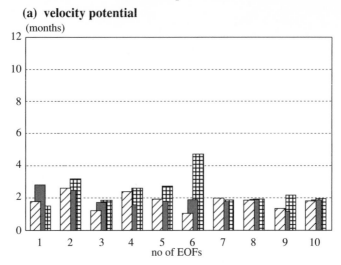

(b) stream function

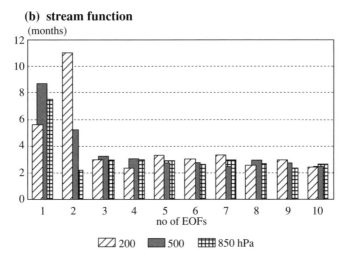

Fig. 4.11. The bending time scales T^* (in months) as functions of the orders of EOFS. The values are derived from the PCs of velocity potential (a) and stream function (b) at 200, 500 and 850 hPa. From von Storch (1997a)

function in Fig. 4.13b. The first EOF, which is absent in section 4.2.2a, may be related to the fact that the combined data set emphasizes variability of the tropical Pacific near the interface of the atmosphere and the ocean. On the other hand, the mode related to EOF1 of the 200-hPa stream function is a global mode of upper troposphere (a more detailed description is given in section 4.4.2).

a) PC1

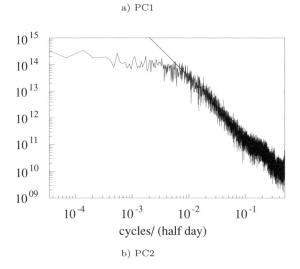

cycles/ (half day)

b) PC2

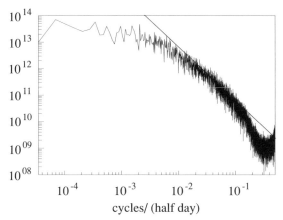

cycles/ (half day)

Fig. 4.12. Spectra of time series of EOF1 and EOF2 of 200-hPa stream function, as derived by projecting 200-year twice daily data onto the EOF patterns. The straight lines indicate the $1/\omega^2$ slope

To precisely determine whether the reddest modes appear in the observation, the 200-hPa stream function from the reanalysis project of the European Center for Median Range Forecast (Gibson et al., 1997) for the period 1981 to 1989 is analyzed. Figure 4.14 shows the first and third EOFs of monthly 200-hPa stream function anomalies of reanalysis data. Note that the order of the EOF has nothing to do with the redness, or the length of the bending time scale. The first EOF and its relation to the global relative angular momentum have been discussed by Kang and Lau (1994). The third EOF

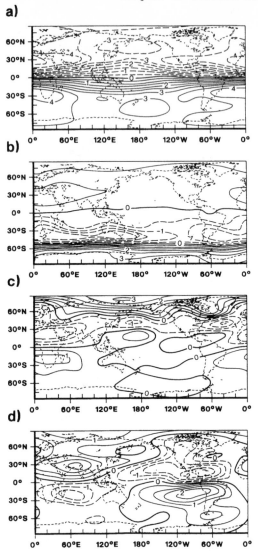

Fig. 4.13. The first four EOFs as derived from 500-year monthly anomalies of 200-hPa stream function. The anomalies are obtained by subtracting the 500-year mean annual cycle. From von Storch (1997a)

describes the same variations as found by Rogers and van Loon (1982) and Trenberth and Christy (1985). Despite the different lengths of data used, the similarity between Fig. 4.13a and b and Fig. 4.14 is remarkable. The blocked pattern with two anticyclonic cells over the central Pacific, which is absent

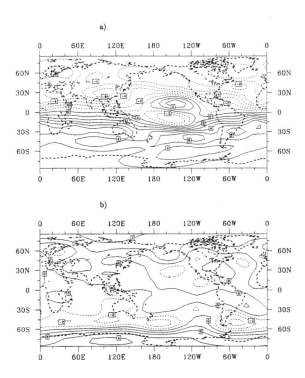

Fig. 4.14. The first and the third EOFs as derived from 9-year monthly anomalies of 200-hPa stream function of the ECWMF reanalysis. The anomalies are obtained by subtracting 9-year mean annual cycle

in Fig. 4.13, reflects the modification of the EOF pattern through ENSO which is essentially absent in the coupled ECHAM/LSG model. Kang and Lau (1994) found that the blocked pattern disappears when EOF analysis is applied to the stream function without ENSO years. We can conclude that the two reddest modes do exist in the observations.

The above discussion suggests that the deficiency of models in producing some aspects of the variability of the real world may introduce certain bias into the results. For instance, the absence of ENSO in the ECHAM1/LSG model leads to a pattern which differs from the observed one, and might also underestimate the power at low frequencies. However, it is difficult to assess the degree of the bias.

Confined to the atmosphere of the ECHAM1/LSG model, the result of EOF decomposition suggests that the most anisotropic patterns have the reddest spectra. In general, the spectra of all large-scale patterns become flat on time scales beyond a year.

c) Summary

The above two examples suggest that the most dominant features of the variability of the present-day climate, as defined by frequency-decomposed covariances, can be characterized by pairs of patterns and spectra. One class of pattern-spectrum pairs represents standing patterns with red spectra. The structures of the patterns seem to be coupled to the redness of the spectra. The reddest spectra are related to the most anisotropic patterns. The other kind of pattern-spectrum pair describes an eastward propagation of anomalies of e.g. sea level in the tropical Pacific, and is associated with a peaked spectrum. As in the univariate case, spectra of large-scale patterns are also smooth and continuous, indicating that the randomness is the most dominant feature of climate variability.

However, the extent to which these estimates reflect the true second moment statistics of the present-day climate is unknown. In particular, one should be aware of two limitations. First, the estimates derived from observational data are likely biased toward variability of those regions of the climate system for which more measurements were carried out. Second, the low-frequency covariances might be underestimated. Due to model deficiencies in producing less-known or unknown low-frequency phenomena, the consideration of model integration does not necessary improve the low-frequency estimates.

In general, even though more information is available for the present-day climate than for the paleo climate, the variability of the present-day climate, as described by frequency-decomposed variances, is not completely known. This is particularly true for the variability in the deep ocean. The derivation of the complete second moment statistics of the present-day climate is and remains a problem of climate research yet to be solved.

4.3 Concepts on the Generation of Temporal Variability

As suggested in the introduction, climate variability should be considered as averaged features of a climate state variable. The most important averaged features are mean $< \mathbf{v}(t) >$ and covariance $< \mathbf{v}'(t)\mathbf{v}'^{\dagger}(t) >$. In the following, concepts concerning temporal variability related to these two quantities are discussed.

There are other concepts based on the consideration of individual time evolution of state variable $\mathbf{v}(t)$. The most prominent one is formulated in terms of low-order models. Finite systems of ordinary nonlinear differential equations are used to describe the behavior of climate state variables. Lorenz (1963, 1982) considered a dissipative system forced with a constant forcing and showed that such a system is able to reveal irregular behaviors. The origin of the irregularity lies in the non-linearity of the governing equations.

However, since low order models are designed to study deterministic features of individual evolution, rather than averaged behaviors of state variables, they will not be discussed here.

4.3.1 Classical Hypotheses and the Concept of Multiple Equilibria: The Origin of Mean $< \mathbf{v}(t) >$

In many concepts concerning the temporal climate variability, the meaning of the term "climate variability" is not explicitly defined, and has to be inferred from the arguments used to formulate the hypothesis. In most cases, variations of the mean climate conditions, i.e. $< v(t) >$, rather than $< v'(t)v'^{\dagger}(t) >$ are considered. Two classes of concepts exist: One suggests that changes in $< v(t) >$ are induced by external forcings of the climate system, the other argues that internal forcings can also be responsible for changes in $< v(t) >$.

The classical hypotheses as summarized e.g. by Huntington and Visher (1922), were focused on identification of possible external forcings of $< v(t) >$. Representative of these hypotheses is Croll's eccentricity theory, later referred to as the Milankovitch theory. It suggested that the variations of the earth's temperature related to the glacial cycles are caused by variations of the geographical distribution of incoming solar radiation produced by periodic orbital perturbations of the earth. The theory became popular again in recent years when new data was introduced to support this idea.

Figure 4.15 shows $\delta^{18}O$ record for planktonic foraminifera from several deep-sea cores and the July insolation at 65°N. The $\delta^{18}O$ record primarily reflects changes in the amount of continental ice in the Northern Hemisphere; the more positive the $\delta^{18}O$ value, the colder the temperature and the larger the ice volume. There are a few periods characterized by long-lasting large positive $\delta^{18}O$ values, indicating long-lasting glacial conditions, followed by rapid decreases of $\delta^{18}O$ to large negative values, indicating rapid disappearances of ice that lead to interglacial conditions. A comparison of the two curves in Fig. 4.15 shows that the rapid disappearances of ice correspond to prominent peaks in summer insolation for the latitudes at which the excess ice was located. This tie between insolation and ice volume leads to the belief that the earth's glacial cycles are driven in part by changes in seasonality brought about by cyclic changes in the earth's orbital elements.

The second class of concepts assumes that variations of $< v(t) >$ result essentially in variations of slowly varying climate variables, such as oceanic variables, and considers an ocean which is separated from the atmosphere. The mean effects of the atmosphere and the ocean, i.e. the mean fluxes of momentum, heat and freshwater, are included in the consideration, whereas the fluctuating fluxes which reflect the weather events are ignored. Without the fluctuating fluxes, which induce the random character, the operator $< >$ can be dropped.

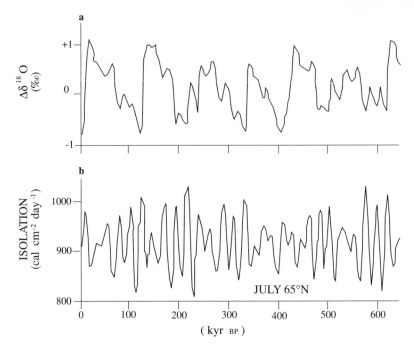

Fig. 4.15. Composite of the $^{18}O/^{16}O$ record for planktonic foraminifera from several deep-sea cores (top panel) and July isolation at 65°N (bottom panel), which reflect changes in seasonality brought about by cyclic changes in Earth's orbital elements. From Broecker et al. (1985)

Representative of this class is the more recent concept of multiple equilibria of the deep ocean circulation. It is thought that the present-day deep ocean circulation, as characterized by the oceanic "Conveyor Belt" with the deep water formation in the North Atlantic, its subsequent spreading as a deep western boundary current in the Atlantic, its entry and upwelling into the other oceans and its return to the Atlantic, determines the present-day climate condition $< \mathbf{v}(t) >$. It is suggested that the transition from one oceanic equilibrium to another may cause drastic changes in $< \mathbf{v}(t) >$. These transitions are thought to be triggered by changes in mean fresh water fluxes induced, for instance, by deglaciation or glaciation.

Studies using general circulation models suggest that the deep ocean circulation indeed possesses different equilibrium states. Figure 4.16 shows the stream function for the Atlantic meridional circulation of two stable equilibria obtained from a global model of the coupled ocean-atmosphere system developed at Geophysical Fluid Dynamics Laboratory (GFDL) in Princeton, USA (Manabe and Stouffer, 1988). Starting from two different initial conditions, "asynchronous" time integrations of the coupled model, under identical boundary conditions, leads to a vigorous thermohaline circulation

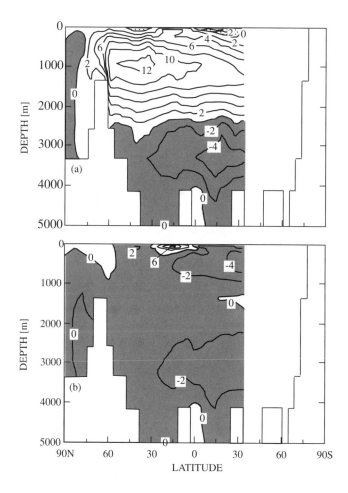

Fig. 4.16. Stream function in Sverdrups illustrating two stable equilibrium meridional circulations in the Atlantic Ocean. From Manabe and Stouffer (1988)

in one equilibrium (the upper diagram in Fig. 4.16), and no thermohaline circulation in the other (the lower diagram in Fig. 4.16). The difference in surface air temperature between these two equilibria is shown in Fig. 4.17. The climatic change induced by the shut-down of deep water formation in the North Atlantic is characterized by large changes in surface air temperature in the Northern Hemisphere, in particular over the northern North Atlantic. If one equilibrium occurs at time t_1 and the other at time t_2, the experiments of Manabe and Stouffer can be used to estimate $< \mathbf{v}(t_1) > - < \mathbf{v}(t_2) >$ for various state variables.

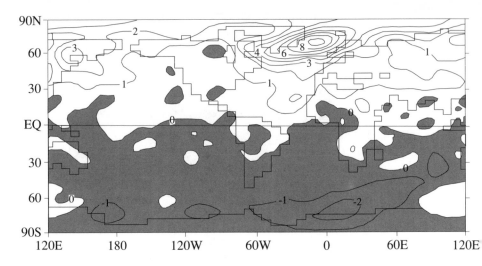

Fig. 4.17. Difference in surface air temperature (degrees C) between two equilibria obtained by Manabe and Stouffer (1988)

The fundamental mechanism, which generates the multiple equilibria, is captured by a variant of Stommel's box model (1961). The model consists of two boxes, one representing low latitudes and other the high latitudes. The thermohaline circulation is represented as a heat and salt exchange term between the boxes. The basic equations are heat and salt conservation,

$$\frac{\partial T}{\partial t} = \lambda(T^* - T) - 2|m|T \tag{4.30}$$

$$\frac{\partial S}{\partial t} = 2F - 2|m|S.$$

The transport m from one box to the other, which characterizes the overturning flow, is parameterized as

$$m = k(\alpha T - \beta S) \tag{4.31}$$

The transport m is proportional to the density difference between the two boxes. For the simplicity, the equation of state is linearized in (4.31), so that the density difference is proportional to temperature and salinity differences T and S between the two boxes.. The first terms on the right-hand sides of the two equations in (4.30) describe the forcing terms in the form of "mixed boundary conditions". The heat flux is parameterized as a restoring term with restoring temperature T^*, whereas the fresh water forcing by a prescribed flux F, which describes the vapor flux from low to high latitudes. The restoring parameter λ is large, so that the surface flux produces nearly fixed temperature difference between the two boxes with $T > 0$.

The equilibrium solution of system (4.30) and (4.31) can be derived analytically. Considering the steady state of (4.30), and combining the second equation of (4.30) with (4.31), one obtains

$$m|m| - \alpha kT|m| + k\beta F = 0 \tag{4.32}$$

The solutions of (4.32) for $m > 0$ are

$$m = \frac{\alpha kT}{2} + \sqrt{\frac{(\alpha kT)^2}{4} - \beta kF} \qquad \text{for } F < F^{crit} \tag{4.33}$$

and

$$m = \frac{\alpha kT}{2} - \sqrt{\frac{(\alpha kT)^2}{4} - \beta kF} \qquad \text{for } 0 < F < F^{crit} \tag{4.34}$$

and for $m < 0$ is

$$m = \frac{\alpha kT}{2} - \sqrt{\frac{(\alpha kT)^2}{4} + \beta kF} \qquad \text{for } F > 0 \tag{4.35}$$

Note that in order to ensure $m > 0$ in (4.34), $F < 0$ is not permitted. For the same reason, $F < 0$ and $m = \frac{\alpha kT}{2} + \sqrt{\frac{(\alpha kT)^2}{4} + \beta kF}$ are not permitted for $m < 0$. Equations (4.33)-(4.35) represent four branches of equilibrium solutions and are shown in Fig. 4.18a. Three of them, indicated by solid lines, are stable, and one of them, indicated by the dashed line, is unstable.

For $F < 0$, (4.33) describes a thermohaline driven flow, since both the temperature difference T and fresh water forcing F work together to force the flow. The larger $|F|$, the larger is the overturning flow m. For $F > 0$, (4.33) is purely thermally driven. Fresh water forcing F brakes the overturning flow m. With increasing F, m decreases. This solution branch ends in a saddle-node bifurcation (point S) at a critical fresh water input of $F^{crit} = (k\alpha^2 T^2)/(4\beta)$. Solution (4.34) is also thermally driven. However, different from the solution obtained from (4.33), the flow is sluggish. At this point, the dependence of the effect of salinity on the flow itself comes into play. A strong overturning flow is more efficient in reducing the salinity contrast, which brakes the thermally dominant circulation. With a sluggish overturning flow, the brake induced by salinity contrast is strong enough to make the solution unstable. This branch of solution is shown by the dashed line in Fig. 4.18a. The final solution (4.35) is purely haline driven. m is zero for $F = 0$ and its amplitude increases with increasing $|F|$.

Even though the governing dynamics of the multiple equilibria, i.e. the overturning flow is proportional to density difference between the two boxes, are extremely simple, the detailed balances of these equilibria are different, depending on the fresh water forcing F, and the internal feedbacks coupled to the fresh water forcing. Rahmstorf (1996) shows that Stommel's box model can be extended to describe the present-day North Atlantic deep water which crosses the equator. He suggested that for the cross equatorial overturning

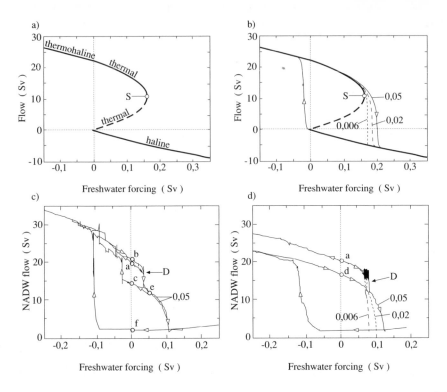

Fig. 4.18. a) Equilibrium flow and hysteresis response of Stommel's box model. b) Same as in a), but obtained from two different methods. The thick lines are the analytical solutions shown in a), whereas the thin curves are the result of time integrations with slowly changing freshwater forcings. c) and d) are hysteresis responses of the North Atlantic overturning circulation to a slowly changing fresh water forcing in the high latitude (left) and in the tropics (right). Arrows mark the direction in which the curve is traced. In the left panel, after one hysteresis loop, starting from a towards the right, the model arrives at equilibrium b. From there the right half of the hysteresis curve was traced a second time. In the right panel, the right half of hysteresis curve was traced at three different rates; the rate of fresh water forcing change is labeled in units $10^{-3} SV yr^{-1}$. For more details see Rahmstorf, 1995

flow, F in Fig. 4.18a should represent the fresh water forcing in the Atlantic catchment rather than the vapor flux from the low to the high latitudes as in the original Stommel's box model.

Figure 4.18a is derived from an extremely simple model. The question now is to what extent the result of the box model can be applied to the real ocean, or to an approximation of the real ocean, i.e. the ocean of a general circulation model. In this case, the situation is so complicated that the equilibrium states can no longer be obtained analytically. One way to estimate these states was suggested by Rahmstorf (1995). He performed time integrations with various fresh water forcings, which change at very slow rates. The slow rate of external forcing changes ensures that the system is almost in equilibrium during the course of integration. How well this technique works is demonstrated in Fig. 4.18b for the box model. The thick lines are the analytically obtained equilibria shown in Fig. 4.18a and the thin lines are obtained by time integration with slowly changing fresh water flux F. They show that, except near the bifurcation point, the equilibrium branches can be tracked closely by slow changes of fresh water forcing.

Applying this technique to the GFDL general circulation model, Rahmstorf (1995) obtained hysteresis responses of the North Atlantic overturning circulation to slowly changing fresh water forcings in the high latitudes (Fig. 4.18c) and in the topics (Fig. 4.18d). The responses are similar to that obtained from the box model (Fig. 4.18b). Furthermore, the resemblance between Fig. 4.18c and Fig. 4.18d confirms the founding of cross equatorial box model, i.e. the equilibrium overturning flow does not depend on the exact location of fresh water forcing, provided it enters the Atlantic catchment area. Rahmstorf's experiments suggest that the deep ocean circulation of a three-dimensional general circulation ocean possesses multiple equilibria, and the mechanisms of these equilibria are identical to those as captured by the simple box model.

The concept of multiple equilibria demonstrates that, besides external forcings, changes in the mean atmospheric fluxes also can produce variations in the mean values of oceanic variables. At this stage, the cause of the flux changes is not yet completely understood.

4.3.2 The Concept of Stochastic Climate Models: The Origin of Covariance $< \mathbf{v}'(t)\mathbf{v}'(t)^\dagger >$

The concept of stochastic climate models is based on the following assumption:

> The climate system consists of a fast component $\mathbf{v}_f$ with short response time scale τ_f and a slow component $\mathbf{v}_s$ with long time scale τ_s, such that
> $$\tau_f \ll \tau_s$$

In the case that the climate system consists of the atmosphere and the ocean, $\mathbf{v}_f$ and $\mathbf{v}_s$ represent the state variables of the atmosphere and the ocean, respectively. The evolution of the coupled system is described by

$$\frac{dv_{f,i}(t)}{dt} = g_i(\mathbf{v}_s, \mathbf{v}_f) \tag{4.36}$$

$$\frac{dv_{s,i}(t)}{dt} = f_i(\mathbf{v}_s, \mathbf{v}_f) \tag{4.37}$$

where $g_i(\mathbf{v}_s, \mathbf{v}_f)$ represents the atmospheric dynamics and the coupling of the atmosphere to the ocean, whereas $f_i(\mathbf{v}_s, \mathbf{v}_f)$ represents the oceanic dynamics and the coupling of the ocean to the atmosphere. g_i and f_i may include external forcing factors, such as those related to variations in incoming solar radiation. In general, (4.36) and (4.37) correspond to a fully coupled GCM of the atmosphere-ocean system.

In the investigations using ocean-only GCMs, such as those discussed in section 4.3.1, it had been argued that only the mean atmospheric forcings are important for the evolution of the ocean. Applying ensemble average on (4.37), one has

$$\frac{d < v_{s,i} >}{dt} = < f_i(\mathbf{v}_s, \mathbf{v}_f) > \tag{4.38}$$

$< f_i(\mathbf{v}_s, \mathbf{v}_f) >$ describes the dynamics of the ocean plus the averaged flux forcings from the atmosphere. (4.38) represents an ocean-only GCM driven by mean atmospheric fluxes. In practice, the averaged flux forcings are obtained from time average $< >_T$. v_s generated by (4.38) varies only on time scale $\tau \gg \tau_s$.

Thus, ocean-only models driven by averaged fluxes can be used to study the time evolution of mean climate $< \mathbf{v}(t) >$. Generally, external forcing or changes in the averaged fluxes are required for the reduced equation system (4.38) to produce changes in the mean climate.

Hasselmann (1976) pointed out that even though (4.38) is valid for the mean $< \mathbf{v}_s(t) >$, it is not appropriate for studying variations around the mean. He further suggested that an equation which is able to describe second moment statistics of the slowly varying climate component should have the form

$$\frac{dv_{s,i}}{dt} = < f_i(\mathbf{v}_s, \mathbf{v}_f) > + f_i'(\mathbf{v}_s, \mathbf{v}_f) \tag{4.39}$$

with

$$< f_i'(\mathbf{v}_s, \mathbf{v}_f) > = 0, \tag{4.40}$$

where $f_i'(\mathbf{v}_s, \mathbf{v}_f)$ is a stochastic forcing. In the formulation of (4.39), the atmospheric forcings are divided into two components. One represents the mean fluxes and the other the deviations from the mean characterizing the fluctuating weather-related fluxes. The former is included in (4.38) and in the first term on the right-hand side of (4.39). The second is given by $f_i'(\mathbf{v}_s, \mathbf{v}_f)$.

Equation (4.39) is a stochastic climate model. The implication is that climate variability is a statistical rather than a deterministic phenomenon. Hasselmann's concept of stochastic climate models is the only concept which explicitly recognizes and deals with the stochastic nature of climate variability.

To demonstrate how $< \mathbf{v}'_s(t)\mathbf{v}_s'^\dagger(t) >$ is generated by (4.39), the equation of deviation $\mathbf{v}'_s(t) = \mathbf{v}_s(t) - < \mathbf{v}_s(t) >$, obtained by subtracting (4.38) from (4.39),

$$\frac{dvp_{s,i}}{dt} = f'_i(\mathbf{v}_s, \mathbf{v}_f) \tag{4.41}$$

is considered. (4.41) can be further simplified and written in vector form as

$$\frac{d\mathbf{v}'_s}{dt} = \mathcal{F}\mathbf{v}'_s + \mathbf{f}'_{<v_s>} \tag{4.42}$$

where the matrix $\mathcal{F}$ represents the linear dynamics around $< \mathbf{v}_s(t) >$ and $\mathbf{f}'_{<v_s>}$ is the stochastic forcing related to the mean state $< \mathbf{v}_s(t) >$.

With the time-scale-separation assumption stated at the beginning of this section, $\mathbf{f}'_{<v_s>}$ can be described by a white noise forcing. (4.42) represents a continuous multivariate first-order Markov process. $\mathbf{f}'_{<v_s>}$ acts to generate the covariance of $\mathbf{v}'_s$. In other words, if $\mathbf{f}'_{<v_s>}$ would not be present, the covariance of v's would be strongly underestimated. On the other hand, in the absence of linear dynamics, such as for a integration time shorter than τ_s, $\mathbf{f}'_{<v_s>}$ in (4.42) would create a non-stationary covariance of $\mathbf{v}'_s$ which increases with increasing time. The multivariate spectra of $\mathbf{v}'_s$

$$\Gamma_{\mathbf{v}'_s}(\omega) = \frac{\Gamma_{f'_{i,<v_s>}}(\omega)}{\omega^2} \tag{4.43}$$

with $\Gamma_{f'_{i,<v_s>}}(\omega)$ being the cross-spectrum of stochastic forcing $\mathbf{f}'_{<v_s>}$ reveal an increasing spectral level with decreasing frequence. In order to obtain stationary variance, $\mathcal{F}$ must be negative definite. In this case, the net effect of the linear dynamics is to dissipate the variance of $\mathbf{v}'_s$. In the univariate case, the spectrum reduces to that of an AR(1) process, which has a finite value at zero frequency and a $1/\omega^2$-slope at high frequencies.

An example of a stochastic climate model was given by Mikolajewicz and Maier-Reimer (1990). They carried out an experiment in which the Hamburg Large-Scale Geostrophic ocean general circulation model (Maier-Reimer et al., 1993) was driven by a spatially correlated white-noise freshwater flux superimposed on the climatological fluxes. A new type of variability characterized by a red noise spectrum, which is not produced by the same model driven by climatological fluxes, is found.

Following the concept of stochastic climate models, second moment statistics of slowly varying climate components are generated by fluctuations of fast varying climate components. Consequently, variability around the mean $< \mathbf{v}_s(t) >$ can be produced either by a fully coupled atmosphere-ocean model which explicitly produces atmospheric fluctuations, or by an ocean-only GCM in which fluctuating fluxes are prescribed as extra stochastic forcings.

In the case that both the mean and fluctuating fluxes change with time, coupled GCMs become the only possibility to describe the variability around the mean, since these models are potentially capable in producing changes in fluxes.

4.4 Concepts on the Generation of Spatial Variability

4.4.1 The General Idea

The concepts discussed in section 4.3 are essentially focused on the temporal climate variability. However, sections 4.1 and 4.2 show that both mean $< \mathbf{v}(t) >$ and covariance $< \mathbf{v}'\mathbf{v}'^\dagger >$ reveal also spatial variability. The major effort in theoretical meteorology and oceanography has been to identify processes which produce the spatial structures of the (present-day) mean climate state, such as the global distribution of surface easterly and westerly winds, or the western intensification of the surface current in the ocean. With increasing information about the variability around the mean, more studies are focused on the processes responsible for spatial features related to variability around the mean state.

These studies rely mainly on simplified equations of motions. The usefulness of this approach comes from the fact that processes responsible for a spatial feature of the considered phenomenon, whether related to the mean or to the variability around the mean, are not dynamically equally important. If one is interested in the *dominant* features of the phenomenon, *simplified* equations of motions, which contain only the dynamically most important processes of this phenomenon, are a powerful approach.

In practice, the following four-step approach is used. First, the spatial features of the considered phenomenon are identified. Second, a simplified equation system is formed. Third, the spatial signature produced by the simplified model is derived. This can be done analytically, numerically or in terms of scaling arguments. Fourth, the signature of the simple system is compared to that found in the first step. A correspondence suggests that the processes isolated in the simple model are responsible for the spatial features of the considered phenomenon.

In the following, this approach is applied to the modes (described by the first two EOFs of 200-hPa stream function shown in Fig. 4.13a and 4.13b) discussed in section 4.2.2b. Section 4.4.2 describes the dominant features of the two modes. Simplified equation systems which produce these features are discussed in section 4.4.2.

4.4.2 An Example: Identification of Processes Responsible for the Dominant Spatial Features of the Two Anisotropic Modes in the ECHAM1/LSG Atmosphere

Dominant spatial features of the modes

The anomalous streamlines shown in Fig. 4.13a and b suggest that motions related to the two reddest modes are anisotropic. In order to further isolate the meridional distributions of these anisotropic features, regressions between the normalized PC1 and PC2 of 200-hPa stream function and anomalies of zonally averaged zonal and meridional wind (Fig. 4.19) and of zonally averaged surface pressure (Fig. 4.20) are calculated. The amplitudes of anomalies shown in Fig. 4.19 and Fig. 4.20 represent one-standard-deviation anomalies. 500-year monthly data are used to produce the regression patterns.

Figure 4.19 shows that EOF1 in Fig. 4.13a is related to a maximum of zonal wind anomalies centered at the equator, whereas EOF2 in Fig. 4.13b is related to a meridional dipole structure of zonal wind anomalies at the mid- and high-latitudes. The meridional extension of these anomalies covers 50-60 degrees of latitude. The amplitudes reach about 1-2 ms^{-1} for both modes. The significance of these wind patterns is suggested by the percentage of the explained variances. The shaded areas in Fig. 4.19 indicate regions where the variance explained by the regression pattern is larger than 20% of the total unfiltered monthly variance. They coincide with the regions of large wind anomalies and have maximum values of about 40-70%. In the following, the mode described by PC1 of 200-hPa stream function is referred to as the tropical mode and that by PC2 as the Southern Hemisphere mode.

The vertical structures of the two modes are distinctly different from each other. The tropical mode is located in the upper troposphere, and disappears almost entirely below 700 hPa. In contrast to that, the Southern Hemisphere mode is barotropic. It exists throughout the whole troposphere.

Consistent with the strong anisotropy already seen in Fig. 4.13a and 4.13b, the anomalies of meridional velocity (Fig. 4.19) are found to be much smaller than those of zonal velocity (Note that the unit of meridional velocity is 0.1 ms^{-1}). This is particularly true for the tropical mode (Fig. 4.19a and b). For the Southern Hemisphere mode (Fig. 4.19c and d), the anomalies of meridional velocity reach about 0.02-0.04 ms^{-1} in the troposphere, and are about 0.3 ms^{-1} at the surface. The ratio of the meridional velocity to zonal velocity is about $O(10^{-2})$ in the troposphere and about $O(10^{-1})$ at the surface. Although the amplitudes are small, the corresponding pattern is significant in terms of the explained variance. Maximal explained variance of about 60% is found near the surface at about $50 - 60^o S$, and about 45% in the upper troposphere at about the same latitudes. It indicates a meridional circulation with an anomalous divergence at lower levels and an anomalous convergence at upper levels.

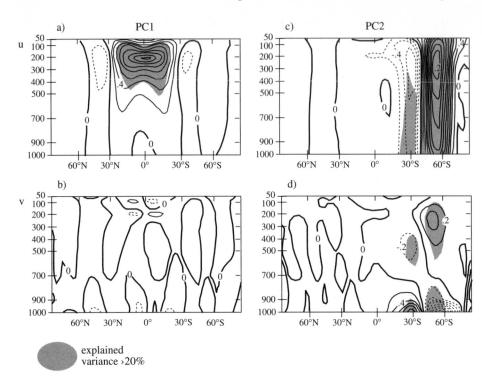

Fig. 4.19. Regression patterns between normalized PCs of 200-hPa stream function and zonally averaged zonal and meridional velocity, with a) and b) on the left for PC1 and c) and d) on the right for PC2. The shaded areas indicate areas where the variance explained by the corresponding regression pattern is larger than 20% of the monthly unfiltered variance. Since PCs are normalized, the anomalies are in physical units with zonal wind in ms^{-1} and meridional wind in $10^{-1}\ ms^{-1}$. From von Storch (1997b)

Figure 4.20 describes mass anomalies related to the tropical and Southern Hemisphere modes. In terms of both amplitude and explained variance (not shown), surface pressure anomalies are not related to the tropical mode (solid line in Fig. 4.20), but are significantly related to the Southern Hemisphere mode (line marked by $*$).

The different involvements of mass and velocity of the two modes can be further quantified by considering the global Ω angular momentum M_Ω,

$$M_\Omega = \frac{2\pi r^4 \Omega}{g} \int_{-\pi/2}^{\pi/2} [p_s] \cos^3 \varphi \; d\varphi \tag{4.44}$$

which is a function of latitudinally weighted mass described by zonally aver-

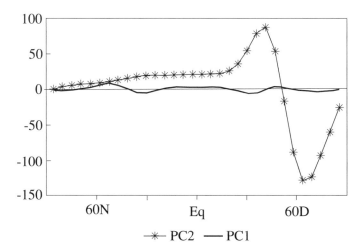

Fig. 4.20. Regression patterns between normalized PC1 and PC2 of 200-hPa stream function and zonally averaged area weighted surface pressure $p_s(\varphi)\cos\varphi$ in Pa. From von Storch (1997b)

aged surface pressure $[p_s]$, and the global relative angular momentum M_r,

$$M_r = \frac{2\pi r^3}{g} \int_{p_s}^{0} \int_{\pi/2}^{-\pi/2} [u] \cos^2\varphi \, d\varphi dp \qquad (4.45)$$

which is the global integral of latitudinally weighted zonally averaged zonal velocity $[u]$.

Denote the relative (Ω) angular momentum related to the tropical and Southern Hemisphere modes by $M_{r,tr}$ and $M_{r,sh}$ ($M_{\Omega,tr}$ and $M_{\Omega,sh}$), respectively. The anomalies in Fig. 4.19 and Fig. 4.20 are used to calculate $M_{r,tr}$, $M_{r,sh}$, $M_{\Omega,tr}$ and $M_{\Omega,sh}$. One obtains $\frac{M_{r,tr}}{M_{\Omega,tr}} = O(10^2)$, and $\frac{M_{\Omega,sh}}{M_{r,sh}} \simeq 3$, which clearly demonstrate the difference between ratios of zonal velocity to mass for the two modes.

The dominant features of the tropical and Southern Hemisphere modes are summarized below.

1. Different involvements of zonal velocity and mass, as characterized by the ratios $\frac{M_{\Omega,sh}}{M_{r,sh}} \simeq 3$ for the Southern Hemisphere mode and $\frac{M_{r,tr}}{M_{\Omega,tr}} \gg 1$ for the tropical mode.

2. Large spatial scales with the meridional scale L_m being comparable to the radius r of the earth.

3. Strong anisotropy with the ratio of amplitude of meridional velocity to that of zonal velocity being much smaller than one.

4. Different horizontal and vertical structures. The tropical mode is characterized by a non-uniform vertical distribution of u and a latitudinal mono-pole of u centered at the equator, whereas the Southern Hemisphere mode is described by a latitudinal dipole of u in the Southern Hemisphere mid- and high-latitudes and a vertical non-uniformly distribution of v indicating a meridional circulation.

Results of simple models

The investigation is carried out in two steps. First (section 4.4.2a), the features concerning the overall scales, i.e. points 1 to 3 as summarized above, are considered. It will be shown that these features are produced by linear quasi-geostrophic potential vorticity dynamics. As the second step (section 4.4.2b), more detailed structures of the modes are considered. Progress is made for the Southern Hemisphere mode. It will be shown that the effect of surface friction, as described by the Ekman velocity, is crucial for determining both the latitudinal distribution of mass and through the geostrophic balance that of zonal velocity, and the vertical-meridional circulation shown in Fig. 4.19d. In both sections, scaling analysis is used to characterize the solutions of simplified equations of motions.

a) The shallow water model: The role of the quasi-geostrophic
and linear quasi-geostrophic potential vorticity dynamics

The linear quasi-geostrophic dynamics are contained in the shallow water system linearized about the state of rest. The scaled linear shallow water equations are given below.

$$
\begin{array}{lll}
\varepsilon_t & & \dfrac{\partial u^*}{\partial t^*} - \delta \sin \varphi v^* + \delta N \dfrac{1}{\cos \varphi} \dfrac{\partial \eta^*}{\partial \theta^*} = 0 \\[3mm]
\varepsilon_t & \delta & \dfrac{\partial v^*}{\partial t^*} + \sin \varphi u^* + N \dfrac{\partial \eta^*}{\partial \varphi^*} = 0 \\[3mm]
\varepsilon_t & FN & \dfrac{\partial \eta^*}{\partial t^*} + \delta \dfrac{1}{\cos \varphi} \left(\dfrac{\partial u^*}{\partial \theta^*} + \dfrac{\partial \cos \varphi v^*}{\partial \varphi^*} \right) = 0
\end{array}
\qquad (4.46)
$$

Variables marked by an asterisk are dimensionless. t^* is time scaled by the typical time scale T of the considered motion. $\partial \theta^*$ and $\partial \varphi^*$ are zonal and meridional displacements $r\partial\theta$ and $r\partial\varphi$ scaled by the typical zonal and meridional length scales L_z and L_m. u^*, v^* are eastward and northward velocity scaled by U and V respectively. The vertical displacement of free surface η from a mean height H of the fluid is scaled by N_o. δ, F, ε_t and N

are dimensionless parameters defined as

$$\delta = \frac{L_m}{L_z} = \frac{V}{U} \tag{4.47}$$

$$F = \frac{L_m^2}{R^2} = \left(\frac{L_m}{\sqrt{(gH)}/2\Omega}\right)^2$$

$$\varepsilon_t = \frac{1}{(2\Omega T)}$$

$$N = \frac{gN_0/L_m}{2\Omega U}$$

where g is gravity, and Ω is the angular velocity of the earth.

The first parameter δ in (4.47) characterizes the degree of anisotropy of the motion. It equals one in a scaling analysis for isotropic motions. Values of δ for the tropical and the Southern Hemisphere modes are much smaller than one, but do not equal zero, i.e. the motions are not strictly zonally symmetric, but clearly prefer the zonal direction. The following three parameters in (4.47) are familiar. ε_t is the local Rossby number. F is the ratio of the square of the meridional length scale to that of the Rossby radius of deformation $R = (gH)^{1/2}/(2\Omega)$. N describes the ratio of the meridional pressure gradient to the typical Coriolis force $2\Omega U$ induced by zonal velocity U.

Using the length scale given in section 4.4.2, and furthermore considering the bending time scale of about a few months as the characteristic time scale, the orders of ε_t and F can be estimated for the tropical and the Southern Hemisphere modes. One has

$$F = O(10) \tag{4.48}$$

$$\varepsilon_t = O(10^{-3})$$

N is not a function of length scale only, its amplitude results from dynamical balance.

From the first two equations of (4.46), the geostrophic balance is given by

$$\sin\varphi = N \equiv \frac{gN_0/L_m}{2\Omega U} \tag{4.49}$$

Equation (4.49) suggests that at the equator where $\sin\varphi$ vanishes, the amplitude of mass anomaly, which is described by the amplitude N_o of surface elevation η, must be zero. This result is in good agreement with Fig. 4.20, where significant mass anomalies are found for the Southern Hemisphere mode, but not for the tropical mode which is centered at the equator.

Using (4.49), the ratio U/N_o, and furthermore the ratio of relative angular momentum to Ω angular momentum can be derived. Denote relative and Ω angular momenta per unit area, that is $HU\rho r\cos\varphi$ and $N_o\Omega\rho r^2\cos^2\varphi$, by m_r and m_Ω respectively, one has

$$\frac{m_r}{m_\Omega} = \frac{HU\rho r\cos\varphi}{N_o\Omega\rho r^2\cos\varphi^2} = \frac{2L_m}{Fr}\frac{1}{\cos\varphi\sin\varphi} = \frac{1}{O(5)\cos\varphi\sin\varphi} \tag{4.50}$$

where $L_m = O(r)$ and $F = O(10)$ are used. (4.50) indicates that $m_\Omega/m_r > 1$ for motions at the mid- and high-latitudes where $O(\sin\varphi) = O(\cos\varphi) = 1$, whereas $m_r/m_\Omega > 1$ for motion centered at the equator. However, the upper limits of m_Ω/m_r in extratropics and m_r/m_Ω in the equatorial area are distinctly different due to the singularity of $\sin\varphi$ at $\varphi = 0$. m_Ω/m_r of extratropical motions cannot exceed $O(5)$, but m_r/m_Ω of motion centered at the equator can have a much larger value. This result suggests that the observed ratios of zonal velocity to mass, as summarized in section 4.4.2, result from geostrophic balance of zonal velocity. The result that quasi-geostrophic motion exists in equatorial areas was also discussed by other authors (e.g. Matsuno, 1966).

It is noted that the expression in (4.50) exists only outside the poles. At the poles where $\cos\varphi = 0$, relative and Ω angular momenta per unit volume (that is, $HU\rho r\cos\varphi$ and $N_o\Omega\rho r^2\cos^2\varphi$) vanish, and m_r/m_Ω is not defined. Consequently, the ratio m_r/m_Ω has only one singularity at $\varphi = 0$. A high-latitude motion which extends to the pole is not related to an extremely large m_r/m_Ω.

Consider now the potential vorticity equation derived from (4.46).

$$\varepsilon_t \frac{\partial}{\partial t^*}(\zeta^* - FN\sin\varphi\eta^*) + \delta\frac{L_m}{r}\cos\varphi v^* = 0 \tag{4.51}$$

where ζ^* is the dimensionless vertical component of relative vorticity. r is the radius of the earth. $L_m/r \cos\varphi$ results from $\partial\sin\varphi/\partial\varphi^* = (L_m/r)\partial\sin\varphi/\partial\varphi$.

Equation (4.51) suggests that the time rate of change of relative potential vorticity (the term proportional to ε_t) must be balanced by the gain/loss of the planetary vorticity due to motion across isolines of planetary vorticity (the term proportional to v^*). The former is of the order of $\varepsilon_t \, max(1, FN\sin\varphi)$, and the latter of the order of $\delta(L_m/r)\cos\varphi$. For planetary-scale motions outside the direct neighborhood of $\varphi = \pm90°$, $O(\delta(L_m/r)\cos\varphi = O(\delta))$. If these motions also satisfy (4.48) and are quasi-geostrophic, one has
$$O(\varepsilon_t \, max(1, FN\sin\varphi)) = O(\varepsilon_t max(1, F\sin^2\varphi)) < O(10^{-2}).$$

The potential vorticity balance then requires

$$O(\delta) = O(\varepsilon_t \, max(1, F\sin^2\varphi)) < O(10^{-2}) \tag{4.52}$$

(4.52) suggests that the motion must be strongly anisotropic. In contrast, a small-scale ($L_m \ll r$) and low-frequency motion with $\varepsilon_t max(1, FN\sin\varphi) \ll 1$ can satisfy (4.51) and at the same time remains isotropic with $\delta = O(1)$. Thus, anisotropy results from the potential vorticity constraint for planetary-scale and low-frequency motions.

(4.52) suggests further that the degree of anisotropy is a function of latitude φ. (4.52) yields $O(\delta) = O(\varepsilon_t F) = O(10\,\varepsilon_t)$ for motions at the mid- and high-latitudes, where $O(\sin\varphi) = 1$; and $O(\delta) = O(\varepsilon_t)$ for motions centered at the equator where $\sin\varphi = 0$. Thus, the extratropical motions can be less anisotropic than the tropical motions. This result is consistent with the much smaller meridional velocity in Fig. 4.19b than in Fig. 4.19d.

The above consideration suggests that the features concerning the relative scales of zonal velocity to mass, and the zonal to meridional scale (i.e. the anisotropy) can be explained by the linear quasi-geostrophic potential vorticity dynamics.

b) The two-layer model:
The role of the lower boundary for the Southern Hemisphere mode

Processes related to properties of the lower boundary, such as surface friction and surface topography, produce vertically varying motions. However, vertical variations are generally related to horizontal features, so that surface processes are potentially capable of determining both vertical and horizontal structures. For the Southern Hemisphere mode, which locates over the southern ocean, the process induced by surface friction is much more important than the process induced to topography.

The simplest model, which takes the vertical dependence of flow into account, is a two layer model with the lower layer being the boundary layer and the upper layer representing the interior fluid. The effect of surface friction on the interior flow on a f-plane is well known. However, von Storch (1997b) pointed out that this effect can be distinctly different from the effect of surface friction on *planetary* interior flow, which has to be formulated on a sphere. In order to demonstrate that, we consider the potential vorticity equation of the interior flow.

For motions on a f-plane, the dimensional potential vorticity equations of the interior flow in the absence and presence of surface friction are respectively given by

$$\frac{\partial}{\partial t^*}(\zeta^* - F\eta^*) = 0 \qquad (4.53)$$

and

$$\frac{\partial}{\partial t^*}(\zeta^* - F\eta^*) = -lw_e^* = -l\zeta^* \qquad (4.54)$$

where $\zeta^* = \nabla^2\eta^*$ for quasi-geostrophic motions. l is depth of Ekman layer scaled by the depth of total fluid. w_e^* is the dimensionless Ekman velocity. It is shown in many textbooks (e.q. Pedlosky, 1987) that the Ekman velocity induced by the interior flow with relative vorticity ζ^* on a f-plane is proportional to ζ^*. When η^* is expressed in terms of harmonic functions, so that $\nabla^2\eta^* = c\eta^*$ holds, the inviscid and viscid solutions of (4.53) and (4.54), η_{inv}^* and η_v^*, can be written as

$$\eta_{inv}^* = \eta_{inv}^*(x^*, y^*) \qquad (4.55)$$

and

$$\eta_v^* = e^{-\alpha t^*}\eta_{inv}^*(x^*, y^*) \qquad (4.56)$$

where $\alpha = cl^*/(c - F)$. $\eta^*_{inv}(x^*, y^*)$ is a function of the coordinates x^* and y^* on a f-plane. (4.55) and (4.56) show that for motions on a f-plane, w^*_e acts only to reduce the amplitude of the interior flow, and does not affect the horizontal structure of the interior flow.

The situation is different for planetary motions whose potential vorticity equation is given by (4.51). For planetary-scale motions at mid- and high-latitudes, (4.51) can be further simplified. For motions at mid and high-latitudes where $O(\sin \varphi) = O(\cos \varphi) = 1$, one has $O(FN \sin \varphi) = O(F) \gg 1$, indicating that relative to the contribution from η^*, the contribution from relative potential vorticity ζ^* to total potential vorticity can be ignored. In this case, the potential vorticity equation of the interior flow, in the absence of surface friction, is given by

$$
\varepsilon_t FN \sin \varphi \, \frac{\partial \eta^*}{\partial t^*} - \delta \frac{L_m}{r} \cos \varphi \, v^* \tag{4.57}
$$
$$
= \varepsilon_t FN \sin \varphi \, \frac{\partial \eta^*}{\partial t^*} - \delta \frac{L_m}{r} N \frac{1}{\sin \varphi} \frac{\partial \eta^*}{\partial \theta^*} = 0
$$

where v^* is expressed in terms of zonal pressure gradient using the geostrophic balance. In the presence of surface friction, its effect on the interior flow must be taken into account. As in the case of motions on a f-plane, this effect is described by Ekman velocity w_e, so that the potential vorticity equation of interior flow becomes

$$
\varepsilon_t FN \sin \varphi \frac{\partial}{\partial t^*} \eta^* - \delta \frac{L_m}{r} N \frac{1}{\sin \varphi} \frac{\partial \eta^*}{\partial \theta^*} = l \sin \varphi w^*_e \tag{4.58}
$$

Different from (4.54), the variation of $\sin \varphi$ with latitude φ introduces the additional term $\delta \frac{L_m}{r} N \frac{1}{\sin \varphi} \frac{\partial \eta^*}{\partial \theta^*}$ into the potential vorticity equation. Furthermore, w^*_e depends not only on the interior relative vorticity, but also on the interior velocity. For strong anisotropic flow, one has

$$
w^*_e = \pm \frac{1}{2\gamma} \zeta^* \pm \frac{\cos \varphi}{4\gamma \sin \varphi} u^* \tag{4.59}
$$
$$
= \pm \frac{N}{2\gamma \cos \varphi \sin \varphi} \frac{\partial}{\partial \varphi^*} (\frac{\partial \eta^*}{\partial \varphi^*} \cos \varphi) \mp \frac{3N \cos \varphi}{4\gamma \sin^2 \varphi} \frac{\partial \eta^*}{\partial \varphi^*}
$$

with $\gamma^2 = \pm \sin \varphi$ for $\varphi > 0$ and $\varphi < 0$ respectively.

Denoting solutions of inviscid and viscid planetary-scale potential vorticity equations (4.57) and (4.58) by η^*_{inv} and η^*_v, respectively, one has

$$
\eta^*_{inv} = \eta^*_{inv}(\alpha_1 t^* + \alpha_2 \theta^*) \tag{4.60}
$$

and

$$
\eta^*_v = \eta^*_v(\theta^*, \varphi^*, t^*) \tag{4.61}
$$

The inviscid solution $\eta^*_{inv}(\alpha_1 t^* + \alpha_2 \theta^*)$ is any function of $\alpha_1 t^* + \alpha_2 \theta^*$ and does not depend on latitude φ^*. On the contrary, the viscid solution (4.61) is latitudinally specified because of the latitudinal derivatives appearing in the expression of w_e^*. Thus, the presence of w_e^* induces a meridional restriction for the meridional distribution of mass and, through geostrophic balance, also for the meridional distribution of zonal velocity. It is suggested that the meridional dipole structure shown in Fig. 4.19c is controlled by Ekman velocity w_e.

A consideration of (4.58) further suggests that the vertical structure, in particular the meridional circulation shown in Fig. 4.19d, also results from the surface friction. In contrast to a quasi-geostrophic motion on a f-plane, which is free of divergence, the divergence of a quasi-geostrophic planetary-scale motion is given by

$$\delta \frac{1}{\cos \varphi} \left(\frac{\partial u^*}{\partial \theta^*} + \frac{\partial \cos \varphi v^*}{\partial \varphi^*} \right) = \delta \frac{L_m}{r} N \frac{1}{\sin \varphi} \frac{\partial \eta^*}{\partial \theta^*} \tag{4.62}$$

Thus, the second term on the left-hand side of (4.58) represents the divergence of quasi-geostrophic planetary-scale interior motion. The term on the right-hand side describes the mass divergence in the boundary layer. The two terms together describe the total meridional mass divergence. Within the two-layer system, this divergence would produce a meridional-vertical circulation of the type of that shown in Fig. 4.19d. Furthermore, the total divergence determines the amplitude of the time rate of change of mass; the smaller the amplitude of the total divergence, the smaller is ε_t and the larger is the time scale T. In order to obtain a small total divergence for the persistent Southern Hemisphere mode, the interior divergence needs to be compensated, to a large extent, by the boundary divergence. This is only possible when v in the boundary layer, which is much thinner than the interior fluid layer, is much larger than v in the interior layer. Such a vertical distribution of the amplitude of v is observed in Fig. 4.19d.

The major difference between (4.53), (4.54) and (4.57), (4.58) results from the fact that for planetary motions with $L_m = O(r)$, the latitudinal variations of $\sin \varphi$ can no longer be ignored. With $\partial \sin \varphi / \partial \varphi \neq 0$, two conditions which lead to (4.55) and (4.56), namely the non-divergence of quasi-geostrophic flow and w_e being a function of ζ only, are no longer valid. Without these two conditions, another type of quasi-geostrophic potential vorticity dynamics which has a three-dimensional nature comes into play. The Ekman velocity w_e induced by a planetary-scale interior motion controls both the latitudinal distribution of mass and zonal velocity, and the meridional-vertical circulation with low level divergence being out of phase with the high-level divergence.

4.5 Conclusions

The study of climate time series suggest that climate variability is a statistical phenomenon, so that expected quantities, such as the mean and the covariance, are required for its description. However, these two quantities contain different types of information. Variability related to the mean becomes only apparent when paleo time scales are involved, whereas variability around the mean can occur on any time scale. For the present-day climate, which is considered to be essentially stationary, the climate variability is described by the covariances or the frequency-decomposed covariances, i.e. cross-spectrum matrices.

For discrete time series, the cross-spectrum matrix is represented by many $M \times M$ matrices, each of them containing spectral information at one frequency. Due to the high dimensionality M of climate state variables, it is difficult to obtain compact information from these matrices. Several techniques which derive interpretable information about the cross-spectrum matrices using different assumptions are discussed.

Two estimates of second moment statistics of present-day climate system are presented. One is derived from the observed and the other from the simulated data. The results suggest that the dominant variability is described by smooth and continuous spectra associated with large-scale patterns. Except for the ENSO mode, which has a multivariate peaked spectrum, the considered atmospheric spectra are flat (i.e. without peak), and can be represented by an AR(1) process. The spatial structures seem to be coupled to the redness of the spectra. The reddest spectra are related to the most anisotropic patterns. The results provide estimates of the variability of the troposphere and the very upper layer of the ocean, for which data are available. Very little is known about second moment statistics of the deep ocean.

The concepts concerning the origin of temporal variability can be divided into two groups. The first deals essentially with changes of the mean climate. It is assumed that for these changes, only the averaged effects of the atmospheric forcings need to be considered. Changes of the mean climate conditions are induced either by external forcings of the climate system, such as variations of incoming solar radiation, or by changes in the mean atmospheric fluxes, such as changes in the mean fresh water fluxes, which lead to multiple equilibria of the deep ocean circulation.

The second group of concepts concerning origin of temporal variability is proposed by Hasselmann. He suggested that a model describing second moments of a slowly varying climate component should include a stochastic forcing stemming from fast varying weather events. Within such a model, the variance of the slowly varying climate variable is excited by the stochastic forcing, whereas the net effect of the internal dynamics of the slow component is to dissipate the variance. The concept of stochastic climate models is the only concept which explicitly considers climate variability as being a statistical phenomenon.

The spatial variability is normally studied using simplified equations of motions. Simple models are used to identify the dominant processes responsible for the spatial features of the two reddest modes found in the ECHAM1/LSG model. The overall scales of the two modes can be explained by the linear quasi-geostrophic potential vorticity dynamics. The Ekman velocity of a planetary motion plays an important role in shaping the latitudinal structure and in forming the meridional-vertical circulation of the Southern Hemisphere mode. This role of the Ekman velocity becomes only apparent for motions with large meridional scale.

The success of simplified equation systems indicates that the averaged spatial structures have a deterministic nature. However, one cannot explain all the features of second moments, in particular those of spectra, by considering simplified equations only. Further studies are required to understand the nature of temporal and spatial variabilities and their relationship.

Part II

Climate Change

Chapter 5

Weather Modification by Cloud Seeding – A Status Report 1989-1997

by William R. Cotton

Abstract

In this chapter the results and status of weather modification research during the period 1989-1997 are examined. I focus on three methods of cloud seeding: the "static mode" and "dynamic mode" of seeding supercooled clouds, and the hygroscopic seeding of warm clouds.

During this period the scientific basis of static seeding has undergone much scrutiny and evaluation resulting in considerable controversy in the literature. While some of the recent work bolsters the early optimism of cloud seeding, overall the image of the scientific credibility of the static seeding concept has been tarnished more than it has been enhanced. One fallout is that scientific research in weather modification has been essentially curtailed in all the developed countries.

While the dynamic seeding concept offers the potential for greater rainfall enhancements than static seeding, it is a much more complex concept. The basic working hypothesis behind static seeding has undergone substantial revision during the period. There is increasing evidence that dynamic seeding can increase rainfall from radar-defined floating targets, but demonstrations of consistent increases in rainfall over fixed ground targets remains elusive. It still remains an unproven candidate for application to water resource management.

While the optimism for enhancing precipitation by glaciogenic seeding of supercooled clouds has waned, optimism for the potential of hygroscopic seeding has grown. This is a result of the refinement of hygroscopic seeding delivery systems, modeling studies, and the results of observations from purposefully- and inadvertently-modified clouds. Nonetheless this work is still very exploratory and is a long way from proving that hygroscopic seeding can result in reliable, significant increases in rainfall on the ground.

We conclude by discussing the parallels between purposeful weather modification and studies of anthropogenic influences on global climate and consider:

- The importance of natural variability,

- The dangers of overselling,

- The capricious administration of science, and

- scientific credibility and advocacy.

5.1 Introduction

The purpose of this paper is to update the findings and concepts summarized in *Human Impacts on Weather and Climate* by Cotton and Pielke (1995). In that book we describe the concepts related to purposeful weather modification by cloud seeding, inadvertent modification of weather and climate on regional scales, and human impacts on global climate change. We also discuss the methods and status for evaluating a cause and effect between human activities and observed weather and climate changes. I will focus specifically on revisions to Part I of that book which we call "The Rise and Fall of the Science of Weather Modification by Cloud Seeding". While this book was first published by Cambridge Press in 1995, prior to that it was self-published by ASTeR Press in 1992. Because Part I was the earliest chapter written, we have updated this part for the period beginning about 1989 through 1997.

In this paper I will focus only on three methods of seeding clouds. The first two are related to supercooled clouds and are called the "static mode" of cloud seeding and the "dynamic mode" of cloud seeding. The third method is the modification of warm clouds by hygroscopic seeding.

5.2 The Static Mode of Cloud Seeding

The main objective of the "static mode" of cloud seeding is to increase the efficiency of precipitation formation by introducing an "optimum" concentration of ice crystals in supercooled clouds. It was originally thought that clouds were deficient in ice nuclei and therefore additions of modest concentrations of ice nuclei should result in a more efficient precipitation-producing cloud system. All that was needed was to introduce seeding material from the ground or at the base of clouds which would then enhance ice crystal concentrations and thereby increase rainfall. Cotton and Pielke (1995) concluded that physical studies and inferences drawn from statistical seeding experiments over the last 50 years suggests that there exists a much more limited window of opportunity for precipitation enhancement by the static-mode of cloud seeding than was originally thought. The window of opportunity for cloud seeding appears to be limited to:

1) clouds which are relatively cold-based and continental;

2) clouds having top temperatures in the range -10 to -25 °C;

3) a time scale limited by the availability of significant supercooled water before depletion by entrainment and natural precipitation processes.

This limited scope of the opportunities for rainfall enhancement by the static-mode of cloud seeding that has emerged in recent years may explain why some cloud seeding experiments have been successful, while other seeding experiments have yielded either inferred reductions in rainfall from seeded clouds or no effect. A successful experiment in one region does not guarantee that seeding in another region will be successful unless all environmental conditions are replicated as well as the methodology of seeding. This, of course, is highly unlikely.

We argued that the success of a cloud seeding experiment or operation, therefore, requires a cloud forecast skill that is far greater than is currently in use. As a result, such experiments or operations are at the mercy of the *natural variability* of clouds. The impact of natural variability may be reduced in some regions where the local climatology favors clouds, which are in the appropriate temperature windows and are more continental. A 'time window' may still exist, however, and this will yield uncertainty to the results unless the field personnel are particularly skillful in selecting suitable clouds.

Furthermore, we concluded by stating that "the 'static' mode of cloud seeding has been shown to cause the expected alterations in cloud microstructure including increased concentrations of ice crystals, reductions of supercooled liquid water content, and more rapid production of precipitation elements in both cumuli (Cooper and Lawson, 1984) and orographic clouds (Reynolds, 1988; Super and Boe, 1988; Super et al., 1988; Super and Heimbach, 1988; Reynolds and Dennis, 1986). The documentation of increases in precipitation on the ground due to static seeding of cumuli, however, has been far more elusive with the Israeli experiment (Gagin and Neumann, 1981) providing the strongest evidence that static seeding of cold-based, continental cumuli can cause significant increases of precipitation on the ground. The evidence that orographic clouds can cause significant increases in snowpack, we argued, is far more compelling, particularly in the more continental and cold-based orographic clouds (Mielke et al., 1981; Super and Heimbach, 1988)."

But even these conclusions have been brought into question in the last 10 years. The Climax I and II wintertime orographic cloud seeding experiments (Grant and Mielke, 1967; Mielke et al., 1971; Chappell et al., 1971; Mielke et al., 1981) are generally acknowledged by the scientific community (National Academy of Sciences, 1975; Sax et al., 1975; Tukey et al., 1978) for providing the strongest evidence that seeding those clouds can significantly increase precipitation. Nonetheless, Rangno and Hobbs (1987; 1993) question both the randomization techniques and the quality of data collected during those experiments and conclude that the Climax II experiment failed to confirm that precipitation can be increased by cloud seeding in the Colorado Rockies. Even so, Rangno and Hobbs (1987) did show that precipitation may have been increased by about 10% in the combined Climax I and II experiments. This should be compared, however, to the original analyses by Grant et al. (1969), Mielke et al. (1970) and Mielke et al. (1971), which indicated greater than 100% increase in precipitation on seeded days for Climax I and 24% for Climax II. Subsequently, Mielke (1995) explained a number of the criticisms made by Rangno and Hobbs in regard to the statistical design of the experiments, in particular the randomization procedures, the quality and selection of target and control data and the use of 500 mb temperature as a partitioning criteria. It is clear that the design, implementation, and analysis of this experiment was a learning process not only for meteorologists but statisticians as well.

The results of the many re-analyses of the Climax I and II experiments have clearly "watered down" the overall magnitude of the possible increases in precipitation in wintertime orographic clouds. Furthermore, they have revealed that many of the concepts that were the basis of the experiments are far too simplified compared to what we know today. Furthermore, many of the cloud systems seeded were not simple "blanket-type orographic clouds", but were part of major wintertime cyclonic storms that pass through the region. As such, there was a greater opportunity for ice multiplication processes and riming processes to operative in those storms, making them less susceptible to cloud seeding.

As noted above, Cotton and Pielke (1995) concluded that the strongest evidence of significant precipitation increases by static seeding of cumulus clouds came from the Israel I and II experiments. It is clear there are no "sacred cows" in the science of weather modification! Even the Israeli experiments (Gagin and Neumann, 1981) have come under attack by Rangno and Hobbs (1995). From their re-analysis of both the Israel I and II experiments, they argue that the appearance of seeding-caused increases in rainfall in the Israel I experiment was due to "lucky draws" or a Type I statistical error. Furthermore, they argued that during Israel II naturally heavy rainfall over a wide region encompassing the north target area gave the appearance that seeding caused increases in rainfall over the north target area. At the same time, lower natural rainfall in the region encompassing the south target area gave the appearance that seeding decreased rainfall over that target area.

Rosenfeld and Farbstein (1992) suggested that the differences in seeding effects between the north and south target areas during Israel II is the result of the incursion of desert dust into the cloud systems. They argue that the desert dust contains more active natural ice nuclei and that they can also serve as coalescence embryos enhancing collision and coalescence among droplets. Together, the dust can make the clouds more efficient rain-producers and less amenable to cloud seeding.

Cotton and Pielke (1995), among others, argued that the "apparent" success of the Israeli seeding experiments was due to the fact that they are more susceptible to precipitation enhancement by cloud seeding. This is because numerous studies (Gagin, 1971; Gagin, 1975; Gagin, 1986; Gagin and Neumann, 1974) have shown that the clouds over Israel are continental, having cloud droplet concentrations of about 1000 cm^{-3} and that ice particle concentrations are generally small until cloud top temperatures are colder than -14 ^{o}C. There is little evidence for ice particle multiplication processes operating in those clouds.

Rangno and Hobbs (1995) also reported on observations of clouds over Israel containing large supercooled droplets and quite high ice crystal concentrations at relatively warm temperatures. In addition, Levin (1994) presented evidence of active ice multiplication processes in Israeli clouds. This further erodes the perception that the clouds over Israel were quite susceptible to seeding. Naturally, Rangno and Hobbs (1995) paper generated quite a large reaction in the weather modification community. The March issue of the Journal of Applied Meteorology contained a series of comments and replies related to their paper (Rosenfeld, 1997; Rangno and Hobbs, 1997a; Dennis and Orville, 1997; Rangno and Hobbs, 1997b; Woodley, 1997; Rangno and Hobbs, 1997c; Ben-Zvi, 1997; Rangno and Hobbs, 1997d). These comments and responses clarify many of the issues raised by Rangno and Hobbs (1995). Nonetheless, the image of what was originally thought as being the best example of the potential for precipitation enhancement of cumulus clouds by static seeding has become considerably tarnished.

Recently, Ryan and King (1997) presented a comprehensive overview of over 47 years of cloud seeding experiments in Australia. These studies almost exclusively focused on the static seeding concept. In this water-limited country, cloud seeding has been considered as a potentially important contributor to water management. As a result, their review included discussions of the overall benefit/cost to various regions.

In spite of having considerable professional contact with many of the scientists in Australia, I was surprised and overwhelmed by the number of cloud seeding experiments that have been carried out there. Over 14 cloud seeding experiments were conducted covering much of southeastern, western, and central Australia, as well as, the island of Tasmania. Ryan and King (1997) concluded that static seeding over the plains of Australia is not effective. They argue that for orographic stratiform clouds, there is strong statistical evidence that cloud seeding increased rainfall, perhaps by as much as 30 % over Tasmania when cloud top temperatures are between -10 and -12 °C in southwesterly airflow. The evidence that cloud seeding had similar effects in orographic clouds over the mainland of southeastern Australia is much weaker. This is somewhat surprising from a physical point of view since the clouds over Tasmania are maritime. As such, one would expect the opportunities for warm-cloud collision and coalescence precipitation processes to be fairly large. Furthermore, in those maritime clouds ice multiplication processes should be operative; especially when embedded cumuliform cloud elements are present. Thus, natural ice crystal concentrations should be competitive with concentrations expected from static seeding, especially in the -10 to -12 °C temperature range. If the results of the Tasmanian experiments are real, benefit/cost analyses suggest that seeding has a gain of about 13/1. This is viewed as a real gain to hydrologic energy production. I guess we'll have to wait and see what Rangno and Hobbs have to say about the Tasmanian experiment.

Another exploratory study of static seeding effects on precipitation that has suggested positive yields was reported by Mather et al. (1996a). They analyzed a total of 127 storms over South Africa using an objective radar-based storm tracking technique. They found that the radar-measured rain flux and storm area from seeded clouds was significantly greater than the control population of clouds. These analyses are for radar-defined floating targets. They do not, however, tell one how effective cloud seeding is in increasing rainfall over fixed target areas on the ground.

Overall, since 1989 the scientific basis of static seeding of supercooled clouds has undergone considerable scrutiny and evaluation. While some of the recent work bolsters the early optimism of the potential of static seeding, overall the image of the scientific credibility of the static seeding concept has been tarnished more than it has been enhanced. Skepticism of its overall potential for a significant cost-effective component to water resource management prevails.

5.3   The Dynamic Mode of Cloud Seeding

While the fundamental concept of the 'static mode' of cloud seeding is that precipitation can be increased in clouds by enhancing their precipitation efficiency, alterations in the dynamics or air motion in clouds due to latent heat release of growing ice particles, redistribution of condensed water, and evaporation of precipitation is also inevitable. Alterations in the dynamics of clouds, however, is not the primary aim of the strategy. By contrast, the focus of the 'dynamic mode' of cloud seeding is to enhance the vertical air currents in clouds and thereby vertically process more water through the clouds resulting in increased precipitation. The main difference in implementation of the strategy is that larger amounts of seeding material are introduced into clouds. A goal in the static mode of seeding is to achieve something like 1 to 10 ice crystals per liter at temperatures warmer than -15 ^{o}C. In the dynamic mode of seeding the target ice crystal concentration is more like 100 to 1000 ice crystals per liter, which corresponds to seeding as much as 200 to 1000 g of silver iodide in flares dropped directly into the high supercooled liquid water content updrafts of cumuli. In the 1960's to the 1980's, the hypothesized chain of physical responses to the insertion of such large quantities of seeding materials as summarized by Woodley et al. (1982) included the following: "(1) the nucleated ice crystals glaciate a large volume of the cloud releasing the latent heat of freezing and vapor deposition, (2) this warms the cloud yielding additional buoyancy in the seeded updrafts, (3)the updrafts with enhanced buoyancy accelerate causing the cloud towers to ascend deeper into the troposphere, (4) pressure falls beneath the seeded cloud towers and convergence of unstable air in the cloud will as a result develop, (5) downdrafts are enhanced, (6) new towers will therefore form, (7) the cloud will widen, (8) the likelihood that the new cloud will merge with neighboring clouds will *therefore increase, and (9) increased moist air is processed by the cloud to form rain. "*

Few of these hypothesized responses to dynamic seeding have been observationally documented in any systematic way. Observations in clouds seeded for dynamic effects showed that seeding did indeed glaciate the clouds, that is, convert the cloud from liquid to primarily ice (Sax, 1976; Sax et al., 1979; Sax and Keller, 1980; Hallett, 1981). Likewise, there is evidence that seeding cumulus clouds in the Caribbean and over Florida results in deeper clouds (Simpson et al., 1967; Simpson and Woodley, 1971). The remainder of the elements of the hypothesized chain of events have not been documented, however.

In recent years the dynamic seeding strategy has been applied to Thailand and West Texas. No results are available yet from Thailand, but some results from exploratory dynamic seeding experiments over west Texas have been reported by Rosenfeld and Woodley (1989; 1993). Analysis of the seeding of 183 convective cells suggests that seeding increased the maximum height of the clouds by 7 %, the areas of the cells by 43 %, the durations by 36 %, and the rain volumes of the cells by 130 %. Overall the results are encouraging, but such small increase in vertical development of the clouds is hardly consistent with earlier exploratory seeding experiments.

As a result of their experience in Texas, Rosenfeld and Woodley (1993) proposed an altered conceptual model of dynamic seeding as follows:

1) NONSEEDED STAGES

(i) *Cumulus growth stage*

The freezing of supercooled raindrops plays a major role in the revised dynamic seeding conceptual model. Therefore, a suitable cloud is one that has a warm base and a vigorous updraft that is strong enough to carry any raindrops that are formed in the updraft above the 0°C isotherm level. Such a cloud has a vast reservoir of latent heat that is available to be tapped by natural processes or by seeding.

(ii) *Supercooled rain stage*

At this stage a significant amount of supercooled cloud and rainwater exists between the 0°C and the -10 °C levels, which is a potential energy source for future cloud growth.

A cloud with active warm rain processes but a weak updraft will lose most of the water from its upper regions in the form of rain before growing into the supercooled region. Therefore, only a small amount of water remains in the supercooled region for the conversion to ice. Such a cloud has no dynamic seeding potential.

(iii) *The cloud-top rain-out stage*

If the updraft is not strong enough to sustain the rain in the supercooled region until it freezes naturally, most of it will fall back toward the warmer parts of the cloud without freezing. The supercooled water that remains will ultimately glaciate. The falling rain will load the updraft and eventually suppress it, cutting off the supply of moisture and heat to the upper regions of the cloud, thus terminating its vertical growth. This is a common occurrence in warm rain showers from cumulus clouds.

(iv) *The downdraft stage*

At this stage, the rain and its associated downdraft reach the surface, resulting in a short-lived rain shower and gust front.

(v) *The dissipation stage*

The rain shower, downdraft, and convergence near the gust front weaken during this stage, lending no support for the continued growth of secondary clouds, which may have been triggered by the downdraft and its gust front.

2) SEEDED STAGES

(i) *Cumulus growth and supercooled rain stages*

These stages are the same for the seeded sequence as they are for natural processes.

(ii) *The glaciation stage*

The freezing of the supercooled rain and cloud water near the cloud top at this stage may occur either naturally or be induced artificially by glaciogenic seeding. This conceptual model is equally valid for both cases.

The required artificial glaciation is accomplished at this stage through intensive, on-top seeding of the updraft region of a vigorous supercooled cloud tower using a glaciogenic agent (e.g., AgI). The seeding rapidly converts most of the supercooled water to ice during the cloud's growth phase. The initial effect is the formation of numerous small ice crystals and frozen raindrops.

This rapid conversion of water to ice releases fusion heat—faster and greater for the freezing of raindrops—which acts to increase tower buoyancy and updraft and, potentially, its top height. (The magnitude of the added buoyancy is modified by the depositional heating or cooling that may occur during the adjustment to ice saturation; see Orville and Hubbard, 1973.) Entrainment is likely enhanced in conjunction with the invigorated cloud circulation.

The frozen water drops continue to grow as graupel as they accrete any remaining supercooled liquid water in the seeded volume and/or when they fall into regions of high supercooled liquid water content. These graupel particles will grow faster and stay aloft longer because their growth rate per unit mass is higher and their terminal fall velocity is lower than water drops of comparable mass. This will cause the tower to retain more precipitation mass in its upper portions. Some or all of the increase cloud buoyancy from seeding will be needed to overcome the increased precipitation load.

If the buoyancy cannot compensate for the increased loading, however, the cloud will be destroyed by the downdraft that contains the ice mass. The downdraft will be augmented further by cooling from the melting of the ice hydrometeors just below the freezing level.

The retention of the precipitation mass in the cloud's upper portions delays the formation of the precipitation-induced downdraft and the resultant disruption of the updraft circulation beneath the precipitation mass. This delay allows more time for the updraft to feed additional moisture into the growing cloud.

(iii) *The unloading stage*

The larger precipitation mass in the upper portion of the tower eventually moves downward along with the evaporatively cooled air that was entrained from the drier environment during the tower's growth phase. When the precipitation descends through the updraft, it suppresses the updraft. If the invigorated pulse of convection has had increased residence time in regions of light to moderate wind shear, however, the precipitation-induced downdraft may form adjacent to the updraft, creating an enhanced updraft-downdraft couplet. This unloading of the updraft may allow the cloud a second surge of growth to cumulonimbus stature.

When the ice mass reaches the melting level, some of the heat released in the updraft during the glaciation process is reclaimed as cooling in the downdraft. This downrush of precipitation and cooled air enhances the downdraft and the resulting outflow beneath the tower.

(iv) *The downdraft and merger stage*

The precipitation beneath the cloud tower is enhanced when the increased water mass reaches the surface. In addition, the enhancement of the downdraft increases the convergence at its gust front.

(v) *The mature cumulonimbus stage*

The enhanced convergence acts to stimulate more neighboring cloud growth, some of which will also produce precipitation, leading to an expansion of the cloud system and its conversion to a fully developed cumulonimbus system.

When this process is applied to one or more suitable towers residing within a convective cell as viewed by radar, greater cell area, duration, and rainfall are the result. Increased echo-top height is a likely but not a necessary outcome of the seeding, depending on how much of the seeding-induced buoyancy is needed

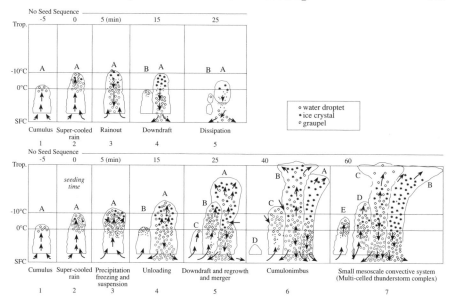

Fig. 5.1. Diagrammatic illustration of the dynamic seeding conceptual model for warm-based supercooled cumuli. Revised as of July 1992. (From Rosenfeld and Woodley, 1993.)

to overcome the increased precipitation loading.

(vi) *The convective complex stage*

When seeding is applied to towers within several neighboring cells, increased cell merging and growth will result, producing a small mesoscale convective system and greater overall rainfall.

This is an idealized sequence of events. Dissipation may follow the glaciation stage or at any subsequent stage if the required conditions are not present."

Figure 5.1 illustrates their revised conceptual model of dynamic seeding. This conceptual model differs from the earlier one in that it emphasizes the conversion of liquid water into graupel particles, which fall slower and grow faster than water drops of comparable mass. The seeding-induced graupel particles will reside in the cloud updraft longer and achieve larger size than a population of water drops in a similar unseeded cloud. They explain the lack of enhanced vertical development of the seeded clouds to increased precipitation mass loading. The enhanced thermal buoyancy of the cloud due to seeding-induced ice phase conversion, they argue, is offset by the increased mass loading, which results in only modest increases in updraft strength and cloud top height.

This new concept emphasizes that rapid conversion of supercooled liquid water into grauple must take place in the seeded plume. As such, it is limited to rather warm-based, maritime clouds having a broad cloud droplet distribution and supercooled raindrops. Numerous modeling studies have shown that the speed of conversion of supercooled liquid water to ice is facilitated by the presence of supercooled raindrops (Cotton, 1972a,b; Koenig and Murray, 1976; Scott and Hobbs, 1977; Lamb et al., 1981). The supercooled raindrops readily collect the ice crystals nucleated by the seeding agent and freeze. The frozen raindrops then collect cloud droplets becoming low-density graupel particles if the liquid water content of the cloud is low or modest, or become high-density hailstones if the liquid water contents are rather large.

Rosenfeld and Woodley (1993) argue that the retention of the increased ice mass in the form of graupel is an important new aspect of their dynamic seeding conceptual model. This may delay the formation of a downdraft and allows more time for further growth of the cloud. The eventual unloading of the enhanced water mass, they argue, is favorable for subsequent regeneration of the cloud by the downdraft-induced gust fronts leading to larger, longer-lived cells.

In summary, the concept of dynamic seeding is a physically plausible hypothesis that offers the opportunity to increase rainfall by much larger amounts than simply enhancing the precipitation efficiency of a cloud. It is a much more complex hypothesis, however, requiring greater quantitative understanding of the behavior of cumulus clouds and their interaction with each other, with larger-scale weather systems, and depends on the details of precipitation evolution. Being a complex, multi-link chain of steps, the hypothesis is very vulnerable to one link of the chain being wrong, or that the full chain works together in rather limited circumstances. Measurements and modeling studies are needed to support this hypothesis since the seeding experiments, while suggestive of being successful, are still vulnerable to type-I statistical errors. This is always a concern with convective storms since the *natural variability* of these storms is so large.

Overall, the dynamic seeding experiments have demonstrated rainfall increases for radar-defined "floating" targets or clusters of convective cells. They have not demonstrated, however, that rainfall can be increased over fixed ground target areas consistently. Thus, the dynamic seeding concept remains as yet an unproven candidate for application to water resource management.

5.4   Hygroscopic Seeding

As noted in Cotton and Pielke (1995) the dominant process for precipitation formation in warm clouds is collision and coalescence. We have seen that this process is very effective in clouds which are warm-based and maritime, or have substantial liquid water contents. The collision and coalescence process among liquid drops is also an important contributor to rain formation in many mixed-phase clouds, and the presence of supercooled drizzle-drops and raindrops enhances the rate of formation of precipitation in supercooled portions of clouds as well.

One method of seeding clouds to enhance precipitation is to introduce hygroscopic particles (salts), which readily take on water by vapor deposition in a supersaturated cloudy environment. The conventional approach is to produce ground salt particles in the size-range of 5-100 μm, and release these particles into the base of clouds. These particles grow by vapor deposition and readily reach sizes of 25 to 30 μm in diameter or greater. They are then large enough to serve as "coalescence" embryos and initiate or participate in rain formation by collision and coalescence.

Cotton and Pielke (1995) reviewed the various physical and statistical experiments that have been carried out over the years. The results of the statistical experiments were generally inconclusive though some suggested positive effects. Observational and modeling studies provide further support for that, at least in some clouds, the addition of hygroscopic seeding material can broaden drop-spectra and at least hasten the onset of precipitation formation. We concluded that there appears to be a real opportunity to enhance rainfall through hygroscopic seeding in some clouds. It has not been determined how open the 'window of opportunity' actually is. In warm-based, maritime clouds the rate of natural production of rainfall may be so great that there is little opportunity to beat nature at its own game. On the other hand, some cold-based continental clouds may have so many small droplets that seeding-produced big drops cannot collect them owing to very small collection efficiencies. Thus, there probably exists a spectrum of clouds between these two extreme types that have enough liquid water to support a warm cloud precipitation process that can be accelerated by hygroscopic seeding. The problem is "to identify those clouds, and deliver the right amount of seeding material to them at the right time."

As optimism for significant precipitation enhancement by static seeding of supercooled clouds has waned, enthusiasm for the potential of hygroscopic seeding has grown. Two ongoing research programs, one in Thailand, the other in South Africa, have contributed to that enthusiasm.

The South African experiment was motivated by a report by Mather (1991), which suggested that large liquid raindrops at $-10°C$ found in a cumulonimbus were the result of active coalescence processes caused by the effluent from a Kraft paper mill. Earlier, Hobbs et al. (1970) found that the effluent from paper mills can be rich in cloud condensation nuclei (CCN). Moreover, Hindman et al. (1977a,b) found paper pulp mill effluent to have high concentrations of large and ultra-giant hygroscopic particles, which is consistent with the idea that the paper pulp mill effectively "seeded" the storm.

Another reason for optimism is that Mather et al. (1996b) applied a pyrotechnic method of delivering salt, based on a fog dispersal method developed by Hindman (1978). This reduced a number of technical difficulties associated with preparing, handling, and delivery of very corrosive salt particles. Seeding with this system is no more difficult than silver iodide flare seeding. Compared to conventional methods of salt delivery, the flares produce smaller-sized particles in the size range of 0.5 to 10 μm. Thus, not as much mass must be carried to obtain a substantial yield of seeding material. The question of effectiveness of this size range will be discussed below. Seeding trials with this system suggested that the pyrotechniques produced a cloud droplet spectrum that was broader and with fewer numbers, which would be expected to increase the chance for initiation of collision and coalescence processes.

Mather et al. (1996b) analyzed radar-defined cells over a period of about an hour to identify the seeding signatures for 48 seeded storms compared to 49 unseeded storms. They showed that after 20 to 30 minutes, the seeded storms developed higher rain masses and maintained those higher rain masses for another 25 to 30 minutes. Bigg (1997) performed an independent evaluation of the South African exploratory hygroscopic seeding experiments and also found that the seeded storms clearly lasted longer than the unseeded storms. Bigg also suggested that there was a clear dynamic signature of seeding. He argued that hygroscopic seeding initiated precipitation lower in the clouds, which, in turn, was not dispersed horizontally as much as the unseeded clouds by vertical wind shears. As a result, Bigg speculated that low-level downdrafts became more intense, which yielded stronger storm regeneration by the downdraft outflows, and longer-lived precipitation cells.

Biggs hypothesis is a plausible scenario that should be examined thoroughly with numerical models and coordinated, high resolution Doppler radars.

Cooper et al. (1997) performed simulations of the low-level evolution of droplet spectra in seeded and unseeded plumes. Following a parcel ascending in the cloud updrafts they calculated the evolution of droplet spectra by vapor deposition and collection. The calculations were designed to emulate the effects of hygroscopic seeding with the South African flares. The calculations showed that introduction of particles in the size-range characteristic of the flares resulted in an acceleration of the collision and coalescence process. If the hygroscopic particles were approximately 10 μm in size, precipitation was initiated faster. But, when more numerous 1 μm hygroscopic particles were inserted, high concentrations of drizzle formed. For a given amount of condensate mass, if the mass is on more numerous drizzle drops than on fewer but larger raindrops, then evaporation rates are greater in the subcloud layer. This could lead to more intense dynamic responses as proposed by Bigg, suggesting that seeding with smaller hygroscopic particles may have some advantages. Keep in mind, however, that this is a very simple model. More comprehensive model calculations should also be performed.

In summary, there are some exciting new results of hygroscopic seeding with flares. This work is still very exploratory and is a long way from proving that such techniques can make significant increases in rainfall on the ground for a variety of weather and climate regimes. It is refreshing for a change, to end an overview of the science of weather modification by cloud seeding on a rather upbeat note!

5.5 Weather Modification Research Funding

Cotton and Pielke (1995) noted that the funding in weather modification research in the United States peaked in the early 1970's at about US$ 19 million per year. By the 1990's that funding level had decreased to less than US$ 5 million a year, with a major part of that funding being a Department of Commerce state/federal cooperative program, which we labeled a "pork barrel" program. We discussed a number of factors that could have contributed to such a collapse in weather modification funding, which I shall not repeat here. Suffice it to say that there are a number of lessons to be learned from what happened, especially in the global climate program.

In 1997, the Department of Commerce State/Federal program is zero budgeted and the identifiable research funding in the U.S. is about US$ 0.5 million. Likewise, the government of Israel decided to terminate funding of weather modification research after 36 years of continuous funding. So funding in weather modification research has continued to slide.

Note that this does not mean that weather modification activities have stopped. Operational weather modification programs exist in something like 22 countries worldwide and in the United States alone there may be as many as 40 operational seeding projects going on in any one year. What has happened is that we have entered the "dark ages" of weather modification where operational cloud seeding projects are, if anything, proliferating without a sound scientific research program supporting them.

5.6 Lessons Learned that the Global Climate Change Community Should Pay Attention To

Cotton and Pielke (1995) examined some of the implications of research on weather modification research to research on global climate change. They considered the following issues:

- The importance of natural variability,

- The dangers of overselling,

- The capricious administration of science and technology, and

- scientific credibility and advocacy.

Let us re-examine each of these common issues more fully, and also discuss the implications of hygroscopic seeding research to the CCN-albedo hypothesis.

5.6.1 Hygroscopic Seeding and the CCN-Albedo Hypothesis

Twomey (1974) first postulated that increased pollution results in greater CCN concentrations and numbers of cloud droplets, which, in turn, increase the reflectance of clouds. Twomey et al. (1984) argued that enhanced cloud albedo has a magnitude comparable to that of greenhouse warming and acts to cool the atmosphere, in opposition to greenhouse warming.

Subsequently Albrecht (1989) proposed that higher droplet concentrations in polluted air will reduce the rate of drizzle formation, resulting in wetter, more reflective clouds. Furthermore, Ackerman et al. (1993; 1994) proposed that heavily drizzling straticumulus clouds can reduce cloud top cooling to such an extent that a stratus-topped boundary layer can collapse. Pollution-caused increases in CCN can thereby suppress the drizzle process leading to the formation of a stratus-topped boundary layer in some boundary layer regimes that may not otherwise be sustainable. Cotton and Pielke (1995) noted, however, the susceptibility of the drizzle process in marine stratocumulus clouds to anthropogenic emissions of CCN may depend on the presence or absence of large and ultra-giant aerosol particles in the subcloud layer. In other words, the drizzle formation process is not solely regulated by the concentrations of CCN and cloud liquid water contents, but possibly also by the details of the spectrum of the hygroscopic aerosol population. This concept has been reinforced by the hygroscopic seeding simulations by Cooper et al. (1997). Their model calculations suggest that high concentrations of hygroscopic particles in the 0.1 μm to 1.0 μm size range can accelerate the drizzle formation process. An implication is that even though pollution may increase the total concentration of CCN particles, if the concentration of particles greater than 0.1 μm is likewise increased, the drizzle formation process may not be suppressed, but instead could be actually enhanced. This could counter the tendency of clouds in polluted air from being more reflective. Again, we emphasize that knowledge of the total size-distribution of hygroscopic aerosol is needed to assess the potential impacts of pollution on global climate!

5.6.2 The Importance of Natural Variability

We have seen that our ability to determine if cloud seeding *causes* some *observed or hypothesized effect*, such as changes in local rainfall in specified target areas, is strongly dependent upon the *natural variability* of the system. However, the same can be said in assessing if anthropogenic greenhouse gas emissions, or deforestation, or release of CCN, have any significant impact on global climate. While the time and space scales are very different, nonetheless the bottom line in examining potential human-caused effects is: *are these effects large enough in magnitude to be extricated from the 'noise' of the natural variability of the system?* There are few, if any, cases in which we can answer this question affirmatively. Ice cores have shown, for example, that a switch from an ice age to a non-ice age climate can occur over only a few decades (La Brecque, 1989a,b) without human intervention.

5.6.3 The Dangers of Overselling

We have seen that funding of the science of weather modification underwent a period of rapid rise, followed by an abrupt crash. One of the leading causes of that crash, as we believe, is that the program was oversold. The claims that only a few more years of research and development will lead to a scientifically-proven technology that will contribute substantially to water management and severe weather abatement, were either great exaggerations, or just false. This is largely because we greatly underestimated the complexity of the scientific and technological problems we were (and still are) faced with.

The same can be said about human impacts on global climate. There are many scientists who are claiming that the short-term (periods of year-to-year, or decades) variations in weather and climate are clear evidence that we are experiencing the effects of anthropogenic greenhouse emissions. Moreover, many claim that the 'forecasts' being made by global climate models represent realistic expectations of global-averaged changes in temperature and rainfall in the next decade or century. In our opinion, both of these claims represent overselling of the climate program. These claims appear and are discussed in the professional literature (e.g., Schneider, 1990; Titus, 1990a,b; IPCC, 1991; Kellogg, 1991) and in the lay press (e.g., Brooks, 1989; Schneider, 1989; Thatcher, 1990; Bello, 1991; Luoma, 1991; UCAR/NOAA, 1991). Titus (1990a), for example, proposes the rerouting of the Mississippi River to save coastal Louisiana! As an example of such extreme claims to mitigate anthropogenically caused global warming, a 1991 National Academy Press report (National Academy of Sciences, 1991) has considered the insertion of 50,000 mirrors of 100 km^2 each in space to reflect incoming sunlight. Such gross global climate engineering represents a close analog to the exaggerated claims in weather modification, which were made in the 1960s and 1970s. Short-term variations of weather and climate are clearly within the *natural variability* of climate to the extent that we can realistically assess it. Moreover, the models are not really 'forecast' models. They are simply research models designed to simulate the responses of hypothesized anthropogenic changes to weather and climate, *other things being the same*. Besides having many limitations in their physical/chemical parameterizations, they are not designed to simulate (or predict) the consequences of many other natural factors affecting climatic change. That is because we simply do not know enough about all the processes of importance to climatic change to include them in any quantitative forecast system. What it amounts to is that many scientists are grossly underestimating the complexity of interactions among the earth's atmosphere, ocean, geosphere, and biosphere. These problems are so complex that it may take many decades, or even centuries, before we have matured enough as a scientific community to make *credible predictions* of long-term climate trends and their corresponding regional impacts. Even then, we may find that the uncertainty level of those predictions due to outside (the earth) influences may be so large that those predictions are not useful for social planning.

5.6.4 The Capricious Administration of Science and Technology

In the United States, as well as in many developed countries, science and technology are often poorly administered. As we have seen in weather modification, administration of many programs is fragmented among a number of basic and mission-oriented agencies, all of which compete for funding at national and state levels. This competition amongst the agencies often leads to the greatly exaggerated claims that many of the scientific and technological issues will be solved in the next five to ten years.

In addition, because many of these agencies are mission-oriented, their job is to examine the impacts of human-induced changes on weather and climate on energy, air quality, water resources, or agriculture. Their job is not to advance the fundamental scientific issues regarding the behavior of the earth system, but to get on with the business of evaluating the impacts of anthropogenic activity on their programs. As a result, they are often looking for shortcuts to bottom line answers that can probably only be obtained through meticulous, often time consuming scientific research.

Moreover, national governing bodies (legislatures, presidents, etc.) all work on time scales of two, four, or six years, and want to be able to identify impacts of their programs on time scales of their tenure. If significant progress is not made on those time scales, then often funding in those programs is reduced, if not curtailed, and new, competing programs are brought to the forefront. This results in shortsighted funding in science and technology in which programs are begun and before they reach maturity they are curtailed, then the rush is on to get on the bandwagon with the latest fad. The scientific and technological problems associated with furthering our understanding of human impacts and natural variability and feedbacks of weather and climate are so complicated and multifaceted that many of the issues will not be resolved on time scales of decades or possibly centuries. *Thus, programs associated with the investigation of human impacts on weather and climate require sustained, stable national funding at a high level.* A view supporting this idea has been recommended in the Policy Statement of the American Meteorological Society on Global Climate Change (1991) and in the report

"Global Climate Change: A New Vision for the 1990s" (Michaels, 1990), which was produced by a group of climate scientists who questioned the overselling and shortsighted perspective of current climate change government policy.

5.6.5 Scientific Credibility and Advocacy

We have seen that, with few exceptions, the scientific evidence is not conclusive that cloud seeding is causing the desired responses. Moreover, the evidence that human activity is "causing" observed changes in weather and climate is also quite tenuous. With that in mind we ask, *should scientists be actively involved in advocating that we apply cloud seeding techniques to enhancing rainfall, or reducing emissions of greenhouse gases to alleviate greenhouse warming?* Certainly the scientists are the best informed with regard to the consequences of human activity and, one could say, that if the informed scientist does not take an advocacy role in recommending that action be taken, then no one else will.

Such a position is not without its dangers, however. If, for example, scientists participate in an operational cloud seeding program or play an obvious role as advocates of applying cloud seeding, they can jeopardize their credibility as truly objective scientists and, therefore, adversely affect both the program and the individual scientists. The same can be said with regard to advocates of major disruptive societal changes with regard to greenhouse emissions.

Some might argue that the risk of losing one's scientific credibility is purely a personal one and must be weighed against the potential societal gains by taking immediate action to relieve drought or reduce greenhouse warming. In fact, the adverse impacts extend far beyond those affecting the individual scientist. Loss of scientific credibility is infectious and can, therefore, propagate through an entire scientific discipline and even to the scientific community as a whole. The fall of the science of weather modification by cloud seeding was almost certainly due, in part, to a loss of scientific credibility. The global climate change community must likewise be careful that a loss of scientific credibility does not propagate through their discipline, or the discipline of atmospheric science as well. Thus, premature advocacy that action be taken now, could, in the long run, destroy the prospects for obtaining solid scientific evidence that human activity is affecting weather and climate.

5.7 Should Society Wait for Hard Scientific Evidence?

Overall there is little hard scientific evidence that anthropogenic activity, either advertently or inadvertently, is causing significant changes in weather and climate, particularly on the global scale. This is certainly true with respect to cloud seeding where there are only a few limited examples of where cloud seeding has been scientifically shown to be effective in enhancing rainfall. Nonetheless, there are many nations which are currently running operational cloud seeding projects. Apparently, the decision has been made in those nations and states that the benefits outweigh the risks of applying the scientifically unproven technology of weather modification by cloud seeding. The major risks, however, are limited to the possibility of creating severe weather or floods, and to increasing rainfall in one local region at the expense of rainfall in a neighboring local region. Often the decision to apply cloud seeding technology in a particular country or state is a prescription of a *political placebo* or a decision that it is better to do something than to sit idly by and do nothing as reservoirs dry up and crops wither and die due to the absence of water.

Again, the situation is not much different with respect to human impacts on global climate. We lack hard scientific evidence that anthropogenic activity is causing, or will cause, changes in global climate. Nonetheless, there is convincing evidence that CO_2 concentrations are increasing at an alarming rate. Clearly, reductions in CO_2 emissions in many of the industrialized countries will have a significant impact on global CO_2 emissions and reduce the chance that human activity will have a significant impact on weather and climate. Certainly there is evidence that the more developed nations are at least causing a leveling off of CO_2 emissions.

But what are the costs? Some of the costs are in terms of reduced industrial productivity which impacts employment and the general standard of living. Another cost is associated with the impacts of using alternative energy resources. One could decide to convert to "cleaner" nuclear power rather than fossil fuels as has been done in France. In this case one is trading off the potential impacts of CO_2 emissions on global warming against the long-term problem of disposing of nuclear waste, as well as, the dangers of inadvertent releases of nuclear materials. Is this a wise tradeoff? Without solid scientific evidence that CO_2 emissions are causing significant changes in climate, one cannot make an objective evaluation of the relative cost of each alternative.

An alarming trend is found in India (Marland, 1991) where a sharp rise in CO_2 emissions in India is occurring, while per capita emissions remain steady. This shows a clear impact of increased population on CO_2 emissions and suggests that one of the most important steps in reducing those emissions is getting the world population stabilized.

One does not need strong scientific evidence that human activity is causing global warming to recognize a reduction in the population on earth will have long-term benefits; common sense is all that is needed!

Indeed, population growth is a problem, which is much more severe than any of the scenarios proposed to occur as a result of greenhouse gas warming. Catastrophic social upheavals are likely to result as the human density continues to increase. Bryson (1989) presents evidence that even today India is at their absolute water limit, such that below average rainfall can cause massive deaths.

5.8 Acknowledgements

I would like to thank constructive comments and input on recent papers from Art Rangno and Harry Orville, Joanne Simpson, William Woodley, Roelof Bruintjes, and Bernie Silverman. Steve Nelson also provided input on the status of NSF funding in weather modification research.

Brenda Thompson is thanked for her assistance in processing this manuscript.

Although this research is not directly supported by any funding agency, the sustained support of basic research on storm dynamics and mesoscale systems under NSF Grant #ATM-9420045 and on the influence of CCN on boundary layer clouds under NSF Grant #ATM-9529321 provided the technical background that I built upon in performing this review.

Chapter 6

The Detection of Climate Change

by Francis W. Zwiers

Abstract

This paper reviews some of the techniques that have been proposed for climate change detection during the past fifteen years and describes several applications. We also briefly consider the question of the attribution of climate change. An attempt is made to use uniform notation and nomenclature throughout so that similarities are more readily apparent.

The methodologies fall into two groups; those that use 'optimal detectors' and those that use 'pattern similarity' statistics. Both approaches require the use of climate models for estimating the signals that are to be searched for, and also for estimating the natural variability of the detectors in the absence of signal. The latter is necessary because the detection statistics are influenced by low frequency climate variations that are not well sampled in the relatively short instrumental record that is currently available. Formally, the use of optimal detectors enhances the prospects for early climate change detection by increasing the signal-to-noise ratio by of the order of 20%. However, real-world differences in performance are likely more strongly affected by how well models represent the climate's response to changes in forcing from human activity and by variations in the intricate processing of the data that precedes the formal detection step. Recent applications of both types of detection methodology have uncovered 'smoking gun' evidence that human activity is having an effect on the recent climate.

6.1   Introduction

There is an extensive climate change detection literature that extends back almost twenty years, beginning perhaps with Hasselmann (1979). It includes descriptions of numerous applications of a seemingly wide range of detection methodologies that are described using a diverse set of terminologies.[1] The purpose of this paper is to review only a selection of papers from the recent literature, comparing the methods that are proposed and some of the results that have been obtained. We try to cut through the Gordian Knot of diverse terminology and notation that inhibits understanding, by using uniform notation and language throughout, and by attempting to convey concepts in as intuitively clear a manner as possible.

The outline of this paper is as follows. The basic ideas that underlie climate change detection studies are reviewed in Section 6.2. Here we review a relatively early contribution to the literature from Bell (1982) and outline the basic theory of optimal climate change detection. Section 6.3 describes some extensions and Section 6.4 reviews the methodology proposed by Hasselmann (1993, 1997) and its recent applications by Hegerl et al. (1996, 1997). Section 6.5 reviews a similar methodology proposed by North et al. (1995) and applied by Stevens and North (1996) and North and Stevens (1997). The somewhat different approach taken by Santer and colleagues (e.g., Santer et al. 1993, 1995, 1996a; Karoly et al. 1994) is described in Section 6.6. The paper concludes with an inter-comparison of the techniques and a brief description of some evolving approaches in Section 6.7.

6.2   Basic Ideas

We begin by supposing that we observe a quantity $\mathbf{T}(x, t)$ such as surface temperature where $x = 1, \ldots, m$ identifies the observing locations and $t = 1, 2, \ldots$ represents the observing times. We suppose also that $\mathbf{T}$ consists of a deterministic signal S and random noise $\mathbf{N}$,

$$\mathbf{T} = S + \mathbf{N}. \tag{6.1}$$

The signal S represents the climate's deterministic response to changes in external forcing. Possible forcings include changes in the amount of energy reaching the earth (i.e., variability in solar output), and changes in the earth's radiative balance caused by variations in the stratospheric aerosol loadings from volcanism, variations in tropospheric aerosol loadings from sources such as fossil fuel burning, and variations in the concentration of greenhouse gases

---

[1]Santer et al. (1996b) give a complete, up to date, overview in the most recent report of the Inter-governmental Panel on Climate Change (IPCC). See also the earlier IPCC overview of Wigley and Barnett (1990). An additional source which contains a number of seminal papers are the proceedings of a 1989 meeting organized by Michael Schlesinger (Schlesinger, 1991).

such as carbon dioxide, methane and the various chlorofluorocarbons. The noise **N**, which is assumed to be a zero mean stochastic process, arises from the natural chaotic variability of the climate system. A characteristic feature of this noise is that there is substantial variability over a broad range of time scales, ranging from the short time scale turbulent variations we recognize as weather, to long time scale variations that result from interaction between the atmosphere and other parts of the climate system, such as the ocean with its very large thermal inertia.

The climate change detection problem is, foremost, the problem of detecting the presence of S in **T**, and secondarily, a problem of attributing all or part of the detected signal to human activities. The former is, formally at least, a problem that can be well posed statistically. The latter is more contentious and difficult to pose.

There are many parallels in the detection problem with the (in some quarters continuing) debate over the relationship between cigarette use and lung cancer. In both cases, it is impossible to conduct direct experiments with the affected entity to confirm that the theoretical agent of change (anthropogenic emission of greenhouse gases in the one case and tobacco use in the other) is in fact responsible. It is impossible to conduct controlled experiments with earth's climate system, and it is unethical to conduct human clinical trials with a deleterious substance such as tobacco. Thus, in both cases we must rely on observations in which cause and effect *can not* be completely disentangled because confounding sources of variation are present, and on proxies that *can* be used for controlled experimentation. In the case of tobacco, those proxies are animals and one can debate whether the results of animal studies can be extrapolated to our species. In the case of climate change, the proxies are global climate models and debate swirls, at least in some circles, about the fidelity of these models to the observed system.

We will focus primarily on the detection question in this paper. Most people who have worked on this problem have proposed some kind of *filtering* over space and time to separate signal from noise. Bell (1982), for example, proposes a detection variable given by a weighted space-time average

$$\mathbf{A}_t = \sum_{\tau=1}^{l} \sum_{x=1}^{m} w(x,\tau) \mathbf{T}(x, t-\tau+1) \tag{6.2}$$

where

$$\sum_{\tau=1}^{l} \sum_{x=1}^{m} w(x,\tau) = 1.$$

To use such a filter it is necessary to select *optimal* weights that maximize the chances of signal detection. This will be described shortly, but first we will introduce some notation which will, hopefully, simplify the task of finding weights and also serve as an aid to understanding. Filter (6.2) can be re-expressed in matrix-vector notation as

$$\mathbf{A}_t = \vec{w}^T \vec{\mathbf{T}}_t \tag{6.3}$$

where $\vec{\mathbf{T}}_t$ is an ml-dimensional random vector constructed so that the first m entries of $\vec{\mathbf{T}}_t$ are $\mathbf{T}(1,t), \ldots, \mathbf{T}(m,t)$, the next m entries of $\vec{\mathbf{T}}_t$ are $\mathbf{T}(1, t-1), \ldots, \mathbf{T}(m, t-1)$, and so on. The ml-dimensional weight vector $\vec{w}$ is constructed similarly, as are the *signal* and *noise* vectors $\vec{S}_t$ and $\vec{\mathbf{N}}_t$. The constraint that the weights should sum to one can be expressed as $\vec{w}^T \vec{e} = 1$ where $\vec{e}$ is the vector of units $\vec{e} = (1, 1, \ldots, 1)^T$.

Note that we have used two kinds of notation to this point. We have used plain characters, such as in $\vec{S}_t$, to identify deterministic quantities.[2] Secondly, we have used bold faced characters, such as $\vec{\mathbf{N}}_t$, to represent quantities that vary stochastically.

Since we observe a system that consists of both signal and noise, it follows that no matter how the weights $\vec{w}$ are chosen, the *detection variable* (i.e., filter output) $\mathbf{A}_t$ will be subject to random variation induced by $\vec{\mathbf{N}}_t$. The problem of designing the optimal filter is, therefore, one of choosing the weight vector $\vec{w}$ so that the amount of signal that passes through the filter is large relative to the amount of noise that is passed. Typically then, $\vec{w}$ is chosen to maximize the *signal to noise ratio* defined as

$$\gamma^2 = \frac{\mathcal{E}(\mathbf{A}_t)^2}{Var(\mathbf{A}_t)}.$$

The mean, or expected value, of $\mathbf{A}_t$ is given by

$$\mathcal{E}(\mathbf{A}_t) = \mathcal{E}\left(\vec{w}^T \vec{\mathbf{T}}_t\right) = \vec{w}^T \mathcal{E}\left(\vec{\mathbf{T}}_t\right) = \vec{w}^T \mathcal{E}\left(\vec{S}_t + \vec{\mathbf{N}}_t\right) = \vec{w}^T \vec{S}_t$$

and its variance is given by

$$Var(\mathbf{A}_t) = Var\left(\vec{w}^T \vec{\mathbf{T}}_t\right) = \vec{w}^T \mathrm{Cov}\left(\vec{\mathbf{T}}_t, \vec{\mathbf{T}}_t\right)\vec{w} = \vec{w}^T \Sigma_{N_t N_t} \vec{w}$$

where $\Sigma_{N_t N_t}$ is the $ml \times ml$ covariance matrix of the noise process $\vec{\mathbf{N}}_t$. Thus, the signal to noise ratio is given by

$$\gamma^2 = \frac{(\vec{w}^T \vec{S}_t)^2}{\vec{w}^T \Sigma_{N_t N_t} \vec{w}}. \tag{6.4}$$

The signal to noise ratio is easily maximized in the usual manner by differentiating γ^2 with respect to $\vec{w}$.[3] Doing so, setting the resulting vector of derivatives $\frac{\partial \gamma^2}{\partial \vec{w}}$ to zero, and solving for $\vec{w}$, yields optimal weights

$$\vec{w} = c \Sigma_{N_t N_t}^{-1} \vec{S}_t \tag{6.5}$$

---

[2]We have implicitly assumed that the climate system will respond to a change in forcing in a deterministic way. This is a reasonable expectation for a given forcing, but the forcing itself varies stochastically in accordance with variations in economic activity, volcanism and perhaps solar output.

where the scaling factor c is chosen so that the weights satisfy the constraint $\vec{w}^T \vec{e} = 1$. The scaling factor is not important in the following discussion, so we will set $c = 1$ for convenience.

This simple derivation results in an *optimal detector* that is given by

$$\mathbf{A}_t = \vec{w}^T \vec{\mathbf{T}}_t = \vec{S}_t^T \Sigma_{N_t N_t}^{-1} \vec{\mathbf{T}}_t \tag{6.6}$$

and has variance

$$Var\left(\mathbf{A}_t\right) = \vec{w}^T \Sigma_{N_t N_t} \vec{w} = \vec{S}_t^T \Sigma_{N_t N_t}^{-1} \vec{S}_t. \tag{6.7}$$

Substituting (6.5) into (6.4), we find that the signal to noise ratio of this detector is

$$\gamma^2 = \frac{((\Sigma_{N_t N_t}^{-1} \vec{S}_t)^T \vec{S}_t)^2}{(\Sigma_{N_t N_t}^{-1} \vec{S}_t)^T \Sigma_{N_t N_t} (\Sigma_{N_t N_t}^{-1} \vec{S}_t)} = \vec{S}_t^T \Sigma_{N_t N_t}^{-1} \vec{S}_t = Var\left(\mathbf{A}_t\right). \tag{6.8}$$

Assuming that the observed process $\vec{\mathbf{T}}_t$ is multivariate Gaussian, the no-signal null hypothesis

$$\mathrm{H}_0 : \vec{S}_t = \vec{0} \tag{6.9}$$

can be rejected when

$$\mathbf{Z}_t = \frac{\vec{S}_t^T \Sigma_{N_t N_t}^{-1} \vec{\mathbf{T}}_t}{(\vec{S}_t^T \Sigma_{N_t N_t}^{-1} \vec{S}_t)^{1/2}} = \frac{\vec{S}_t^T \Sigma_{N_t N_t}^{-1} \vec{\mathbf{T}}_t}{\gamma} \geq \sim 2. \tag{6.10}$$

This provides a rough and ready criterion for deciding whether the signal has been seen in the observed data.

At this point it is useful to consider a simple example. Suppose that $\vec{S}_t$ and $\vec{\mathbf{N}}_t$ are two dimensional, and that $\vec{\mathbf{N}}_t$ is bi-variate Gaussian with mean $\vec{0}$ and covariance matrix

$$\Sigma_{N_t N_t} = \begin{pmatrix} 4 & 0 \\ 0 & 1 \end{pmatrix}.$$

A contour plot of the probability density function of $\vec{\mathbf{N}}_t$ is displayed in the lower left hand corner of Fig. 6.1 (the ellipsoids centered on the origin). Suppose also that we have a constant signal $\vec{S}_t = (5,5)^T$. The contour plot in the upper right hand corner of the diagram displays the probability density of function of $\vec{\mathbf{T}}_t = \vec{S}_t + \vec{\mathbf{N}}_t$, and the vector joining the centers of the two density functions depicts the signal itself. The weight vector for the optimal detector in this example, which is given by

$$\vec{w} = \Sigma_{N_t N_t}^{-1} \vec{S}_t = (\frac{5}{4}, 5)^T,$$

lies to the left of the signal. The optimal detector looks for the signal in a direction that is rotated away from the horizontal axis because this is the direction in which the noise predominates. The signal to noise ratio of the optimal detector is

$$\gamma_{opt}^2 = \begin{pmatrix} 5 & 5 \end{pmatrix} \begin{pmatrix} \frac{1}{4} & 0 \\ 0 & 1 \end{pmatrix} \begin{pmatrix} 5 \\ 5 \end{pmatrix} = 31.25.$$

In contrast, using (6.4) we find that the signal to noise ratio of the detector that looks in the direction of the signal (i.e., it uses weights $\vec{w} = \vec{S}_t$) is

$$\gamma_{signal}^2 = \frac{(\vec{S}_t^T \vec{S}_t)^2}{\vec{S}_t^T \Sigma_{N_t N_t} \vec{S}_t} = \frac{50^2}{125} = 20,$$

which is substantially less than the signal to noise ratio of the optimal detector.[4]

We now look at the optimal detector in some more detail. First recall that the noise covariance matrix can be decomposed as $\Sigma_{N_t N_t} = \mathcal{P} \Lambda \mathcal{P}^T$ where the columns of $\mathcal{P}$ are the EOFs of $\vec{N}_t$ and where Λ is a diagonal matrix that contains the corresponding eigenvalues.[5] Using the EOFs, the noise process can be decomposed as

$$\vec{N}_t = \sum_{i=1}^{ml} \alpha_{i,t} \vec{p}_i \tag{6.11}$$

where $\vec{p}_i$ is the ith EOF and $\alpha_{i,t} = \vec{N}_t^T \vec{p}_i$ is the ith *principal component* (PC). The PCs are uncorrelated and have variance $Var(\alpha_{i,t}) = \lambda_i$ where λ_i is the eigenvalue that corresponds to the ith EOF. The signal can also be decomposed with the EOFs as

$$\vec{S}_t = \sum_{i=1}^{ml} \beta_{i,t} \vec{p}_i \tag{6.12}$$

---

[4]A diagram similar to Fig. 6.1 appears in Hasselmann (1979; see Fig. 1, page 258) who describes signal to noise problems in more generality than we do here. The concluding section of Hasselmann (1979), which deals with applications in atmospheric problems, is revealing of the progress that has been made in the intervening period. He says that estimation of the covariance matrix will 'generally present no basic difficulty'. However, he says this will be more difficult for simulated climates because 'continuous model time series of comparable length to analyzed global or hemispheric real data are not available'. Fifteen years later we have essentially the opposite view. It is now generally recognized that observed data sets are not long enough to provide reliable information about the variability of the climate system on the time scales needed to make inferences about climate change. Instead, we have come to rely heavily upon very long ($\sim$1000 years) simulations from coupled climate models which would have been unimaginable given the computing power available in the late 70's.

[5]exaatrix $\mathcal{P}$ is *orthonormal*. That is, $\mathcal{P}\mathcal{P}^T = \mathcal{P}^T\mathcal{P} = \mathcal{I}$ where $\mathcal{I}$ is the $ml \times ml$ identity matrix.

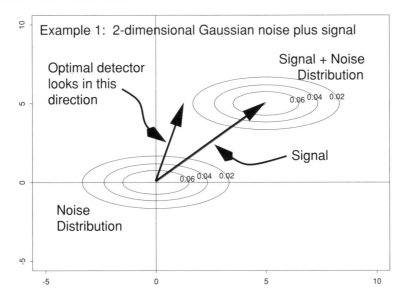

Fig. 6.1. A simple example illustrating that the optimal detector will generally look in a direction different from that of the signal

where $\beta_{i,t} = \vec{S}_t^T \vec{p}_i$. With these decompositions the variance of the optimal detector can be expressed as

$$Var\left(\mathbf{A}_t\right) = \sum_{i=1}^{ml} \frac{\beta_{i,t}^2}{\lambda_i}. \tag{6.13}$$

Thus, the total signal to noise ratio can be expressed as a sum of signal to noise ratios in the direction of each of the EOFs.

The optimal detector itself can also be decomposed with the EOFs. We have

$$\mathbf{A}_t = \vec{S}_t^T \Sigma_{N_t N_t}^{-1} \vec{\mathbf{T}}_t = \vec{S}_t^T \mathcal{P}\Lambda^{-1}\mathcal{P}^T \vec{\mathbf{T}}_t = \sum_{i=1}^{ml} \frac{\beta_{i,t}\delta_{i,t}}{\lambda_i} = \sum_{i=1}^{ml} \frac{\beta_{i,t}}{\lambda_i^{1/2}} \cdot \frac{\delta_{i,t}}{\lambda_i^{1/2}} \tag{6.14}$$

where $\delta_{i,t} = \vec{\mathbf{T}}_t^T \vec{p}_i$ is the projection of the observed process onto the ith EOF. The term on the left-hand side of each product is the strength of the signal in the direction of the ith EOF expressed in standard deviations of the noise in that direction. Similarly, the term on the right is the value of the observed field in the direction of the ith EOF, again expressed in standard deviations of the noise. Clearly, less attention is paid to the components of the signal and observed field that correspond to directions in which there is a lot of noise. This is why the detection variable in the example (see Fig. 6.1) searches for a signal in a direction closer to the vertical axis.

Bell (1982) discussed a number of practical issues in conjunction with the application of the optimal detector. For example, he advised that the dimensionality of $\vec{\mathbf{T}}_t$ should be kept small so that $\Sigma_{N_t N_t}$ can be estimated from data. Hasselmann (1993) explains that sampling error in the estimated value of $\Sigma_{N_t N_t}$ leads to error in the estimated EOFs, and especially to error in the estimates of the corresponding eigenvalues. The high order eigenvalues, which are also the smallest eigenvalues, are strongly underestimated when samples are small. Their inclusion can lead to large error in the optimal detector because they appear in the denominator in (6.14).

Bell also discussed the length l of the detection window. He advised that the window should be long enough to ensure that successive detectors $\mathbf{A}_t$ are approximately independent so that standard methods of statistic inference can be used to test the no signal null hypothesis. Also, selection of a longer window leads to a greater signal to noise ratio. But the window can not be too large because the noise covariance matrix must also be estimated from the available data record.

The basic ingredients of the detection procedure described above are shared with others that we will consider in the following Sections. They are:

- Observations in which to look for the signal. That is, it is necessary to characterize $\vec{\mathbf{T}}_t$ by determining the variable(s), locations and/or indices, and time window that will be used.

- A signal $\vec{S}_t$ estimated by using a climate model to simulate the response to a prescribed change in forcing.

- An estimate of the variability $\Sigma_{N_t N_t}$ of the noise process $\vec{\mathbf{N}}_t$.

Given these three ingredients, it becomes possible to construct an 'optimal' detector $\mathbf{A}_t = \vec{S}_t^T \Sigma_{N_t N_t}^{-1} \vec{\mathbf{T}}_t$. However, to *interpret* the detector, we also need

- a method for deciding when a signal has been detected. That is, we need a means by which to test $\mathrm{H}_0 : \mathcal{E}(\mathbf{A}_t) = 0$.

In the simple paradigm (ansatz) discussed above, where sampling variability in the estimate of $\Sigma_{N_t N_t}$ is not taken into account, the decision making tool is given by equation (6.10).

Bell (1982) attempts a simple detection exercise using information that was available at the time. He starts with a 3-dimensional vector $\vec{\mathbf{T}}_t$ that consists of seasonal mean zonally averaged 1000 hPa temperature (T_{1000}) at $15°N$, $45°N$ and $75°N$.

Estimation of the covariance matrix is simplified by assuming that the covariance between latitudes is zero (Weare, 1979) so that the off-diagonal part of $\Sigma_{N_t N_t}$ can be set to zero. The variances on the diagonal are estimated by combining Oort's (1982) estimates of the variance of monthly mean temperatures with a simple red noise model to infer the variance of the seasonal mean. Details can be found in Bell (1982).

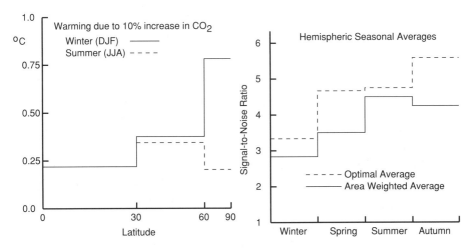

Fig. 6.2. Left: Surface warming in degrees Celsius in DJF and JJA due to a 10% increase in CO_2 as a function of latitude and season, obtained by scaling the results of Manabe and Stouffer's (1980) calculation by a factor of $\log 1.1/\log 4$. Corresponding warming values expected for high-, middle- and low-latitude zones in MAM and SON (not shown) are $0.49°$, $0.35°$, and $0.21°$ (for MAM) and $0.56°$, $0.35°$, and $0.24°C$ (for SON) respectively.

Right: Signal to noise ratio as a function of season for seasonally, hemispherically averaged temperature and for seasonally averaged temperature optimally weighted geographically.

After Bell (1982)

The signal is obtained from an early equilibrium climate change experiment performed at GFDL (Manabe and Stouffer, 1980) that simulated the effect of changing the atmospheric concentration of CO_2 from 300 to 1200 ppm.[6] The observed concentration of CO_2 during the 1970's was about 30 ppm ($\sim$10%) greater than the pre-industrial concentration. To first order, the radiative forcing resulting from the greenhouse gas effect is proportional to the log of the greenhouse gas concentration. Thus, Bell inferred that the expected equilibrium warming corresponding to the 10% change in CO_2 concentration would be approximately $\log 1.1/\log 4 \approx 0.07$ times the model's $4 \times CO_2$ response. Figure 6.2 (left) displays the resulting signal vector $\vec{S}_t$ used by Bell.

---

[6]Equilibrium climate change experiments are conducted by coupling an atmospheric general circulation model to a mixed-layer ocean model and a sea-ice model. The coupled system is first brought to equilibrium under control conditions. The climate simulation is then extended beyond this point so that natural variations about the equilibrium state can be sampled. Sampling typically continues for 10-20 simulated years. Separately, the coupled system is also brought to equilibrium under experimental conditions, and is again sampled for an extended period after equilibrium has been achieved.

This approach was to estimate a signal to noise ratio in each season. The estimates, which are displayed as the dashed line in Fig. 6.2 (right), range from about 3.4 in Northern Hemisphere winter (DJF) to about 5.3 in Northern Hemisphere autumn (SON). The signal to noise ratio of the optimal detector was greater in all seasons than that of the simple detector (solid line) that uses the area weighted hemispheric average temperature.

Bell then extends his detector into a weighted annual average of T_{1000} in the three latitude bands in all of four seasons of the year. This results in a 12-dimensional temperature vector $\vec{\mathbf{T}}_t$ and a corresponding 12×12 covariance matrix $\Sigma_{N_t N_t}$. He shows that optimal weighting yields a signal to noise ratio of 14, which is about three times the performance of the detector built from the seasonal mean, and about 25% greater than the performance that would be obtained by using a non-optimal detector, which computes an annual average of the area weighted seasonal means.

Given the power of the optimal detector, Bell speculates on why the warming had not been detected at the time of his article. He discusses factors such as oceanic delay (thought to be about 10 years), external forcing factors that had not been accounted for (such as volcanism), sampling variability that this study did not take into account, and the possibility that the natural variability of the system, which is represented by $\Sigma_{N_t N_t}$, had been underestimated.

6.3 More Elaborate Ideas

Bell revisited the detection issue in 1986, considering various aspects of the problem in more depth. Three things were added in this paper:

1. He considered how the detection strategy might change if the natural variability, represented by $\Sigma_{N_t N_t}$, changed with time.

2. He considered also the distribution of the decision making statistic $\mathbf{Z}_t$ (see equation (6.10)) when $\Sigma_{N_t N_t}$ is estimated, as it must be in reality.

3. Finally, he examined the power (i.e., detection ability) of this decision making tool.

We will briefly examine each of these extensions in the following subsections.

6.3.1 Changing Natural Variability

Suppose that the variability of the system depends upon the presence or absence of the signal so that

$$\Sigma_{N_t N_t} = \begin{cases} \Sigma_0 & \text{when } \vec{S}_t = \vec{0} \\ \Sigma_t & \text{when } \vec{S}_t \neq \vec{0} \end{cases}$$

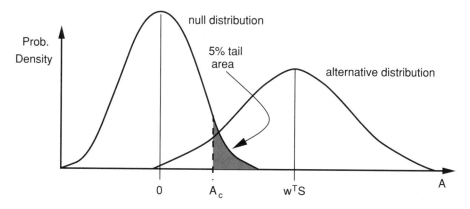

Fig. 6.3. A schematic diagram of the distribution of the detector $\mathbf{A}_t$ under the no signal null hypothesis (left curve) and under an alternative distribution (right hand curve). The critical value for rejecting the null hypothesis at the 5% significance level is labelled A_c

and suppose also that $\vec{\mathbf{T}}_t$ is multivariate Gaussian. Then the optimal detector $\mathbf{A}_t$ will be distributed

$$\mathbf{A}_t \sim \begin{cases} \mathcal{N}(\vec{0}, \vec{w}^T \Sigma_0 \vec{w}) & \text{when } \vec{S}_t = \vec{0} \\ \mathcal{N}(\vec{w}^T \vec{S}_t, \vec{w}^T \Sigma_t \vec{w}) & \text{when } \vec{S}_t \neq \vec{0} \end{cases}$$

where the notation $\mathbf{X} \sim \mathcal{N}(\mu, \sigma^2)$ indicates that $\mathbf{X}$ has a Gaussian distribution with mean μ and variance σ^2. The change in distribution is illustrated schematically in Fig. 6.3. We see that a change in covariance changes the width of the detector distribution, which in turn affects the power, or sensitivity, of the detector. This suggests that $\vec{w}$ should be chosen so that the power of the signal detection test is maximized.

The power, or probability of detecting the signal when it is present, is given by

$$prob\left(\mathbf{A}_t > A_c \mid \vec{S}_t \neq 0\right) = \frac{1}{\sqrt{2\pi}\sigma_t} \int_{A_c}^{\infty} \exp^{-\frac{1}{2}(A - \vec{w}^T \vec{S}_t)^2 / \sigma_t^2} dA \qquad (6.15)$$

where $\sigma_t^2 = \vec{w}^T \Sigma_t \vec{w}$ and where A_c is the critical value for a *one sided* test of the no signal null hypothesis.[7] When the test is conducted at the $p \times 100\%$ significance level, the critical value is given by $A_c = Z_{1-p}(\vec{w}^T \Sigma_0 \vec{w})^{1/2}$ where Z_{1-p} is the $(1-p)$ quantile of the standard Gaussian distribution.[8]

---

[7]That is, the null hypothesis is rejected when values of the detector greater than A_c are observed. One sided tests have greater power than that two-sided tests when it is known, a priori, that the mean of the detector increases when a signal is present. Detector $\mathbf{A}_t$ has this property by design.

[8]$Z_{1-p} = 1.645, 1.96, 2.326, 2.576$, and 3.090 for tests conducted at the 5%, 2.5%, 1%, 0.5% and 0.1% significance levels respectively.

The weight vector $\vec{w}$ that maximizes the power (6.15) is given by

$$\vec{w} = \left[\frac{Z_{1-p}}{\sigma_0} \Sigma_0 + (1 - \frac{Z_{1-p}\sigma_0}{\sigma_t^2})\Sigma_t \right]^{-1} \vec{S}_t \qquad (6.16)$$

where $\sigma_0^2 = \vec{w}^T \Sigma_0 \vec{w}$ and $\sigma_t^2 = \vec{w}^T \Sigma_t \vec{w}$. Thus, the most powerful detector has weights $\vec{w} = \Sigma_0^{-1} \vec{S}_t$ if the covariance matrix does not change when the signal appears, that is, when $\Sigma_t = \Sigma_0$. But if $\Sigma_t \neq \Sigma_0$, the optimal weights depend upon the initial covariance matrix Σ_0, the covariance matrix Σ_t that prevails at the time of the test, the signal $\vec{S}_t$ *and* the significance level of the test. That is, the direction in which you look for the signal depends upon the risk of incorrectly rejecting the null hypothesis that you are willing to take!

The effect of the choice of significance level is illustrated in the following simple example. Suppose that we observe a bivariate Gaussian temperature vector $\vec{\mathbf{T}}_t$ with pre- and post-industrial means $\vec{S}_0 = \vec{0}$ and $\vec{S}_t = (5,5)^T$ respectively. Also suppose that the pre- and post-industrial covariance matrices are given by

$$\Sigma_0 = \begin{pmatrix} 4 & 0 \\ 0 & 1 \end{pmatrix} \quad \text{and} \quad \Sigma_t = \begin{pmatrix} 4 & 0 \\ 0 & 4 \end{pmatrix}.$$

Contour diagrams of the pre- and post-industrial distributions of $\vec{\mathbf{T}}_t$ are displayed in Fig. 6.4. The vector that connects the centers of the two distributions is the signal $\vec{S}_t$. Vectors labelled 5%, 2.5% and 0.1% indicate the direction in which the most powerful detector looks when decision making is performed at the 5%, 2.5% and 0.1% significance level. In this example, the most powerful detector rotates towards the vertical axis as decision making becomes more conservative so that it avoids more of the noise under the null hypothesis. Since there is more noise in the horizontal direction, the conservative decision maker will preferentially look for the signal in a direction close to the vertical axis. The ultra-conservative limiting behaviour is to avoid all noise in the horizontal direction. The detector that assumes $\Sigma_t = \Sigma_0$ is also displayed in Fig. 6.4. This less sophisticated detector operates very conservatively in this particular example.

6.3.2   Taking Sampling Variability Into Account

In practice, the covariance matrix $\Sigma_{N_t N_t}$ must be estimated, and thus, it is natural to ask how sampling uncertainty in the estimated $\widehat{\Sigma}_{N_t N_t}$ affects the 'optimal' detector. The word 'optimal' is put in quotation marks because sampling variability in $\widehat{\Sigma}_{N_t N_t}$ makes the optimal weighting problem mathematically intractable. The practical solution is to assume that replacing $\Sigma_{N_t N_t}$ with an estimator $\widehat{\Sigma}_{N_t N_t}$ in (6.6) leads to a nearly optimal detector. Similarly, it is necessary to replace $\vec{\mathbf{T}}_t$ with $\vec{\mathbf{T}}_t - \overline{\overline{\mathbf{T}}}$ where $\overline{\overline{\mathbf{T}}}$ is an estimate of the mean of $\vec{\mathbf{T}}_t$ during a suitable reference period. The resulting ad-hoc

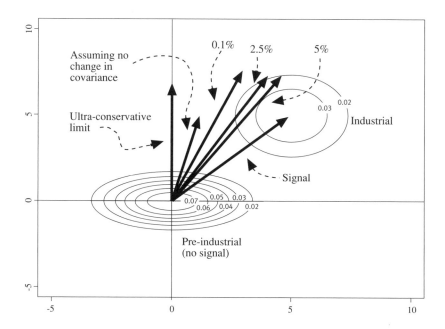

Fig. 6.4. A simple example illustrating that the optimal detector changes with the level of risk taken when testing for the presence of a signal if the variability changes when the signal appears

detector has the form $\tilde{\mathbf{A}}_t = \vec{S}_t^T \widehat{\Sigma}_{N_t N_t}^{-1} (\vec{\mathbf{T}}_t - \overline{\vec{\mathbf{T}}})$. This detector is *asymptotically* optimal provided both $\widehat{\Sigma}_{N_t N_t}$ and $\overline{\vec{\mathbf{T}}}$ are consistent estimators.[9] A corresponding estimator of the variance of $\tilde{\mathbf{A}}_t$ is $\widehat{Var}\left(\tilde{\mathbf{A}}_t\right) = \vec{S}_t^T \widehat{\Sigma}_{N_t N_t}^{-1} \vec{S}_t$. Together, they can be used to form a decision making statistic

$$\zeta_t = \frac{\vec{S}_t^T \widehat{\Sigma}_{N_t N_t} (\vec{\mathbf{T}}_t - \overline{\vec{\mathbf{T}}})}{(\vec{S}_t^T \widehat{\Sigma}_{N_t N_t}^{-1} \vec{S}_t)^{1/2}}. \tag{6.17}$$

Bell (1986) shows that ζ_t has a distribution similar to the Student's *t*-distribution under the no signal null hypothesis $H_0 : \mathcal{E}\left(\vec{\mathbf{T}}_t - \overline{\vec{\mathbf{T}}}\right) = 0$. The distribution has the pleasant property that its form does not depend upon the specifics of either the signal $\vec{S}_t$ or the covariance $\Sigma_{N_t N_t}$.

---

[9]Estimator $\hat{\mu}$ of parameter μ is *consistent* if $\lim_{n \to \infty} \mathcal{E}\left((\hat{\mu} - \mu)^2\right) = 0$ where n is the size of the sample from which $\hat{\mu}$ is computed. Note however, that asymptotic optimality is of little practical use in the context of climate change detection since it is conceptually impossible to increase the length of the reference period used to estimate $\overline{\vec{\mathbf{T}}}$ and $\widehat{\Sigma}_{N_t N_t}$ in an unbounded manner.

6.3.3   The Power of the Optimal Detector

Finally, Bell discusses a strategic problem in his 1986 paper. Specifically, suppose that the climate change signal has components throughout a high dimensional space. This might be the case, for example, if $\vec{S}_t$ has a complex spatial structure. Should we also search for the signal in a high dimensional space? The answer can be obtained by deriving the power, or sensitivity, of the decision making test that is based on the optimal detector.

Bell (1986) considers a pedagogical example in which the signal has p components

$$\vec{S}_t = \begin{cases} \vec{0} & \text{for } t \leq t_0 \\ (2, 0.75, 0.75, \ldots, 0.75)^T & \text{for } t > t_0 \end{cases}$$

and the noise vector has p independent univariate Gaussian elements so that $\Sigma_{N_t N_t} = \mathcal{I}$. He also supposes that the reference mean and covariance can be estimated from a sample of $n = 26$ independent observations taken at times $t \leq t_0$. Bell then considers the power of the optimal detector when $\Sigma_{N_t N_t}$ is known and when it is estimated, and compares these with the power of Hotelling's T^2 test. The latter, given by $T^2 = (\vec{\mathbf{T}}_t - \overline{\vec{\mathbf{T}}})^T \widehat{\Sigma}_{N_t N_t}^{-1} (\vec{\mathbf{T}}_t - \overline{\vec{\mathbf{T}}})$, is a detector that looks in all directions rather than just the single direction used by the optimal detector.

The result of the calculation is shown in Fig. 6.5. The power of the optimal detector increases steadily with the dimension p, if we have perfect knowledge of the covariance structure of the noise. On the other hand, the sensitivity of the detector increases only gradually until $p \approx 10$ when $\Sigma_{N_t N_t}$ is estimated, beyond which it degrades. This shows that is it necessary to keep the dimensionality of the detection problem small when the amount of data available to estimate the natural variability of the process is small. The third curve depicts the sensitivity of the 'brute force' detector that looks in all directions at the same time. This detector decreases in sensitivity with increasing dimension because the broad search incorporates noise, as well as, signal from all directions.

6.4   Signal Patterns and Spaces

The Battelle Pacific Northwest Laboratory was given a contract in 1992 to develop a proposal for a United States Department of Energy early climate change detection program. They enlisted a small panel of eminent scientists, including people like Tim Barnett, Klaus Hasselmann, Gerry North and Ben Santer and closeted these people in two intense two-day meetings to thrash out a detection strategy. Discussion was heated and lively. A small report was ultimately produced (Pennell, et al. 1993), but more importantly, the meetings seemed to have stimulated an intensified effort on climate change detection amongst those present. It may be coincidence, but

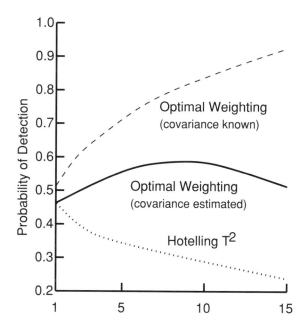

Fig. 6.5. Probability of detecting climate change $\vec{S}_t = (2, 0.75, 0.75, \ldots, 0.75)^T$ at the 2.5% significance level when the noise vector $\vec{N}_t$ consists of p uncorrelated Gaussian random variables $N_i, i = 1, \ldots, p$, with unit variance. The dashed curve shows the detection probability of the optimal detector with the covariance is known. The solid curve gives the detection probability when the covariance is estimated from $n = 26$ observations. The dotted curve shows the detection probability that Hotelling's T^2 test would yield. After Bell (1986)

papers on methodology (Hasselmann, 1993, 1997; North et al. 1995; North and Kim, 1995; Santer et al. 1993) and applications (Hegerl et al. 1996, 1997; Stevens and North, 1996; North and Stevens, 1997; Santer et al., 1995, 1996a) soon followed.

The first of these methodological papers was published by Klaus Hasselmann (1993). It generalized on an earlier paper (Hasselmann, 1979) that, like Bell (1982, 1986), discussed what Hasselmann called the *signal pattern* problem. That is, optimal detection when the response pattern $\vec{S}_t$ is known. Hasselmann (1993) generalized this to the *signal space* problem in which it is known a priori that the signal lies somewhere in a space spanned by several *guess* patterns. For example, one might use the signals estimated by several different coupled climate models, or one might use the ensemble of signals produced by an ensemble of climate change simulations.[10]

---

[10]See, for example, Boer et al. (1998), who describe an ensemble of three transient

6.4.1   The Signal Pattern Problem

We now briefly review the main points of Hasselmann's analysis of the *signal pattern* problem, which is similar to Bell's analysis, albeit somewhat more general. The basic ideas, which predate Bell, were first presented in Hasselmann (1979).

- Hasselmann derives the same optimal detection statistic ζ_t (recall equation (6.17)), but notes that $\vec{\mathbf{T}}_t$ could be an *extended* vector that spans space, time *and* multiple variables.

- He suggests the use of Monte Carlo simulation to find the null distribution of ζ_t. This was also suggested by Bell (1986), who performed simulations and compared with the exact distribution under idealized conditions.

- He calls $\Sigma_{N_t N_t}^{-1} \vec{S}_t$ the *optimal fingerprint* and notes that it is usually not parallel to $\vec{S}_t$.

- He points out that $\vec{\mathbf{T}}_t$ and $\vec{S}_t$ should be truncated to a finite number of EOFs if $\Sigma_{N_t N_t}$ is estimated to avoid poorly estimated high-index eigenvalues and EOFs (see equation (6.14) in Section 6.2 and the subsequent discussion).

6.4.2   The Signal Space Problem

The more general approach considered in Hasselmann (1993, 1997) is to suppose that we know only that the signal resides in a p-dimensional space spanned by p *guess patterns* $\vec{S}_{1,t}, \dots, \vec{S}_{p,t}$. As noted above, these might be obtained from p different climate models or from p independent climate change simulations performed with the same model. They might also be estimates of the response to p different perturbations in forcing. The detection problem can be solved independently for each guess pattern, in which case we find that the optimal detector in the ith direction is given by $\mathbf{A}_{i,t} = \vec{S}_{i,t} \Sigma_{N_t N_t}^{-1} \vec{\mathbf{T}}_t$. Since we do not have prior information about the direction, which is most likely to carry the signal, we must perform a test that searches in all directions simultaneously.[11] The no signal null hypothesis (6.9) now becomes

$$H_0 : \vec{S}_{i,t} = \vec{0} \ \text{ for all } \ i = 1, \dots, p \tag{6.18}$$

and decisions must be made with the *Hotelling* T^2 statistic

$$\mathbf{T}^2 = \vec{\mathbf{A}}_t^T \Sigma_{A_t A_t}^{-1} \vec{\mathbf{A}}_t \tag{6.19}$$

---

climate change simulations with the CCCMA coupled model (Flato et al. 1998).

[11] A Bayesian approach, such as that described by Leroy (1998) or Hasselmann (1998), avoids this problem because subjective information about the likely direction of the signal can be incorporated via the *prior* distribution. Bayesian statisticians use an approach to decision making that is philosophically different from the frequentist approach that has commonly been used in climatology.

where $\vec{\mathbf{A}}_t$ is the vector of optimal detectors $\vec{\mathbf{A}}_t = (\mathbf{A}_{1,t}, \ldots, \mathbf{A}_{p,t})^T$ and $\Sigma_{A_t A_t}$ is the matrix of covariances between the elements of $\vec{\mathbf{A}}_t$.

It is worth taking note of the algebraic similarity between the signal pattern and signal space detectors. In the signal space case, the vector of optimal detectors can be written

$$\vec{\mathbf{A}}_t = \begin{pmatrix} \mathbf{A}_{1,t} \\ \vdots \\ \mathbf{A}_{p,t} \end{pmatrix} = \begin{pmatrix} \vec{S}_{1,t}^T \\ \vdots \\ \vec{S}_{p,t}^T \end{pmatrix} \Sigma_{N_t N_t}^{-1} \vec{\mathbf{T}}_t = \mathcal{S}_t^T \Sigma_{N_t N_t}^{-1} \vec{\mathbf{T}}_t \qquad (6.20)$$

where $\mathcal{S}$ is the matrix composed of the p guess patterns. Using this expression it is easy to demonstrate that the covariance matrix required for the signal space decision maker (6.19) is $\Sigma_{A_t A_t} = \mathcal{S}_t^T \Sigma_{N_t N_t}^{-1} \mathcal{S}_t$.

An important piece of advice given by Hasselmann, which he reiterates from Hasselmann (1979), is that the signal space *should be kept small*. Hasselmann's point is easily illustrated with a simple example. Assume, for simplicity, that the elements of the detection vector $\vec{\mathbf{A}}_t$ are independent Gaussians with unit variance. This means that $\Sigma_{A_t A_t} = \mathcal{I}$. Also, suppose that only one of the guess patterns carries the signal. Specially, let's assume that $\mathcal{E}\left(\vec{\mathbf{A}}_t\right) = (2, 0, 0, \ldots, 0)^T$. The power of the Hotelling T^2 test, which is easily computed in this situation, is displayed as a function of p in Fig. 6.6. For each significance level considered, the power is high for the $p = 1$ (signal pattern) case, drops off very quickly as p increases and then decays gradually to the significance level of the test as p becomes very large.

This example is similar to that considered in Section 6.3.3 (see Fig. 6.5) except that it is assumed here that the first guess pattern captures the full signal. More generally, we would expect to capture the signal with increasing fidelity as the number of guess patters increases. However, more noise will also be captured by the decision making statistic, with the result that we may have both a better representation of the signal and lower power.

One other thing to note about Fig. 6.6 is that the test conducted at the 5% significance level is considerably more powerful than the one conducted at the 1% level. This is always the case with statistical hypothesis testing procedures. The least conservative decision maker will be the first to detect the climate change signal on average, but he or she will also incur more frequent (and perhaps larger) losses on average from actions taken on the basis of incorrect reject decisions.

6.4.3 Attribution of Climate Change

The question of attributing the causes of climate change has received less attention than that of detection, but some authors (e.g., Santer et al. 1993, 1995, 1996a,b; Hasselmann, 1997 and Hegerl et al. 1997) have given this some thought. We briefly describe Hasselmann's (1997) proposal for making

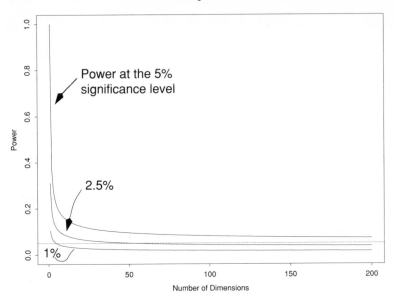

Fig. 6.6. An example illustrating the power of the signal space detector for a p-dimensional detector $\vec{\mathbf{A}}_t$ with mean $\mathcal{E}(\vec{\mathbf{A}}_t) = (2, 0, 0, \ldots, 0)^T$ and covariance $\Sigma_{A_t A_t} = \mathcal{I}$. It is assumed that $\vec{\mathbf{A}}_t$ is multivariate Gaussian. The curves display the probability of detection as a function of dimension for decision makers that operate at the 5%, 2.5% and 1% significance levels. For reference, the horizontal line indicates a probability of 5%

attribution in this subsection and will briefly describe an application in Section 6.4.5. Santer's approach to the question is briefly described in Section 6.6.

We begin by noting that the vector of optimal detectors $\vec{\mathbf{A}}_t$ given by (6.20) can be re-expressed as a vector of least squares estimates of the amplitudes θ_i of the signals $\vec{S}_{i,t}$ in the observed record. In fact, if we suppose that the observed process can be represented as a linear combination of the signals plus noise,

$$\vec{\mathbf{T}}_t = \sum_{i=1}^{p} \theta_i \vec{S}_{i,t} + \vec{\mathbf{N}}_t,$$

then the usual algebra leads to

$$\widehat{\vec{\theta}} = (\mathcal{S}_t^T \Sigma_{N_t N_t}^{-1} \mathcal{S}_t)^{-1} \vec{\mathbf{A}}_t \tag{6.21}$$

where the signal matrix $\mathcal{S}_t$ is as defined in (6.20).

This vector of amplitudes can be estimated from the observations (say $\widehat{\vec{\theta}}_{obs}$) and also from a transient climate change simulation (say $\widehat{\vec{\theta}}_{mod}$). If the signals are well represented in the model, and if they are truly the cause of observed climatic change, then we should expect the estimates to be equal to within sampling variability. Hasselmann (1997) and Hegerl et al. (1997) therefore propose the *non-rejection* of

$$\text{H}_0 : \vec{\theta}_{obs} = \vec{\theta}_{mod} \tag{6.22}$$

as a criterion for attribution. That is, the proposal is that we should only claim attribution if we detect climate change (i.e., reject (6.18)) and *fail* to find significant evidence that the signal amplitudes in the observed and simulated climates are unequal.

At first glance, this would appear to be a somewhat awkward approach because of the asymmetrical roles of the null and alternative hypotheses in statistical hypothesis testing. Tests are set up to look for evidence of the type suggested by the alternative hypothesis that is contrary to the null hypothesis.[12] Failure to reject, at say the 1% significance level, simply means that there was not enough evidence to safely reject the null hypothesis.[13] This is not the same as the common, but incorrect, interpretation that non-rejection at the 1% level is equivalent to acceptance at the 99% confidence level.

However, despite its awkward nature, the approach does bring us closer to real attribution by subjecting a detection finding to a second objective criterion that measures the similarity between the observed and model simulated signal amplitudes. In practice, the awkwardness is at least partly ameliorated by using the attribution test to eliminate contending explanations for detected climate change. This is the approach taken in Hegerl, et al. (1997; see Section 6.4.5). *Multiplicity* is not likely to be a great concern in the attribution problem because the test will be used to screen a limited number of contending mechanisms for climate change. Moreover, the general expectation is that the the probability of rejecting the null hypothesis will be large for all but one of these mechanisms (i.e., the correct one).[14]

The choice of the significance level for the attribution test is guided by two considerations. We want the probability of rejecting incorrect signals

---

[12]For example, when testing $\text{H}_0 : \mathcal{E}(\mathbf{X}) = 0$ against $\text{H}_1 : \mathcal{E}(\mathbf{X}) > 0$, we reject the null when $\mathbf{X}$ is large and positive. Unusually large negative values of $\mathbf{X}$ are ignored.

[13]That is, the evidence was not strong enough to warrant the 1% risk of incorrectly rejecting the null hypothesis when it is true.

[14]*Multiplicity* is a problem that arises when conducting multiple tests of hypothesis. The significance level of the combined outcome is difficult to assess as the number of tests performed increases because some tests will reject the null hypothesis by random chance even if all null hypotheses are true. If all tests are independent of each other, and all are performed at say the 1% significance level, then the probability of making one or more reject decisions in n is given by $1 - 0.99^n$. For example, there is a 4.9%, 9.6% or 18.2% chance of making at least one reject decision in 5, 10 or 20 independent tests conducted at just the 1% level.

(i.e., not attributing to the wrong climate change mechanism) to be as large as possible. This suggests that we should operate with a fairly large risk, say 10%, of rejecting the null hypothesis. On the other hand, we also would like to avoid rejection in the presence of the correct signal, suggesting that we should operate at a much smaller significance level, say 1%. However, operating conservatively increases the chances of equivocal results, that is, failing to reject the attribution test for more than one signal. Thus, given the economic stakes in the detection and attribution game, it seems prudent to increase the power of the attribution test at the expense of increasing the risk of failing to attribute the correct mechanism.

6.4.4 Applying the Signal Pattern Approach

Hegerl et al. (1996) describe one of the first large scale applications of the signal pattern approach to the detection of the anticipated global warming signal. They focused their efforts on surface temperature because this variable is relatively well observed and has a long instrumental record, and because model studies suggest that the signal-to-noise ratio of the enhanced greenhouse gas effect will be greater in temperature than in most other variables (see, for example, Barnett et al. 1991, Cubasch et al. 1992, Zwiers and Kharin, 1998). Hegerl et al. used the well known Jones and Briffa (1992) surface temperature data set, which is constructed by averaging selected station and ship data into $5° \times 5°$ latitude-longitude boxes.[15]

Hegerl et al. keep the size of their detection problem manageable by using a temperature vector $\vec{\mathbf{T}}_t$ constructed from a two dimensional latitude-longitude field. They do not explicitly include time in the vector by including temperature fields observed at different times. Instead, they incorporate the time derivative of the response into the problem by using linear trend coefficients from a moving window trend analysis as the primary detection variable. Three separate detection analyses are performed using trend coefficients from 15-, 20- and 30-year moving windows. Relatively short windows are used so that accelerating changes can be detected, and also so that an attempt can be made to estimate the natural variability of the trend coefficients from the observed record.

Figure 6.7 displays 30-year trend estimates from two periods in this century. The most recent period (1965-94) shows warming in the Northern Hemisphere over northwestern North America and Siberia of up to $1°C$ per decade, and cooling of a similar magnitude over the eastern North America / Labrador Sea region. The (1916-1945) period also shows warming over western North America, but it does not show large trends in other regions.

The blank areas in Fig. 6.7 identify regions in which there is insufficient data to estimate the 30-year trends. These gaps pose a problem for detection

---

[15]The Jones and Briffa (1992) data set is maintained at the University of East Anglia and is available on the Internet from `http://www.cru.uea.ac.uk/advance10k/climate.htm`.

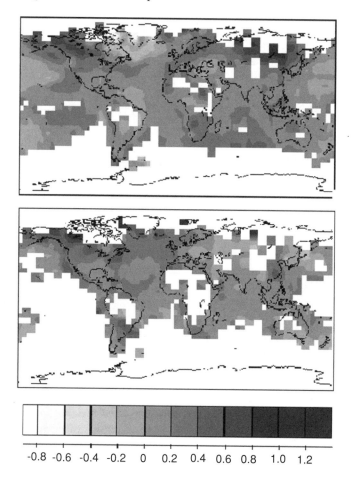

Fig. 6.7. The thirty year trend in observed (Jones and Briffa, 1992) surface temperature in $°C$ per decade for a) 1965-94 and b) 1916-45. Regions with insufficient data are not shaded. Courtesy G. Hegerl

efforts because missing data increases the sampling variability of the detection statistics and, therefore, makes identification of the signal more difficult. Hegerl et al., therefore, construct $\vec{\mathbf{T}}_t$ from the $5° \times 5°$ grid boxes for which trends can be fitted from 1949 onwards. This means that the detector will be influenced by enhanced sampling variability caused by missing data prior to 1949, but it will have uniform sampling properties during the important recent period. This grid corresponds to the shaded areas Fig. 6.8.

Hegerl et al. used the *Early Industrialization* (EIN) transient climate change experiment (Cubasch et al. 1995) performed the ECHAM/LSG coupled model (Cubasch et al. 1992) to estimate the signal pattern $\vec{S}_t$. This experiment specifies equivalent CO_2 concentrations from observations be-

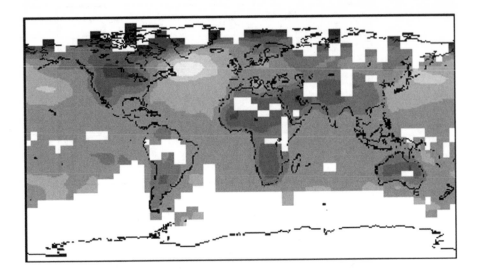

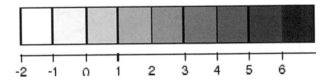

Fig. 6.8. The signal pattern $\vec{S}_t$ used by Hegerl et al. (1996). The pattern was computed from the Max Planck Institute *Early Industrialization* transient climate change experiment (Cubasch et al. 1995) by subtracting the mean temperature field simulated for 1986-1995 from the mean field simulated for 2076-2085. Courtesy G. Hegerl

tween 1935 and 1985 and increases the equivalent CO_2 concentration at the rate of 1% per year beyond 1985. The signal vector is defined as the difference between the mean surface temperature simulated for the last decade in the EIN simulation (2076-2085) and the mean temperature simulated for the 1986-1995 decade at the locations at which temperature trends can be computed from the observed record for 1949 onwards.

The resulting signal pattern estimate is displayed in Fig. 6.8. As Hegerl et al. point out, the detector constructed with this pattern will be optimal *only if* this signal pattern is correct.[16] The features we see in Fig. 6.8 are similar to those obtained with other models as well. There is more warming over the continents than over the ocean, warming at high latitudes that is amplified by ice-albedo feedback, and some cooling in the northwest Atlantic. Comparing with Fig. 6.7, we see that there is at least some similarity with

---

[16]This is obvious, of course, but worth stating because it is easily forgotten.

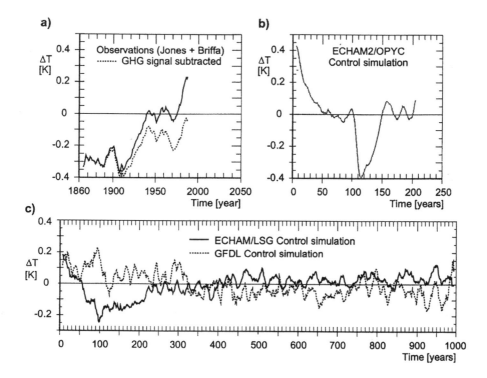

Fig. 6.9. Time evolution of global mean near-surface temperature for (a) the observations, (b) the ECHAM/OPYC (Lunkeit et al. 1995) control simulation, and (c) for two 1000-year control simulations with the GFDL (Manabe and Stouffer, 1996) and ECHAM/LSG models. All three have been smoothed with an 11-year running mean. Note the different scales on the time axis. The dashed line in (a) shows the observed global mean temperature after the greenhouse gas signal has been subtracted. All global mean temperatures are computed on the 'observed' grid (see text). From Hegerl et al. (1996)

the changes that have occurred during the last 30-years.

At this point we have two of the four ingredients required for a detection effort. The next item that is required is an estimate $\widehat{\Sigma}_{N_t N_t}$ of the natural variability of $\vec{\mathbf{T}}_t$. Dimension reduction is needed to achieve this, both to avoid the poorly estimated high index eigenvalues and to avoid the poorly sampled part of the physical domain. Hegerl et al. point out that the high index EOFs will tend to rotate the guess pattern into these regions. On the other hand, it is necessary to retain enough EOFs to represent $\vec{S}_t$ well in the reduced space and to insure that the space is large enough to allow effective optimization. Hegerl et al. settled on $k = 8$ EOFs. The EOFs were obtained from the EIN simulation on the grid shown in Fig. 6.8.

Four data sets were available for estimating natural variability. These where i) 1000 years from the ECHAM/LSG simulation (von Storch, 1994; von Storch et al. 1997), ii) 210 years from the ECHAM2/OPYC simulation (Lunkeit et al. 1995), iii) 1000 years from the GFDL transient climate change simulation (Manabe and Stouffer, 1996), and iv) 1854-1994 observations.[17] In each case the data were put on the 5° × 5° grid shown in Fig. 6.8. Figure 6.9 displays the global means of these data sets. There are large slow variations in the observed means with a large warming between 1910 and 1940, and another from 1970 onwards. The EIN signal accounts for some of this variation. The ECHAM2/OPYC simulation shows an initial adjustment period, followed by a sudden cooling of about 0.4° and then another period of adjustment. The ECHAM/LSG simulation cools for the first 100 years and then gradually adjusts to an equilibrium state at about year 400. The global mean seems to behave as a stationary process beyond that point. The GFDL simulation cools gradually for about 350 years before fluctuating about a steady state. It seems to generate more variability than the ECHAM/LSG model on the decadal and longer time scales displayed in Fig. 6.9.

Figure 6.10 displays a comparison of the variability of Northern Hemisphere summer land temperatures in these these data sets. Note that 'raw' observations, which include external signals, were used for this calculation. It appears that the GFDL simulation has about as much variability as the observations on the decadal and shorter time scale, while the ECHAM/LSG simulation has less variability than is observed. At long time scales both the ECHAM/LSG and GFDL simulations appear to have less variability than observed. The paleo data has less variability than the raw observations at long time scales, perhaps because it does not contain the warming signal. However, precise assessments are difficult because the observed record is short and subject to increased sampling variability earlier in the record (Jones et al. 1997). The paleo data are also subject to various uncertainties.

Before estimating the natural variability, the data were reduced by computing the 15-, 20-, and 30-year moving window trend coefficients at each grid point. These fields were subsequently projected onto the EOFs. In the case of the observations, for which many early trend coefficients are missing, the EOF coefficients were estimated by fitting the EOFs to the available trend coefficients by least squares.

Next, the reduced ECHAM/LSG data were used to estimate the 8×8 covariance matrix $\Sigma_{N_t N_t}$. We denote this matrix $\widehat{\Sigma}_{E/L}$. *Given* this matrix, the other three 'natural variability' data sets were used to estimate the natural variance of the detector $\mathbf{A}_t = \vec{S}_t^T \widehat{\Sigma}_{E/L}^{-1} \vec{\mathbf{T}}_t$. Why use the ECHAM/LSG simulation to estimate $\Sigma_{N_t N_t}$? First, the covariance matrix should be well estimated which suggests that one of the two 1000-year control simulations

---

[17]The signal pattern $\vec{S}_t$ obtained from the EIN simulation was removed from the observations before estimating the covariance matrix.

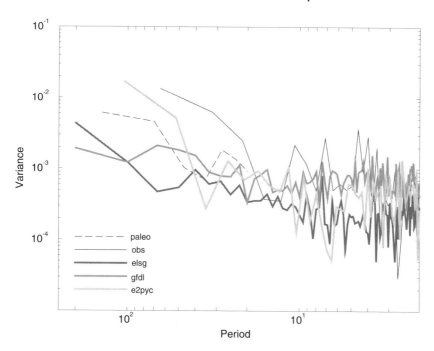

Fig. 6.10. Power spectra of Northern Hemisphere mean summer land temperatures estimated from observations (Jones and Briffa, 1992; solid black line), a paleo climate time series (Bradley and Jones, 1993; dashed black line), the 1000-year ECHAM/LSG simulation, the 1000-year GFDL simulation, and the ECHAM2/OPYC simulation. Courtesy G. Hegerl

should be used. Second, the decision making test should be conservative.[18] This argues that the simulation with the smaller variability should be used. To understand this, recall that the detector can be written

$$\mathbf{A}_t = \sum_{i=1}^{8} \frac{\beta_{i,t}\delta_{i,t}}{\lambda_i}$$

where $\beta_{i,t}$ and $\delta_{i,t}$ are the projections of the reduced signal and observations respectively onto the ith eigenvector of $\widehat{\Sigma}_{E/L}$, and λ_i is the corresponding eigenvalue. Smaller eigenvalues lead to a detector $\mathbf{A}_t$ with greater natural variance, and thus a test that is inherently more conservative.

Once the signal vector and covariance matrix have been determined, the optimal fingerprint (i.e., the weight vector $\vec{w}_t$) in the reduced space is given by

---

[18] A conservative test rejects the null hypothesis less frequently than the nominal significance level when the hypothesis is in fact true.

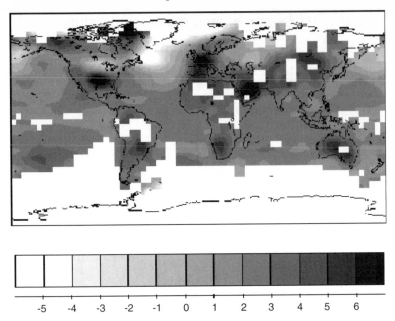

Fig. 6.11. The optimal fingerprint $\vec{w}_t$ for the 30-year temperature trend derived from the EIN signal and the ECHAM/LSG natural variability. Units are $^\circ C$ per decade. Courtesy G. Hegerl

$\widehat{\Sigma}_{E/L}^{-1}\vec{S}_t$. The optimal fingerprint in physical space is easily recovered from this vector by treating its components as EOF coefficients. The resulting fingerprint obtained by Hegerl et al. for the 30-year temperature trends is displayed in Fig. 6.11. We see a pattern that looks for more warming over land than ocean and for cooling in the northwest Atlantic.

Figure 6.12 displays the projection of the reduced observations onto the optimal fingerprint (i.e., $\mathbf{A}_t$) for the 20-year trend coefficients (solid curve). A projection of the EIN simulation surface temperature onto the optimal fingerprint is also displayed (dotted curve). The shaded regions indicate the range of $\mathbf{A}_t$ values that are consistent with the no signal null hypothesis at the 5% significance level. That is, we expect the $\mathbf{A}_t$ time series to wander outside the these bands about 5% of the time when no climate change signal is present. The narrow region is obtained when the GFDL control simulation is used to estimate the natural variability of the detector, and the wide band is obtained when observations (with the signal removed) are used. The solid curve tentatively suggests that the EIN signal is detectable in the observations beginning perhaps in the late 1980's. The dashed curve shows that the signal actually emerges from the EIN simulation itself about a decade later. Results for the 15-year trends (not shown) do not yet show a significant signal, presumably because natural variability is greater on this time

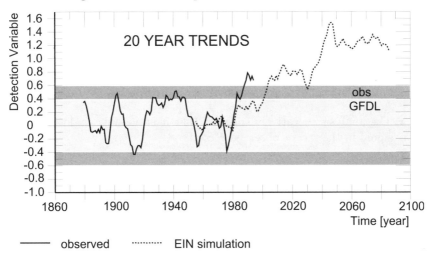

Fig. 6.12. The projection A_t of the 20-year trends onto the optimal fingerprint $\vec{w}_t$ for the observations (solid line) and EIN simulation (dotted line). The shaded bands are expected to contain A_t 95% of the time when no signal is present and when the natural variability of A_t is estimated correctly. The narrow band labelled 'GFDL' uses the variability estimated from the GFDL simulation while the wider band labelled 'obs' uses the Jones and Briffa (1992) data (with signal removed). Courtesy G. Hegerl

scale. Results for the 30-year trends (not shown) are similar to those for the 20-year trends.

We complete this section by summarizing the components of this application of the signal pattern approach.

- The observed surface air temperature data are obtained from Jones and Briffa (1992).

- All data are put on the observed grid, and only those grid points at which trends can be reliably estimated for the modern instrumental period (1949 onwards) are used.

- The signal and the EOFs needed for dimension reduction are obtained from a transient climate change simulation.

- The analysis is performed on 15-, 20- and 30-year moving window trend coefficients that have been projected onto the first 8 EOFs. Hegerl et al. felt that 8 were a reasonable comprise between a) keeping the problem small so that the covariance can be well estimated, b) representing the signal well, and c) retaining a space that is large enough to permit useful optimization. It was found that the optimal fingerprint improves the signal to noise ratio by about 20% for the 30-year trends.

- The covariance matrix $\Sigma_{N_t N_t}$ is estimated from the ECHAM/LSG simulation because it has lower variability than the other 'natural variability' data sets.

- The natural variance of the detector was estimated from the GFDL experiment and the observed record (with the signal removed).

- The analysis tentatively suggests that it is possible to detect the EIN signal in the recent observed record.

6.4.5   A Signal Space Application

Hegerl et al. (1997) update their 1996 study by considering several signals, using two additional long control runs to better assess natural variability, and attempting an application of the multi-pattern attribution methodology discussed Section 6.4.3.

Signals associated with changing greenhouse gas forcing alone ($\vec{S}_{GHG}$), the combined effect of changing greenhouse gas forcing and sulphate aerosol distributions ($\vec{S}_{GHG+A}$) and variations in solar output ($\vec{S}_{SOL}$) were estimated from recent transient climate change simulations with the ECHAM3/LSG model (Hasselmann et al. 1995; Cubasch et al., 1997) for the period 1880 to 2049. The greenhouse gas signal was obtained by computing the first EOF of the annual mean surface temperature in the greenhouse gas only experiment. Hegerl et al. found that the signal is well represented by the first EOF, and argue that this estimate is subject to less sampling variability than the estimate obtained by taking the difference between long period means at the beginning and end of the simulation. The greenhouse gas plus aerosol signal is taken to be the first EOF of the ensemble mean of a pair of independent simulations with identical greenhouse gas and aerosol forcings. The solar forcing signal is obtained similarly from a pair of simulations in which estimated historical changes in solar forcing were prescribed (Cubasch et al. 1997).

Data for estimating the natural variability was obtained from the 1000-year ECHAM/LSG and GFDL control simulations used previously, a 700-year control simulation performed with ECHAM3/LSG and a 611-year control simulation performed with HADCM2 (Johns et al. 1997). Observations were also used to estimate natural variability.

Hegerl et al. (1997) first repeated the signal pattern detection exercise of their previous paper. They searched for the signals in 30-year moving window trends computed from the Jones and Briffa (1992) data set. Dimensionality was reduced by projecting all fields onto 10 EOFs. The EOFs were estimated from one of the two greenhouse gas plus aerosol simulations. The noise covariance matrix needed to derive the optimal detectors was estimated from the 30-year trends observed in the 1000 year ECHAM/LSG simulation. Results (not shown) are similar to those described above (see Fig. 6.12).

Hegerl et al. (1997) demonstrate the pattern space approach to detection and attribution by considering two signals, the greenhouse gas only signal and the greenhouse gas plus aerosol signal. Together, they span a two dimensional vector space. An orthogonal basis (with respect to the noise) is easily obtained by choosing the normalized version of $\vec{S}_{GHG}$ as one of the basis vectors and setting the second basis vector to

$$\vec{S}_A = c(\vec{S}_{GHG+A} - \vec{S}_{GHG+A}^T \Sigma_{N_t N_t}^{-1} \vec{S}_{GHG}) \tag{6.23}$$

where c is a normalizing constant. The elements of the resulting 2-dimensional vector of estimated signal amplitudes, say $(\widehat{\theta}_{GHG}, \widehat{\theta}_A)^T$, should by uncorrelated if $\mathbf{T}_t$ contains only noise with covariance proportional to $\Sigma_{N_t N_t}$.

Figure 6.13 shows the $(\widehat{\theta}_{GHG}, \widehat{\theta}_A)^T$ time series obtained from the 30-year moving window trends in the observations. The upper panel displays the time series alone for clarity while the lower panel also displays three swarms of points that represent natural variability as simulated by the ECHAM3/LSG, GFDL and HADCM2 models. The lower panel also displays the evolution of the observed time series after the signal has been removed. The ellipsoids are 95% confidence regions that should contain 95% of the noise points if the detectors are jointly Gaussian. The fact that these ellipsoids are not circular shows that the natural variability in the ECHAM3/LSG, GFDL and HADCM2 control simulations differs somewhat from that simulated in the ECHAM/LSG control simulation. The differences are not profound.

The observed time series oscillates in and out of the swarm of points in Fig. 6.13 (bottom). Consistent with the signal pattern detection results (not shown), the detectors emerge from the noise for the 30-year windows ending in the mid-40's and at the end of the record when they take their most extreme values. It is encouraging that the observed variations from which the signal has been removed (labelled 'vobs' in Fig. 6.13 (bottom)) are largely confined to the region occupied by the control simulations.

Figure 6.14 displays an *attribution diagram* for 50-year moving window trends of the Northern Hemisphere summer (JJA) mean temperature. The black trace represents the time series of signal amplitude estimates computed from the observations. Detection can be claimed at the 5% significance level for the last 5 points in this time series when any of the three control simulations (ECHAM3/LSG, GFDL or HADCM2) is used to estimate natural variability (not shown). According to Hasselmann's (1997) criterion, attribution of the cause of the warming can only be claimed if null hypothesis (6.22) can not be rejected. This hypothesis can be tested for the, say 1946-95 trend, by computing a statistic of the form

$$\begin{aligned}
\mathbf{D}^2 &= (\widehat{\vec{\theta}}_{obs,46-95} - \widehat{\vec{\theta}}_{mod,46-95})^T (\Sigma_{\theta_o \theta_o} + \Sigma_{\theta_m \theta_m})^{-1} \\
&\times (\widehat{\vec{\theta}}_{obs,46-95} - \widehat{\vec{\theta}}_{mod,46-95})
\end{aligned} \tag{6.24}$$

where $\widehat{\vec{\theta}}_{obs,46-95}$ is the vector of estimated signal amplitudes computed from

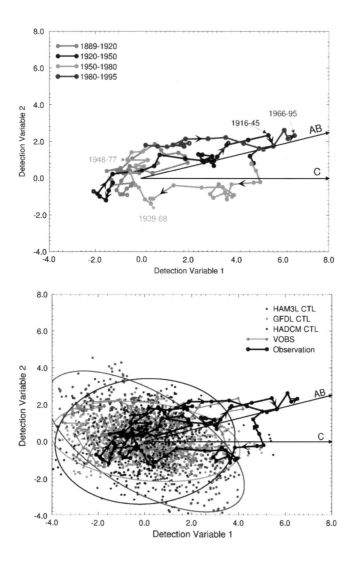

Fig. 6.13. Top: Signal amplitudes estimated from observed 30-year moving window trends.

Bottom: Signal amplitudes together with corresponding amplitudes estimated from 4 natural variability data sets (the ECHAM3/LSG, GFDL and HADCM2 control simulations, and observations from which a signal estimate has been removed). Ellipsoids are estimated 95% confidence regions for the natural variation of the signal amplitudes. Courtesy G. Hegerl

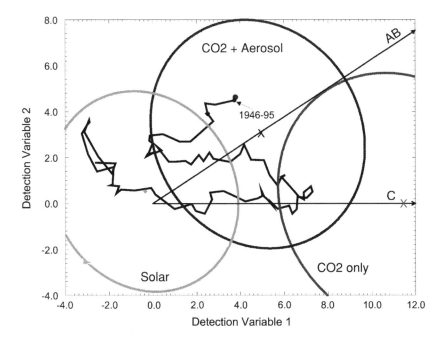

Fig. 6.14. Attribution diagram for observed 50-year trends in JJA mean temperature. The ellipsoids enclose non-rejection regions for testing the null hypothesis that the 2-dimensional vector of signal amplitudes estimated from observations has the same distribution as the corresponding signal amplitudes estimated from the simulated 1946-95 trends in the greenhouse gas, greenhouse gas plus aerosol and solar forcing experiments. Courtesy G. Hegerl

the observed 1946-95 trends, $\widehat{\vec{\theta}}_{mod,46-95}$ is the same quantity estimated for the same period in a forced simulation, and $\Sigma_{\theta_o\theta_o}$ and $\Sigma_{\theta_m\theta_m}$ are estimates of the covariance matrices of $\widehat{\vec{\theta}}_{obs,46-95}$ and $\widehat{\vec{\theta}}_{mod,46-95}$ respectively. These covariances are estimated from the observations with the signal removed and from the ECHAM3/LSG control simulation respectively.

$\mathbf{D}^2$ has a chi-square distribution with 2 degrees of freedom under the null hypothesis when the covariances matrices are well estimated and when the signal amplitudes are Gaussian.[19] A test can be conducted by rejecting null hypothesis (6.22) when $\mathbf{D}^2 > D_c^2$ where D_c^2 is an appropriate critical value.[20] Equivalently, we will not be able to reject when

---

[19]$\mathbf{D}^2$ will be chi-squared distributed with p degrees of freedom when $\vec{\theta}$ has p elements

[20]Critical values of the chi-squared distribution with 2 degrees of freedom for testing (6.22) at the 10%, 5%, 2.5%, 1% and 0.5% significance level are 4.61, 5.99, 7.38, 9.21 and 10.6 respectively.

$$(\widehat{\vec{\theta}}_{obs,46-95} - \widehat{\vec{\theta}}_{mod,46-95})^T (\Sigma_{\theta_o \theta_o} + \Sigma_{\theta_m \theta_m})^{-1}$$
$$\times (\widehat{\vec{\theta}}_{obs,46-95} - \widehat{\vec{\theta}}_{mod,46-95}) < D_c^2.$$

This last relation gives an ellipsoidal region of possible values for $\widehat{\vec{\theta}}_{obs,46-95}$ that are consistent with the null hypothesis that the observed and simulated signal amplitudes are equal. We call this the *non-rejection* region.

   Figure 6.14 displays regions like this for greenhouse gas only simulation, the greenhouse gas plus aerosol simulation, and the solar forcing simulation. The most recent trends appear to be consistent with the simulated greenhouse gas plus aerosol forcing response, and thus give some modest support for the attribution of the observed warming to the combined effects of these two forcings. Unequivocal attribution is still difficult at this time because there remains substantial overlap of the non-rejection regions for the various signals considered. We can anticipate that these regions will diverge in the future as the greenhouse gas and aerosol forcings strengthen.

6.5   Space-Time Filter Theory

Another participant in the Battelle meetings was Gerry North. He and his colleagues have approached the detection problem using a 'filter theory' formalism that puts the problem in a very elegant mathematical framework. However, while the language is different, North's approach ultimately results in optimal detectors that are formally equivalent to those described by Hasselmann (1979, 1993) and Bell (1982, 1986). However, North and colleagues do bring some interesting and useful additional insight to the detection problem by taking a substantially different approach to the application of the optimal detector.

6.5.1   The Basic Filter Theory

North et al. (1995) sets out the basic filter theory. The approach is much the same as that set out by Hasselmann and Bell, but now the quantities involved are continuous in space and time. Retaining as much of our original notation as possible, we let $\mathbf{T}(\vec{r}, t)$ represent the observed field at a location $\vec{r}$ on the surface of the globe and time t. As before, we have in mind that $\mathbf{T}$ consists of deterministic signal plus random noise

$$\mathbf{T}(\vec{r}, t) = S(\vec{r}, t) + \mathbf{N}(\vec{r}, t). \tag{6.25}$$

The goal of the filter theory approach is to construct a linear *smoothing filter* that extracts the signal from $\mathbf{T}$ as efficiently as possible. That is, the approach seeks to find a *filter kernel* $\Gamma(\vec{r}, t, \vec{r}', t')$ such that

$$\widehat{S}(\vec{r}, t) = \int_{\mathcal{D}} \int_{\mathcal{T}} \Gamma(\vec{r}, t, \vec{r}', t') \mathbf{T}(\vec{r}', t') d\vec{r}' dt' \tag{6.26}$$

is an unbiased estimator of S with minimum expected square error

$$\epsilon^2(\vec{r}, t) = \mathcal{E}\left(\left[\widehat{S}(\vec{r}, t) - S(\vec{r}, t)\right]^2\right)$$

for all $\vec{r}$ and t.

North et al. show that the optimal filter kernel is given by

$$\Gamma(\vec{r}, t, \vec{r}', t') = \frac{S(\vec{r}, t)}{\gamma^2} \sum_{i=1}^{\infty} \frac{\beta_i \Psi_i(\vec{r}', t')}{\lambda_i} \qquad (6.27)$$

where $\Psi_i(\vec{r}, t)$ is the ith eigenfunction of $\rho(\vec{r}, t, \vec{r}', t') = \text{Cov}(\mathbf{T}(\vec{r}, t), \mathbf{T}(\vec{r}', t'))$, λ_i is the corresponding eigenvalue, β_i is the projection of the signal $S(\vec{r}, t)$ onto $\Psi_i(\vec{r}, t)$ and $\gamma^2 = \sum_{i=1}^{\infty} \beta_i^2/\lambda_i$.[21] Substituting kernel (6.27) into (6.26) results in an unbiased, minimum squared error estimator for $S(\vec{r}, t)$ that is given by

$$\widehat{S}(\vec{r}, t) = \frac{S(\vec{r}, t)}{\gamma^2} \mathbf{A} \qquad (6.28)$$

where

$$\mathbf{A} = \sum_{i=1}^{\infty} \frac{\beta_i \delta_i}{\lambda_i} \qquad (6.29)$$

and δ_i is the projection of $\mathbf{T}(\vec{r}, t)$ onto the ith eigenfunction $\Psi(\vec{r}, t)$.

The important thing to notice about (6.28) and (6.29) is the similarity between (6.29) and (6.14). Specifically $\mathbf{A}$ has exactly the same form as the optimal detector (6.14) derived in Section 6.2.[22] Indeed, $\mathbf{A}$ *is* the optimal detector when the observed process is continuous rather than discrete. Also, it is worth noting that (6.28) is the continuous space-time analogue of the discrete space-time signal estimator described by Hasselmann (1993; see equation (23), page 1962). The conclusion to be drawn is that despite the differences in language and approach, Hasselmann, Bell, North (and others not mentioned in this review) all arrive at the same optimal detection solution.

North et al. (1995) complete their paper by considering some simple examples. A companion paper (North and Kim, 1995) further embellishes the technique by designing filters that can look for oscillatory signals in a given frequency band. Using an energy balance model, they design filters that might be used to detect the climate's response to solar and greenhouse gas forcing.

---

[21]That is, $\Psi_i(\vec{r}, t)$ and λ_i satisfy an integral equation of the form

$$\int_D \int_T \rho(\vec{r}, t, \vec{r}', t') \Psi_i(\vec{r}', t') d\vec{r}' dt' = \lambda_i \Psi_i(\vec{r}, t).$$

Just as with the familiar EOFs, these eigenfunctions are orthonormal and complete under suitable regularity conditions.

[22]Actually, as stated here, (6.29) is slightly less general than (6.14) because it has been implicitly assumed that the covariance function is independent of the time origin.

6.5.2 An Application - Detecting the Solar Signal

Stevens and North (1996) demonstrate the filter theory (i.e., optimal detector) approach by trying to detect the climate's response to a weak periodic variation in solar output. Satellite observations suggest that solar output varies with the sunspot cycle with an amplitude of about $0.6 \ Wm^{-2}$ (Willson and Hudson, 1991; Lee et al. 1995). Stevens and North point out that the long term variability of the solar output is still highly uncertain since the available satellite record spans only about one and one half solar cycles. They also point out that the signal, if it is present, is very weak. They note that the observed standard deviation of global annual mean temperature is about $0.18°C$ while recent estimates of the response to the solar cycle range from $0.01°C$ to $0.08°C$.

It is instructive to examine this detection study because it takes a substantially different approach to the preparation of the data and the estimation of the detector's natural variability. Also, the evaluation of the optimal detector (i.e., deciding whether a signal has been detected) is dealt with differently here than in studies described above.

Stevens and North cast the problem of finding the optimal filter in the familiar discretized form rather than in the original continuous space-time form. As before, let $\vec{\mathbf{T}}_t$, $\vec{S}_t$ and $\vec{\mathbf{N}}_t$ be ml-dimensional vectors of observed temperature, signal and noise sampled at m locations on the surface of the earth and at l different times. The derivation of the optimal filter proceeds in the usual manner and results in an unbiased, minimum squared error estimate of the signal that is given by

$$\widehat{\vec{S}}_t = \vec{S}_t \frac{\vec{w}_t^T \vec{\mathbf{T}}_t}{\vec{S}_t^T \Sigma_{N_t N_t}^{-1} \vec{S}_t} \tag{6.30}$$

where $\vec{w}_t$ are the optimal weights given by (6.5) and where $\Sigma_{N_t N_t}$ is the covariance matrix of the noise. Equation (6.30) can be re-written in the same form as (6.28), that is, as

$$\widehat{\vec{S}}_t = \vec{S}_t \frac{\mathbf{A}_t}{\gamma^2} \tag{6.31}$$

where $\mathbf{A}_t$ is the optimal detector (6.6) and γ^2 is the signal to noise ratio (6.8). Note that the filter estimates only the amplitude of the signal; *it does not modify its shape.*

As with many other recent examples, the Jones and Briffa (1992) data are used in this study. Thirty-six *detection boxes*, each covering a $10° \times 10°$ region with a 100-year time series of monthly mean near-surface temperature data, are selected. The geographical distribution of the boxes is illustrated in Fig. 6.15. As is to be expected, there is little coverage over the oceans.[23]

---

[23]See Zwiers and Shen (1997), and references therein, for an analysis of the sampling error that results from estimating the global mean and higher order spherical harmonics from a sparse observing network.

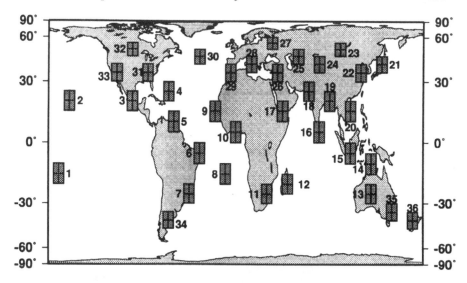

Fig. 6.15. Geographical distribution of the thirty-six $10° \times 10°$ detection boxes

An interesting aspect of the Stevens and North detection strategy is that they put the annual mean temperature in the tropical detection boxes (i.e., those boxes located between $30°N$ and $30°S$), but put the summer mean temperature (where summer is defined as the 6 warm months of the year) in the extra-tropical boxes. As can be seen from Fig. 6.16, the variability of the temperature in the detection boxes is roughly constant throughout the year in the tropics, but it increases sharply during the cold season at higher latitudes. Thus, their strategy effectively increases the signal to noise ratio by avoiding the high variance part of the record.

The thirty-six 100-year times series that result from this temporal sampling and box averaging process are then Fourier transformed, and the sine and cosine coefficients that resolve the 'solar band' are retained. That is, the coefficients at the 9 frequencies $\omega = 0.06, 0.07, \ldots, 0.14 \text{ yr}^{-1}$ located near the solar frequency ($1/11 \text{ yr}^{-1}$) are retained for each detection box. The ultimate result is a single temperature coefficient vector $\vec{T}$ of dimension $36 \times 2 \times 9 = 648$.

The next step in the detection exercise is to estimate the variability Σ_{NN} of the noise present in $\vec{T}$. This can not be done with observations because the entire record has been reduced to a signal realization of a high dimensional random vector. Stevens and North, therefore, turn to the GFDL and ECHAM/LSG 1000-year climate simulations. Their first step in estimating Σ_{NN} is to assume that the cross-spectral density is approximately constant

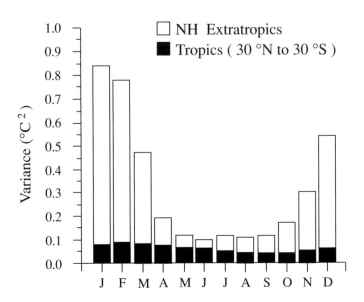

Fig. 6.16. Area-weighted monthly mean variances ($^\circ C^2$) of the observed surface temperature anomalies for the 13 detection boxes in the Northern Hemisphere extratropics and the 20 boxes in the Tropics. Monthly means for the period 1984-1993. From Stevens and North (1996)

for frequencies $\omega \in [0.06, 0.14]$.[24] This makes it possible to estimate Σ_{NN} from a 100-year record by estimating the cross-spectral densities at the center of the solar band with a spectral estimator having an equivalent bandwidth of approximately 0.09 yr^{-1}. They do this 10 times with each climate simulation, once for each 100-year chunk of the 1000-year run, and then average the 10 covariance matrix estimates to arrive at a final estimate of Σ_{NN} for each simulation.

The third ingredient required in order to search for the solar cycle signal is an estimate of the signal. This was achieved in a number of steps. The first step was to linearly correlate irradiance data from the Nimbus-7 satellite with sunspot numbers over a one and one half cycle period (Fig. 6.17, left). Stevens and North found a correlation of 0.725 by lagging the satellite data by one month. It is clear from Fig. 6.17 (left) that the estimated correlation is subject to some large uncertainties. In any case, they used this correlation to estimate the solar output from historical sunspot numbers via a simple linear

[24]The cross-spectral density function can be thought of as the distribution of covariance across time scales in much the same way that the spectral density function can be thought of as the distribution of variance across time scales. See von Storch and Zwiers (1998) for an introduction to time series and spectral analysis.

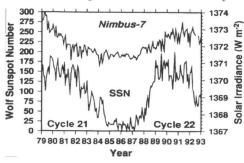 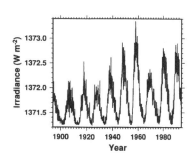

Fig. 6.17. Left: The monthly mean Wolfer sunspot number and total solar irradiance from *Nimbus-7* for the period January, 1979 - December, 1992. Right: Monthly solar-cycle irradiance forcing function for the period 1894-1993. From Stevens and North (1996)

regression. The resulting estimate of the solar output for the period 1894-1993 is shown in Fig. 6.17 (right). Note that the variation in estimated solar forcing (peak to trough) is of the order of $1\ Wm^{-2}$. This forcing function was used to perturb the usual solar forcing of a two-dimensional energy balance model of the climate. The signal $\vec{S}$ was derived from the model's response. The response, which is typically a few one-hundreds of a degree C, is larger over land than over ocean (see Stevens and North, Figure 7).

All the ingredients needed to estimate the signal amplitude

$$\hat{\theta} = \frac{\vec{S}^T \Sigma_{NN}^{-1} \vec{T}}{\vec{S}^T \Sigma_{NN}^{-1} \vec{S}} = \frac{1}{\gamma^2} \mathbf{A}$$

are now available. However, before making this calculation, Stevens and North decompose γ^2 as a function of frequency and eigen-index. Not surprisingly, they find that most of the signal is concentrated near $\omega = 1/11\ \mathrm{yr}^{-1}$. They suspect that there are some spuriously large contributions to γ^2 from high index eigenvalues, and therefore truncate at EOF 21.

The final results obtained after EOF truncation are given in the following table.

Result	Σ_{NN} estimated from:	GFDL	ECHAM/LSG
Signal to noise ratio γ^2		4.75	21.4
Signal amplitude $\hat{\theta}$		1.30	0.93
Standardized detection statistic $\mathbf{Z} = \gamma\hat{\theta}$		3.68	4.30
Modified detection statistic $\mathbf{Z}'$		1.94	1.76

The statistic $\mathbf{Z}$ given in the third line of the table is the standardized optimal detector given in equation (6.10). Note that it is not strongly affected by the choice of climate model used for estimating the natural variability Σ_{NN}.

Statistic $\mathbf{Z}$ can be used to test the no signal null hypothesis (6.9) if the natural variability of $\vec{T}$ (or equivalently $\vec{N}$) is the only source of uncertainty in

$\widehat{\theta}$. Stevens and North call this *filter pass-through* error because it is caused by the noise that remains after $\vec{\mathbf{T}}$ has been filtered. Assuming $\mathbf{Z}$ is Gaussian, the no signal null hypothesis can be rejected at less than the 0.01% significance level with either variability estimate, if pass-through error is the only source of uncertainty.

However, Stevens and North note that error in the estimate of Σ_{NN} (which they call *filter sampling error*) and in the estimate of the signal (*bias error*) also contribute to the uncertainty of $\widehat{\theta}$. They attempt to estimate both of these sources of error with the aid of their energy balance model (EBM). To estimate the filter sampling error, they ran the EBM with noise forcing for 10000 years. This run was broken into 1000 year chunks, each of which was used to estimate Σ_{NN} and thus a signal amplitude θ. The resulting sample of 10 signal amplitudes has mean $\overline{\theta} \approx 1.1$ and variance $\widehat{\sigma}^2_{samp} = 0.23$. An attempt was made to estimate the bias error by varying the depth of the ocean mixed layer in the EBM. This produced some variation in the signal $\vec{S}$ that partially accounts for uncertainty in the EBM's ability to respond correctly to the prescribed solar forcing signal. Keeping Σ_{NN} fixed, the bias error was estimated to be $\widehat{\sigma}^2_{bias} = 0.009$ and $\widehat{\sigma}^2_{bias} = 0.002$ when Σ_{NN} is obtained from the GFDL and ECHAM/LSG simulations respectively. Combining the three sources of uncertainty, they obtain a 'total uncertainty estimate' for $\widehat{\theta}$ of

$$
\begin{aligned}
\widehat{\sigma}^2_{tot} &= \frac{1}{\gamma^2} + \widehat{\sigma}^2_{samp} + \widehat{\sigma}^2_{bias} \\[2mm]
&= \begin{cases} 0.21 + 0.23 + 0.009 \approx 0.45 & \Sigma_{NN} \text{ from GFDL} \\ 0.05 + 0.23 + 0.002 \approx 0.28 & \Sigma_{NN} \text{ from ECHAM/LSG} \end{cases}
\end{aligned}
$$

This results in the modified decision making statistics $\mathbf{Z}' = \mathbf{Z}/(\gamma\widehat{\sigma}_{tot})$ given in the table, which are significant at the 5.2% and 8.0% significance levels respectively. Given that the highly uncertain nature of the forcing itself (recall Fig. 6.17) has not yet been taking into account, and that the full uncertainty associated with the use of the EBM has not been taken into account, the evidence for the detection of a solar signal in the observed records becomes highly tenuous.[25]

---

[25]This ad-hoc method of taking uncertainty into account is used because the *frequentist* approach to statistics makes it difficult to quantify the effects of, for example, uncertain physical models. Bayesians would argue that they have a more satisfactory paradigm for statistical inference in this respect. Formally at least, they can can use a prior distribution to express subjective assessments of the uncertainty of the physical model (the EBM in this case), and consequently propagate this through to uncertainty about the signal. Significant strides have been made in the application of Bayesian reasoning during the past decade. This has become possible because of the revolution in computing power and the concurrent introduction of numerical algorithms such as the *Gibbs Sampler* (Smith and Roberts, 1993) for conducting Bayesian calculations. Leroy (1998) has shown that the optimal detection technique is a special case of the Bayesian approach in which signal pattern is assumed to be completely known.

6.5.3 An Extension - More Signals

North and Stevens (1996) extend the approach described above by considering four types of forcing (solar, volcanic, sulphate aerosols, and greenhouse gases) in various combinations. Signals were estimated as the response of the EBM to the prescribed forcing changes. Four climate models were used to estimate the covariance Σ_{NN}. The following table summarizes some of their results.

Signal	$\overline{\hat{\theta}}$	$\hat{\sigma}_{tot}$	$\mathbf{Z}' = \overline{\hat{\theta}}/\hat{\sigma}_{tot}$	Significance Level
Solar	0.19	0.68	0.28	39%
Volcanic	1.13	0.33	3.42	0.03%
Greenhouse Gas	1.78	0.38	4.68	< 0.005%
Aerosols	3.81	0.95	4.01	< 0.005%
Combined	0.81	0.26	3.12	0.09%

The column labelled $\overline{\hat{\theta}}$ contains the mean signal amplitude that is obtained with the four covariance estimates. The column labelled $\hat{\sigma}_{tot}$ is an estimate of the combined pass-through, bias and sampling error crudely estimated as $\hat{\sigma}_{tot}^2 = \frac{1}{\gamma^2} + 0.05$. North and Stevens argue that 0.05 is a reasonable bound for the bias and sampling errors because most of the latter is avoided by their EOF truncation. They acknowledge that this estimate is uncertain and advise that it be interpreted as a lower bound. The column labelled 'significance level' indicates the maximum level at which the null hypothesis can be rejected in a one-sided test. The signal labelled 'Solar' is the part of the EBM's response to the solar forcing function that is orthogonal to its response to the sum of the other three forcings. The other three signals are defined analogously. The combined signal is the EBM's response to the sum of the four forcings.

There is apparently strong evidence for the presence of the volcanic, greenhouse gas and sulphate aerosol signals in the observed record of the last century. The level of trust that can be placed on these inferences is highly contingent upon the quality of the estimate of uncertainty $\hat{\sigma}_{tot}$. One thing that does seem fairly certain is that there really is very little evidence for a solar forcing signal when other signals have been taken into account.

6.6 Detection with Pattern Similarity Measures

None of the detection studies reviewed to this point (Bell, 1982; Hegerl et al. 1996, 1997; Stevens and North, 1996; North and Stevens, 1997) uses a procedure that is truly optimal. It is true that they apply the same optimal detection methodology to a reduced data set, but as we have seen, considerable intuition, art and ingenuity are needed to perform the data reduction.

Indeed, it is probably the case that the success of a detection study depends as much upon these preliminary steps as it does upon the formal optimized detection step itself. Therefore, while an optimal detection procedure is certainly a useful adjunct to success, it is likely not the crucial determining factor. Santer and colleagues have taken an approach that is simple and direct. Given an estimate of the signal, say $\vec{S}$, and a suitable observed vector $\vec{T}_t$, they construct their detection strategy around the projection $\vec{S}^T \vec{T}_t$ of the observations onto the signal. This approach is generally referred to as the *pattern correlation* approach. Their statistics are not optimal, but they avoid complexities such as dimension reduction and the estimation and inversion of the covariance matrix Σ_{NN}. The main features of the methodology are laid out in Santer et al. (1993) and the application we describe here is documented in Santer et al. (1995).

The signals $\vec{S}$ are estimated from four equilibrium climate change simulations performed with a version of the NCAR CCM1 climate model, which has been coupled to a 50 m mixed layer ocean model (Taylor and Ghan, 1992) and the GRANTOUR tropospheric chemistry model (Walton et al. 1988). All simulations were allowed to run for at least a decade to allow the simulated climate to come into equilibrium with the specified forcing and were then sampled for 20-years. The four experiments (Taylor and Penner, 1994) were

- A control run with pre-industrial CO_2 concentrations.

- A sulphate aerosol run with pre-industrial CO_2 concentrations and present day SO_2 emissions.

- A CO_2 run with present day CO_2 concentrations.

- A CO_2/SO_2 run with present day CO_2 concentrations and SO_2 emissions.

Three signals were defined as

$$\vec{S} = \overline{\vec{T}}_{exp}^{\,20 \; years} - \overline{\vec{T}}_{cont}^{\,20 \; years}$$

where the over bar indicates a time average, *cont* indicates the control run, *exp* indicates one three experimental simulations and $\vec{T}$ is near surface temperature.

Santer et al. use the familiar Jones and Briffa (1992) data set. It was processed by computing annual mean anomalies relative to a fixed reference period (1950-1979) and then low pass filtering these anomalies with a 13-point Gaussian filter to remove high frequency variability such as that associated with El-Niño. They then search for the signals in the departures $\vec{T}_t = \vec{T}_{f,t} - \vec{T}_{f,1954}$ of the filtered anomalies from their 1954 levels.

Two measures of the similarity between the shapes of the signal and the processed temperature fields $\vec{\mathbf{T}}_t$ are considered for detection purposes. The first is the so-called *uncentered* spatial correlation coefficient

$$\mathbf{C}_t = \frac{\sum_{i=1}^{m} S_i \mathbf{T}_{t,i}}{(\sum_{i=1}^{m} S_i^2)^{1/2}} \tag{6.32}$$

that was first proposed by Barnett and Schlesinger (1987). The summations are taken over all points in the field. The uncentered statistic is similar to the optimal detector (6.6). This can be seen by writing $\mathbf{C}_t$ in matrix-vector form as

$$\mathbf{C}_t = \frac{\vec{S}^T \vec{\mathbf{T}}_t}{(\vec{S}^T \vec{S})^{1/2}}$$

and comparing with the standardized optimal detector

$$\mathbf{Z}_t = \frac{\vec{S}_t^T \Sigma_{N_t N_t}^{-1} \vec{\mathbf{T}}_t}{(\vec{S}_t^T \Sigma_{N_t N_t}^{-1} \vec{S}_t)^{1/2}}$$

that we first encountered in Section 6.2 (see equation (6.10)). These detection statistics are equivalent only when the noise in the system is spatially white (i.e., when $\Sigma_{N_t N_t} \propto \mathcal{I}$).

The second pattern similarity measure considered by Santer et al. (1995) is the so-called *centered* statistic

$$\mathbf{R}_t = \frac{\sum_{i=1}^{m}(\mathbf{T}_{t,i} - \overline{\mathbf{T}}_{t,\circ})(S_i - \overline{S}_\circ)}{\left[\sum_{i=1}^{m}(\mathbf{T}_{t,i} - \overline{\mathbf{T}}_{t,\circ})^2 \sum_{i=1}^{m}(S_i - \overline{S}_\circ)^2\right]^{1/2}}. \tag{6.33}$$

The notation $\overline{\mathbf{T}}_{t,\circ}$ indicates that a simple arithmetic average has been computed across the elements of $\vec{\mathbf{T}}_t$. This statistic is also sometimes referred to as the *anomaly correlation* coefficient. A common variant on $\mathbf{R}_t$ area weights the summations in (6.33). For completeness, we note that the centered statistic can also be written in vector-matrix notation as

$$\mathbf{R}_t = \frac{\vec{S}^T(\mathcal{I} - \mathcal{U})^2 \vec{\mathbf{T}}_t}{\left[\vec{S}^T(\mathcal{I} - \mathcal{U})^2 \vec{S}\right]^{1/2} \left[\vec{\mathbf{T}}_t^T(\mathcal{I} - \mathcal{U})^2 \vec{\mathbf{T}}_t\right]^{1/2}}$$

where $\mathcal{I}$ is the $m \times m$ identity matrix and $\mathcal{U}$ is the $m \times m$ matrix with elements $u_{i,j} = 1/m$ for all i and j. However, the centered statistic can not be cast in a form similar to the optimal detector.

Neither statistic is an optimal detector. As noted above, this may not be a serious impediment to successful signal detection. Santer et al. point out that there are various reasons to prefer one statistic over the other. The uncentered statistic $\mathbf{C}_t$ includes the global mean response to forcing changes, and can, therefore, grow in an unbounded manner. This type of behaviour is clearly desirable from a detection perspective. However, it is argued that

$\mathbf{C}_t$ is not well suited for *attribution* studies because it can not distinguish effectively between forcings that lead to similar trends in the global mean. In contrast, the centered statistic is bounded. This makes it less suitable for trend detection, but it is argued that it is more suitable for attribution since it measures the similarity of the patterns that remain after the global mean has been removed.[26] Santer et al. (1995) compute both types of statistics, but focus most of their attention on the centered statistic.

The $\mathbf{R}_t$ time series derived from the observations are analyzed by fitting linear trends to the most recent P-year periods for $P = 10, 20, \ldots, 50$. We will denote these estimated trend coefficients with $\widehat{\Delta}_P$. An upward trend in $\mathbf{R}_t$ during the most recent epoch is an indication that the signal is emerging. However, an estimate of natural variability is needed to assess the trend by testing the null hypothesis $H_0 : \Delta_P = 0$ that the signal is absent during the most recent P-year epoch.

As in the other studies we have examined, the natural variability of the trend coefficient is determined from long control simulations with coupled climate models. In this case, the 1000 year GFDL simulation and a 600 year segment of the ECHAM/LSG simulation are used. The method for estimating the natural variability of $\widehat{\Delta}_P$ is simple. The model data was interpolated to the observed grid, sampled only at grids points that were observed in 1954, and processed as the observations by computing anomalies relative to a reference period and low pass filtering. The centered statistic $\mathbf{R}_t$ was then computed for each year in the simulation. A sample of L estimated P-year trend coefficients $\widehat{\Delta}_{P,mod,l}, l = 1, \ldots, L$ was obtained by moving a P-year window along the model $\mathbf{R}_t$ time series. The sample represents the natural variability of $\widehat{\Delta}_P$ under the null hypothesis. Santer et al. stress that this procedure only accounts for the internal variability of the coupled system as it is represented by the model.

The significance of the observed $\widehat{\Delta}_P$ is estimated from the sample by counting the number of model simulated trend coefficients in the L coefficient sample that are greater than $\widehat{\Delta}_P$. If there are k such occurrences, then the significance of the observed $\widehat{\Delta}_P$ is estimated as $\frac{k}{L} \times 100\%$. The null hypothesis is rejected when this number is small (e.g., less than 5%) because this indicates that the observed value of $\widehat{\Delta}_P$ is highly unusual in the simulated control climate. The following results were obtained:

- The most recent $20, 30, \ldots, 50$-year trends in the uncentered statistic $\mathbf{C}_t$ are highly significant relative to the natural variability in either of the long control simulations for both the CO_2 and SO_2/CO_2 signals.

- The most recent trends in the centered statistic $\mathbf{R}_t$ are not significant for the CO_2 only signal.

---

[26] Hegerl et al. (1997) argue that the centered statistic can give ambiguous attribution results in practice (see their Figure 7 and related discussion). They prefer the multi-fingerprint approach described in Sections 6.4.3 and 6.4.5.

- The most recent 50-year trends in $\mathbf{R}_t$ are significant at the 3% level in northern summer (JJA) and at the 1% level in northern autumn (SON) for the combined SO_2/CO_2 signal when compared to the natural variability in either of the long control simulations. Apparently a signal with spatial structure similar to that of the combined signal is slowly emerging from the data. The important finding here is that the combined SO_2/CO_2 signal is needed to tease this pattern out of the data.

- With one exception, none of the recent trends in $\mathbf{C}_t$ and $\mathbf{R}_t$ are significant for the SO_2 signal. The exception is the 50-year trend in SON, which is significant when compared to the natural variability of both models.

The presence of trend at a high significance level does not imply perfect correspondence between the observed temperature pattern and the combined SO_2/CO_2 signal. This can be seen from the $\mathbf{R}_t$ time series for SON that is displayed in Fig. 6.18. The trend does not evolve smoothly and the spatial correlation coefficient attains values no greater than 0.3 during the last 25 years or so. Thus, while trend may be present, the match between the observed warming pattern and the combined signal is certainly not profound. The 25-year decrease in correlation at the beginning of the record is also somewhat disquieting.

6.7 Summary

Certainly, this tour of the detection literature has not been exhaustive. Many studies have not been mentioned and the reference list is far from complete. Instead, we have tried to cover the basic ideas and present several examples in a tutorial manner.

The basic formalisms used in detection studies are straight forward. We have seen that the ideas proposed by Hasselmann, Bell and North are equivalent and 'optimal'. Hegerl and North (1997) draw much the same conclusion. Recent applications of the optimal detection approach include Hegerl et al. (1996, 1997), Stevens and North (1996), North and Stevens (1997) and Tett et al. (1996). See also Santer et al. (1996b) for a comprehensive review. Leroy (1998) has shown that the optimal techniques are a special case of a Bayesian analysis. It can be argued that the Bayesian approach, which accounts for subjectivity in a probabilistic manner, is the next natural step in the evolution of the methodology. This line of investigation is only just emerging, with participants such as Leroy (1998), Hasselmann (1998) and Levine and Berliner (1998). Santer and colleagues use pattern correlation techniques (Santer et al. 1993) that are sub-optimal, but we have argued that the choice of a sub-optimal or optimal technique is not likely to be the critical determinant of success in a detection study.

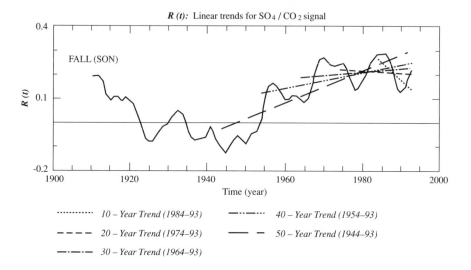

Fig. 6.18. Linear trends for the final 10-50 years of the $\mathbf{R}_t$ time series for the SON near-surface temperature signal from the combined SO_2/CO_2 experiment. The 50-year trend is significant at the 1% significance level when compared to trends obtained from either the 600-year sample from the ECHAM/LSG control simulation or the 1000-year GFDL control simulation. The other trends are not significant. From Santer et al. (1995)

Less attention seems to have been paid to the attribution question. Hasselmann (1997) proposed a formal test that Hegerl et al. (1997) have applied. The procedure is somewhat awkward because the attribution decision is made by *not rejecting* the null hypothesis that the forced signals have the same amplitude in the observed and simulated climate. This problem is ameliorated somewhat by using the procedure to screen competing mechanisms for climate change. Bayesian analysis may be better suited to the attribution question because it treats the null and alternative hypotheses more symmetrically. Santer et al. (1993, 1995, 1996a,b) have also given thought to the question of attribution, arguing that so called centered pattern correlation statistics are better suited for this purpose than the uncentered statistics, and demonstrating that detection is possible with the centered statistics only when a combination of signals are considered.

Our tour of detection studies demonstrates that the application of the detection formalisms requires many intricate and inter-related operations. These include

- Selection of the data that is to be used. The choice is limited because it is generally felt that long records are needed of variables that are sensitive to changes in the composition of the atmosphere to achieve the signal to noise ratios that are required for early detection.

- Dimension reduction. Bell (1982) used zonal mean temperature in three zones; Hegerl et al.(1996, 1997), Stevens and North (1996), and North and Stevens (1997) all used EOF reductions. The approach of Santer et al. (1995) does not require dimension reduction.

- A pre-filtering step of some sort is usually performed to remove noise. For example, Hegerl et al. (1996) compute 15-, 20- and 30-year trends and Santer, et al. (1995) used a 13-year Gaussian filter. Stevens and North (1996) built this in by restricting their analysis to a particular frequency band.

- Estimation of a signal from a climate model. Long transient climate change simulations with fully coupled models are often required. Karoly et al. (1994) and Santer et al. (1995, 1996a) have used shorter equilibrium climate change simulations.

- A detection statistic is required. Bell (1982), Hegerl et al. (1996, 1997), Stevens and North (1996) and North and Stevens (1997) use an optimal detector. Santer et al. (1995, 1996a) use the estimated trend in a time series of pattern correlation statistics as their detector.

- An assessment of the natural variability of the detector is required to determine if the signal has been seen. Hegerl, et al. (1996, 1997), Santer et al. (1995, 1996a), Stevens and North (1996) and North and Stevens (1997) use long control simulations with coupled climate models to estimate the variability that is internal to the climate system. Stevens and North (1996) and North and Stevens (1997) also try to account for signal uncertainty and sampling error.

- Finally, some consideration of attribution is desirable. Hasselmann (1997), Hegerl et al. (1997), Karoly et al. (1994) and Santer et al. (1993, 1995, 1996a, 1996b), amongst others, have taken some initial steps in this direction.

Interpretation of the results of the detection studies is complicated by the caveats that must be applied. First, there is uncertainty about the specifics and magnitude of the signals because models are imperfect. Bear in mind, however, that coupled climate models are built from first principles. It seems unlikely that the general features of the signals generated by modern models will be wrong unless something fundamental and completely unforeseen has been overlooked in their construction. Secondly, we must cope with the uncertainty that results from an observed record that is not complete and free of error. Third, since the observed record is small, we must use models to estimate the natural variability of the observed system. The first two caveats do not seriously compromise detection efforts. Use of an incorrect signal simply results in a somewhat less efficient detector. Missing and erroneous observations result in greater sampling variability in the detector that can largely be accounted for by 'sampling' model output in manner that mimics our sampling of the observed climate. But we must be cautious that models may underestimate the natural variability of our detectors. While the recent long coupled simulations seem to reproduce the variability of the observed record with reasonable fidelity on the time scales that are relevant to detection studies, we need to be alert to the possibility that not all sources or modes of variability are accounted for.

A final caveat to take into consideration is that we can only ever have one realization of the observed climate. We do not have the luxury of restarting humanity's experiment with the climate in 1850 and observing a second, independent evolution of the last one and one half centuries. Consequently, the studies described above have all searched for the global warming signal in the same surface data set. Independent confirmation is required, but this is only possible with the surface data if we are willing to wait a substantial period for nature to generate additional independent data.

One way around this is to search for the signal in other parts of the climate system. One possibility is to extend the search upwards into the free atmosphere by using a historical radiosonde record such as that compiled by Oort and Liu (1993). Doubled CO_2 experiments, such as Boer et al. (1992), show that a characteristic climatic response to enhanced greenhouse gas forcing is a hemispherically symmetric pattern of cooling in the stratosphere and warming in the troposphere with a maximum in the upper troposphere. Karoly et al. (1994) and Santer et al. (1996a) have both searched for this signal in radiosonde record with some success.[27] They conclude that it becomes increasingly apparent in the short (1963-1987) radiosonde record that is available.

Acknowledgments

I am grateful to Gabi Hegerl for useful discussions and for generously making figures available. I thank Hans von Storch and GKSS for travel support which enabled me to participate in the First GKSS Spring School on Environmental Research at which this overview was first presented as an extended lecture. Mike Berkeley generously assisted with the preparation of the figures.

---

[27]Santer et al. (1996a) also account for changes in the vertical structure of the atmosphere that are related to changes in the distribution of sulphate aerosols and to the depletion of stratospheric ozone.

Part III

Implications

Chapter 7

Cooperative and Non-Cooperative Multi-Actor Strategies of Optimizing Greenhouse Gas Emissions

by Klaus Hasselmann

Abstract

A simple Structural Integrated Assessment Model (SIAM) consisting of a linearized impulse response climate model coupled to an economic model is applied to the determination of optimal CO_2 emission paths that minimize the time integrated sum of climate change impact and mitigation costs. The impulse response climate model is calibrated against state of the art three-dimensional carbon cycle and coupled ocean-atmosphere general circulation models. The economic module consists of simple expressions for the climate change impact and emission abatement costs, with some additional cost terms parameterizing the inertia of the economic system.

Application of SIAM to the single-actor case (all economic actors agree on a common abatement policy) yields emission curves that rise for one or two decades before falling monotonically to the asymptotic value of zero after many centuries. Removal of the economic inertia terms yields solutions with an immediate draw-down of emissions, but with little impact on the long-term levels of emissions or climate change. Important for an effective climate mitigation policy are the long-term rather than the short-term emission reductions. The optimal emissions path depends critically on the intertemporal relation assumed for the climate damage costs. Solutions with limited climate change are obtained only if the discount rate for climate damage costs is set at a significantly lower level than the discount rate for mitigation costs. Application of standard discount rates for both cost terms yields optimal emission paths leading to a long term global warming of the order of 10 degrees Centigrade.

The SIAM model is applied also to three non-cooperative multi-actor cases: n identical mitigating actors; a single mitigating actor facing a majority of non-mitigating actors; and a single world-fossil-fuel user interacting with a single world-fossil-fuel producer. In the first two cases the n-actor non-cooperative solutions are found to be less efficient than the cooperative solutions, as expected, but less so than may have been anticipated intuitively from free-rider considerations. In the last case, the fossil fuel supplier can effectively counteract the mitigation efforts of the fossil fuel user by lowering the fuel price. The Nash equilibrium relations derived for these examples can be readily applied to more general multi-actor cases.

PART I: THE SINGLE-ACTOR PROBLEM

7.1 Introduction

A basic challenge facing humankind today is the development of strategies for the mitigation of global warming due to increasing emissions of greenhouse gases. The dominant contribution to greenhouse warming, about 60%, stems from CO_2 emissions, followed by methane (20%), chlorofluorocarbons (CFCs, 15%) and other gases (IPCC, 1996a). In this paper, we consider the problem of determining optimal paths of CO_2 emissions that minimize the net impacts of global warming, including both the direct impacts of climate change and the abatement efforts undertaken to reduce global warming. The following notes represent a summary of more detailed analyses in Hasselmann et al., 1997 and Hasselmann and Hasselmann, 1997, referred to in the following as HHGOS and HH, respectively. The same material is presented also in a lecture given in a symposium in Hawaii in January 1997 (Hasselmann, 1997).

The optimization problem involves a combination of dynamical control theory and game theory. Global CO_2 emissions are produced by many actors with different interests pursuing different goals. The actors are coupled through the global economic system and through climate change, which they produce jointly but experience separately. Because of their different levels of CO_2 emissions, varying trade dependencies and diverse vulnerabilities to climate change, the views of individual actors on the most effective mitigation policy will necessarily diverge. The net outcome of these divergent interests will depend strongly on the game theoretical strategies pursued by the actors individually and in various alliance combinations.

Figures 7.1 and 7.2 (from HHGOS and HH, respectively) indicate schematically the individual modules and interactions that need to be considered in a coupled climate-socio-economic system model designed for such integrated assessment studies. Figure 7.1 represents a single-actor Global Environment and Society (GES) model. It is assumed that all actors participating in the global socio-economic system have agreed on a cooperative climate mitigation policy that maximizes a hypothetical, jointly accepted, global welfare expression. The optimization task in this case is to control the global CO_2 emissions, through application of appropriate regulatory mechanisms, such that the global welfare, defined as an integral over time, is maximized. The arrows in the diagram indicate the various feedbacks that need to be considered in the coupled model, dark arrows represent interactions that are reasonably well understood, in contrast to the poorly known interactions represented by light arrows.

Figure 7.2 represents a decomposition of the GES model of Fig. 7.1 into the individual actors of the global climate-socio-economic system, indicating explicitly the coupling between actors through the global climate, negotiations and trade.

GLOBAL ENVIRONMENT AND SOCIETY (GES) MODEL

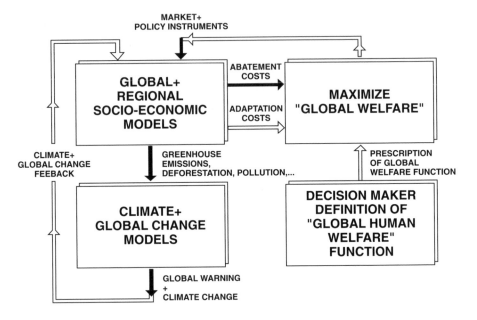

Fig. 7.1. Interactions and sub-systems of an integrated Global Environment and Society (GES) model. Darker and lighter arrows represent better and less well understood interactions, respectively (from HHGOS)

We consider in the first part of this paper the cooperative single-actor scenario, turning then in the second part (section 7.5) to the general multi-actor case.

7.2   The Single-Actor SIAM Model

The goal of the present investigation is not to provide quantitative data on optimal emission paths that can be used for practical political decisions, but rather to reveal the structure of the trade-offs that determine the optimal path, and to identify the principal sensitivities of the solution to the input assumptions. For this purpose, we reduce the GES model to the simplest structural form that is still qualitatively compatible with the basic dynamical properties of the coupled climate-socio-economic system. The resultant Structural Integrated Assessment Model (SIAM), described in detail in HH-GOS, consists of a linearized impulse response model for the climate system

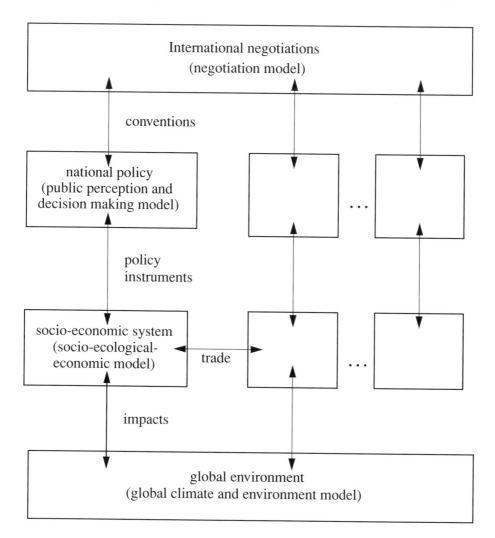

Fig. 7.2. Interactions and sub-systems of a GES model, as in Fig. 7.1, but decomposed into separate interacting players (from HH)

and simple mitigation and climate damage costs expressions for the socio-economic model. Although highly idealized, the model illustrates the central role of time scales and inter-temporal cost relations in the determination of the optimal emissions path.

7.2.1   The Climate Model

The climate module is based on a linearization of complex three-dimensional nonlinear climate models. Although the climate system is strongly nonlinear, for small deviations $\mathbf{x}(t)$ of the climate state vector (representing, in a discretized model representation, the perturbation vector of all climate variables at all model gridpoints) from a reference equilibrium state, the evolution of the climate perturbation in response to an arbitrary, sufficiently small CO_2 emission function $e(t)$ can be represented in the general linear response form

$$\mathbf{x}(t) = \int_{t_0}^{t} \mathbf{R}(t - t')e(t')dt', \qquad (7.1)$$

where the impulse-response function $\mathbf{R}(t - t')$ denotes the climate response at time t to a unit δ-function emission at time t'. It is assumed that the forcing and climate perturbation are zero prior and up to the initial time t_0 : $e(t) = \mathbf{x}(t) = 0$ for $t \leq t_0$.

The linearized form (7.1) is generally applicable for CO_2 concentrations less than about twice the pre-industrial level and an increase in the global mean temperature of less than 3K (which is still small compared to the reference global mean temparature of 287K). Note that, within the linear response regime, the linearized system represents no loss of information compared to the full nonlinear system: eq.(7.1) represents the response of the climate perturbation vector in the full phase space of the nonlinear system. However, in our applications we shall consider only a highly reduced impulse response climate model for the global mean temperature.

The impulse response function can be determined by fitting the linear model to the response computed for the perturbation of a fully nonlinear model, consisting, for example, of a three-dimensional global carbon cycle model combined with a coupled ocean-atmosphere general circulation model (CGCM), cf. Maier-Reimer and Hasselmann (1987), Maier-Reimer (1993), Hasselmann et al (1993). The net impulse response function $R(t)$ for the combined carbon cycle-CGCM sytem is obtained as a convolution

$$R(t) = R_T(t) + \int_{0}^{t} R_T(t - t')\dot{R}_w(t')dt'. \qquad (7.2)$$

of the independently calibrated response functions $R_w(t)$ and $R_T(t)$ for the carbon cycle model and the CGCM, with

$$w(t) = \int_{t_0}^{t} R_w(t - t')e(t')dt', \qquad (7.3)$$

$$T(t) = \int_{t_0}^{t} R_T(t - t')\dot{w}(t')dt', \qquad (7.4)$$

where $w(t)$ is the change in atmospheric carbon content in GtC (Gigatons carbon) relative to the equilibrium pre-industrial state, $e(t)$ denotes the carbon emissions in GtC/year, $T(t)$ is the change in global mean temperature relative to the pre-industrial equilibrium level, and the response function $R_T(t - t')$ denotes the change in the global mean temperature produced at time t (measured in years) by a unit step-function increase in the atmospheric CO_2 concentration at time t'. Details of the derivation are given in HHGOS. We shall chose $t = t_0$ in our examples as the pre-industrial date 1800.

To retain the same carbon units GtC for w and the emissions e (GtC/yr), the atmospheric CO_2 concentration is expressed in eq.(7.3) in terms of the total carbon content of the atmosphere. However, we shall present results for w later in the usual units of ppm. The conversion factor is $w\,[GtC] = 2.123\,w\,[ppm]$. The present atmospheric CO_2 concentration is about $360\,ppm$, corresponding to an atmospheric carbon content of $764\,GtC$, while the preindustrial concentration was $.w_0 = 280\,ppm = 594\,GtC$.

Figure 7.3 shows the response functions R_w, R_T and the resulting net response function R obtained by fitting the response relations (7.3) and (7.4) to experiments with the Hamburg global carbon cycle model and CGCM (adapted from Maier-Reimer, 1993, and Hasselmann et al., 1993, cf. HHGOS). The temperature response functions $R'_T = R_T w_0$, $R' = R w_0$ shown in the figure have been normalized by multiplication with the pre-industrial concentration w_0, thereby representing the response to a step-function doubling of the atmospheric CO_2 concentration at time $t = 0$. The concentration is then either retained at a constant level (in the case of R'_T), or (in the case of R') decreases to an asymptotic value of 7% of the initial level, corresponding to the chemical equilibrium partitioning of the initial CO_2 input between the atmosphere and the ocean (the loss of CO_2 from the ocean through sedimentation on time scales of several thousand years has been ignored). Alternative versions of the response functions are discussed in HHGOS; the uncertainties are not critical for our applications.

The response curves demonstrate that the net climate response to CO_2 emissions cannot be characterized by a single time constant. After a rapid temperature rise in the first few years as the upper mixed layer of the ocean warms, the global mean temperature increases more slowly as the warming penetrates into the main ocean thermocline, reaching its maximum value of about $1°C - 1.5°C$ after one or two decades (compared with the asymptotic temperature response of $2.5°C$ for a CO_2 doubling without CO_2 transfer from the atmosphere to the ocean). Subsequently, the temperature gradually relaxes back over a period of several hundred years to its asymptotic equilibrium value of $2.5 \times 0.07 = 0.175°C$. The initial fast response is governed by the temperature response of the ocean-atmosphere system, while the later relaxation stages are determined by slow response terms in both the carbon cycle and the climate system.

For the optimization of greenhouse-gas emission paths, both the near-time and far-time climate response characteristics must be considered. If the mandate of sustainable development is taken seriously, the socio-economic impact of the long-term climate response over several hundred years cannot

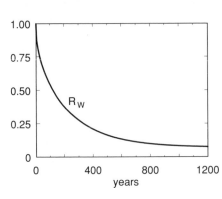
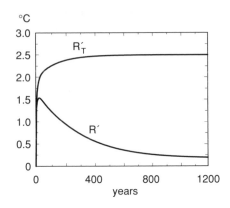

Fig. 7.3. Left panel: Response function R_W representing the atmospheric retention factor for a unit δ-function emission of CO_2 at time $t = 0$. Right panel: Temperature response functions $R'_T = w_0 R_T$ and $R' = w_0 R$ for a step-function doubling of the CO_2 concentration at time $t = 0$

be ignored. We shall demonstrate in our examples that the application of the usual exponential discount factors designed to model economics or intertemporal social preferences over the short or medium term is inappropriate for the representation of the long-range intertemporal values associated by society with the concept of sustainable development. In general, the ecological, sociological and economic response to climatic change, including the associated social preferences, should be modelled (in analogy with the multi-time scale description of the dynamical climate system) in terms of a hierarchy of processes with different characteristic time scales. We shall not attempt to develop a realistic ecological-socio-economic model on these principles in this paper, but will illustrate some of the consequences of the multi-time scale nature of climate change and the choice of intertemporal relations for the mitigation problem.

7.2.2 Cost Functions

As a socio-economic model, we introduce simply an expression for the total climate related costs C, consisting of the sum

$$C = C_a + C_d \tag{7.5}$$

of the CO_2 abatement costs C_a and the climate damage costs. Both costs are defined as the additional costs relative to some business-as-usual (BAU) economic evolution path, for which all climate related costs are ignored. The optimal abatement strategy is then defined as the CO_2 emissions path that minimizes the total costs. We do not consider the regulatory mechanisms (carbon tax, emission certificates, etc.) required to realize the optimal emission path (which would require a more sophisticated economic model), but regard simply the emissions themselves as control variable.

The term 'cost' is used here as a synonym for loss of welfare, including not only normal economic costs, but also quality-of-life factors such as the state of the environment, the maintenance of species, impacts on health and life expectancy, etc. Although often termed non-monetary, these values must be included in any realistic cost-benefit trade-off analysis, whereby they automatically acquire monetary values.

The costs C_a, C_d are expressed as time integrals over the *specific* costs $c_a(t)$, $c_d(t)$:

$$C_a = \int_{t_0}^{\infty} c_a(e(t), \dot{e}(t), \ddot{e}(t), t) dt, \tag{7.6}$$

$$C_d = \int_{t_0}^{\infty} c_d(T(t), \dot{T}(t), t) dt. \tag{7.7}$$

Costs and discount factors are assumed to be inflation adjusted. Time has nevertheless been included explicitly as a separate variable in the specific cost functions c_a, c_d in order to describe the inter-temporal relations between present and future costs. We shall follow the traditional approach and express these in terms of exponential discount factors. In the present context, however, this is controversial, since the relevant climate impact time scales far exceed the typical economic planning horizons of a few years or decades for which discount factors are normally applied. Thus, the relevant inter-temporal relations cannot be inferred directly from market transactions, but must be assessed indirectly, for example, as polled "willingness-to-pay" relations. Moreover, the type of costs and values involved cover a very wide span, and the relevant inter-temporal relations must be expected to vary accordingly. Thus, the discount rate appropriate for technical-economic costs such as reducing CO_2 emissions or increasing the heights of dikes in response to rising sea levels will differ substantially from the inter-temporal relations appropriate for the assessing the long-term value of the environment, the diversity of species or the future costs of health care (cf. discussion in HHGOS and Nordhaus, 1997). As a first-order attempt to differentiate between different inter-temporal relations for different catagories of costs and values we shall introduce different discount factors for abatement and climate damage costs. The optimal CO_2 emission paths will be found to depend sensitively on the assumed ratio of the discount factors.

We shall be concerned only with the ratios of abatement and climate-damage costs. Thus, all costs are defined only to within an arbitrary constant scaling factor. We make no attempt to introduce an absolute scaling with respect to, say, GDP (Gross Domestic Product). Our interest lies in establishing the form of the optimal emission paths for various input assumptions concerning the relative magnitudes and forms of the cost functions. For this analysis the absolute cost values are irrelevant. However, most quantitative cost estimates suggest that the mitigation and damage costs for optimal emission paths are generally of the same order and lie in the range of one to a few percent of GDP.

In addition to the emissions e, first and second time derivatives $\dot{e}$ and $\ddot{e}$ of e are included in the specific abatement-cost function in order to penalize rapid changes in the emissions (in a more sophisticated economic model, these inertia effects would be expressed by capital investments).

As the simplest mathematical expression which captures the principal properties of the abatement costs that may be anticipated from a more detailed economic model, we set

$$c_a = \left\{ (\frac{1}{r} - r)^2 + \tau_1^2 \dot{r}^2 + \tau_2^4 \ddot{r}^2 \right\} D_a(t) \tag{7.8}$$

where $r = e/e_A$ is the abatement factor relative to the BAU reference emissions path e_A, τ_1 and τ_2 are time constants, and

$$D_a(t) = \exp(-t/\tau_a) \tag{7.9}$$

is the abatement-cost discount factor, characterized by an abatement-cost discount time constant τ_a (inverse annual discount factor).

The first term in (7.8) has the property that any positive or negative departure from the reference BAU emission path e_A incurs costs that are quadratic in the deviations $\delta r = r - 1$ for small δr, $(\frac{1}{r} - r)^2 \approx 4(\delta r)^2$, while the term approaches infinity for either $r \to 0$ or $r \to \infty$. The quadratic dependence on the first and second derivatives of $e(t)$ parameterizes the economic inertia; it suppresses discontinuities in the emissions and the rate of change of emissions.

For the specific climate-damage costs we take a similarly simple form:

$$c_d = \left\{ \left(\frac{T}{T_c}\right)^2 + \left(\frac{\dot{T}}{\dot{T}_c}\right)^2 \right\} D_d(t), \tag{7.10}$$

where T_c, $\dot{T}_c$ are scaling constants and

$$D_d(t) = \exp(-t/\tau_d) \tag{7.11}$$

is the climate-damage costs discount factor, with a discount time constant τ_d.

Climate damages are assumed to occur not only through a change in the temperature itself, but also through the rate at which the temperature changes. The quadratic dependencies imply that the incurred climate damages are independent of the sign of the temperature change and reflect the general view that climate damage costs increase nonlinearly with climate change.

We have made use of the freedom to choose an arbitrary common normalization constant in the definition of the cost functions by setting the coefficient of the first term of the abatement cost function (7.8) equal to unity. The parameters T_c and $\dot{T}_c$ define a critical (soft shouldered) elliptical window or corridor in the climate phase space $T_c, \dot{T}_c$ within which the climate-damage costs are less than or of the same order as the mitigation costs at an abatement level of order $r = O(0.5)$. Outside the corridor the climate damage costs are greater than the mitigation costs at this abatement level.

The minimal-cost solution can be found numerically by a method of steepest descent (e.g. a conjugate gradient technique, cf. Press et al., 1986). This requires computing the gradient of the cost with respect to the control function, i.e the emissions $e(t)$. For a climate model expressed in integral response form, the gradient can be computed explicitly (cf. Hasselmann *et al*, 1996). However, in the numerical results presented below the gradient was computed automatically using a general numerical functional derivative compiler developed by Giering (Giering and Kaminsky, 1996). This had the advantage of immediately providing the gradient whenever the climate model was modified.

7.3 Optimal CO$_2$ Emission Paths for the Single-Actor Case

As reference BAU emission scenario $e_A(t)$ for the computation of the abatement costs we have assumed simply a linear increase for the first 205 years, from 1995 until 2200, growing from 6.3 GtC/yr in 1995 at an initial growth rate of 2.5 %/year to 38 GtC/yr in 2200. This is consistent with the upper and lower bounds of the emission projections by different energy models (cf. summary in Table 2.1 of Cline, 1992) and with the range of BAU scenarios considered by IPCC (1990, 1992, 1996a). After 205 years, the emissions have simply been frozen at the 38 GtC/yr level. A constant long-term emissions level will clearly not be attainable indefinitely because of limited fossil fuel resources. However, our optimal emission scenarios are found to be insensitive to the form of $e_A(t)$ beyond a few hundred years, provided a modest discount factor, with a time constant of the order of 50 or 100 years, is applied to the abatement costs. To test the sensitivity of our results with respect to the long term properties of e_A, we considered also a (clearly unrealistic) extension of the linear growth of e_A to 800 years. The differences in the computed optimal emission paths were minimal, since the change in the BAU path occurred only at a late time when the the abatement costs were already strongly attenuated by the exponential discount factor.

Prior to 1995 we have introduced a spin-up period, beginning with the pre-industrial state at time $t_0 = 1800$. For the spin-up period we assumed an exponential emissions growth function

$$e_A(t) = 6.3 \exp\left[(t - t_0 - 195)/t_s\right] \tag{7.12}$$

where $195 = t(\text{today}) - t_0 = 1995 - 1800$ corresponds to the length of the spin-up period. The emissions spin-up time constant was determined as $t_s = 35$ years from the condition that the carbon cycle model must reproduce the 1995 CO_2 concentration $w(1995) = 358\,ppm$ for the given pre-industrial concentration $w_0 = w(1800) = 280\,ppm$. By coincidence, this also almost satisfies the condition for a continuous derivative in the transition from exponential to linear growth in 1995, which would require $t_s = 40$ years.

All computations were carried out with a discretization time step of $\Delta t = 5$ years from the year 1800 over a period of 1200 years, up to the year 3000. However, the emissions were allowed to adjust freely only over 805 years, from 1995 to 2800, and were then frozen at the level $e(2800)$ for the last 200 years. The time span is clearly unrealistically long for economic predictions, but, as is apparent from Fig. 7.3 and the results shown later, is nevertheless appropriate for assessing long-term climate impacts relevant for a sustainable development policy.

7.3.1   Prescribed Scenarios

As reference for the interpretation of the optimal emission solutions, it is useful to consider first the climate evolution for some prescribed emission scenarios. Figure 7.4 shows the CO_2 emissions, atmospheric CO_2 concentrations and resultant global warming for: the BAU emission scenario e_A (scenario SA); a modified BAU scenario SB, which is more consistent with the estimated fossil fuel reserves and assumes that, after achieving a maximal value of 38 GtC in 2200, the emissions decrease linearly to zero in the year 3000; and two frozen emission scenarios SF and SG, in which the emissions are assumed to be stabilized after the year 2000 at the 1990 or 80% of the 1990 levels, respectively.

The upper panels show the evolution over the full 1000 year period of the integration, plus the spin-up period, while the lower panels depict the same curves for the period 1995-2200 only. A comparison of the upper and lower panels illustrates the dangers of restricting the discussion of global warming to time scales of one or two centuries or even few a decades: the impact of CO_2 emissions over longer time scales relevant for the response of the climate system can be dangerously underestimated.

The scenarios SA and SB can be interpreted quantitatively only for the first 100-150 years. After this period, the CO_2 concentrations and temperatures rapidly increase beyond the limits of our linear response model. However, the order-of-magnitude prediction that the CO_2 concentrations will grow to some ten times the present value in the course of several hundred years is, if anything, an underestimate, since it ignores the positive feedbacks of the decreasing solubility of CO_2 in the ocean with increasing temperature

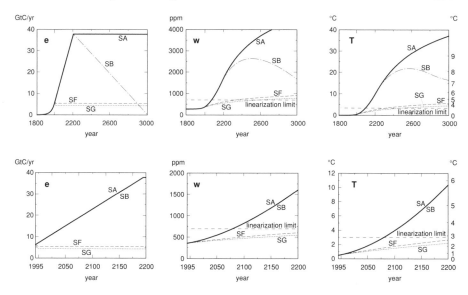

Fig. 7.4. CO₂ emissions, computed CO₂ concentrations and global warming (from left to right) for the time periods 1800-3000 (top) and 1995-2200 (bottom) for the BAU scenario (SA, full curves), modified BAU scenario (SB, dashed-dotted curves), frozen emissions at 1990 levels after the year 2000 (SF, dashed curves) and 20% reduced emissions relative to the 1990 level after 2000 (SG, dotted curves). The linear model is not applicable above the indicated dashed levels. A correction allowing for the logarithmic rather than linear dependence of the climate response on the CO₂ forcing yields the logarithmic temperature scale shown on the right ordinate axis of the top-right panel. However, this temperature reduction is partially offset by the nonlinearity of the dissolution of CO₂ in the ocean. (From HHGOS)

and increasing CO₂ concentrations. The linearized temperature response, on the other hand, is strongly exaggerated for higher temperature values. Application of the normal logarithmic dependence of the radiative forcing on changes in the CO₂ concentration instead of our linear response relation yields a temperature response for a ten-fold increase in the CO₂ level of the order of 8°C (cf. logarithmic temperature scale on the right side of the top-right panel of Fig. 7.4). However, at these temperatures other nonlinearities besides the radiative forcing dependence on the CO₂ concentration become important – including possible instabilities, for example through a breakdown of the North Atlantic circulation. Reliable predictions cannot be made for these extreme climate changes even with complex nonlinear three-dimensional carbon cycle and coupled atmosphere-ocean general circulation models, since there exist no data for model evaluation in this range.

The full severity of the climate-change impact for scenarios SA and SB becomes apparent only in the long-term perspective over several hundred years. While the monotonic increase in the second half of the next millennium

Table 7.1. Emission scenarios (from HHGOS)

SA	business-as-usual (BAU), with constant emissions after 2200
SB	modified business-as-usual, with linearly decreasing emissions after 2200
SF	frozen emissions at 1990 level after 2000
SG	reduced emissions frozen at 80% of 1990 level after 2000
$S0$	baseline reduced-emissions run, cost-function parameters: $T_c = 1°C$, $\dot{T}_c = 0.02°C/yr$, $\tau_1 = \tau_2 = 100$yrs, $\tau_a = 50$ yrs, $\tau_d = \infty$ yrs
$S1a, b$	same as $S0$ but with reduced abatement-cost inertial terms (run $S1a$: $\tau_1 = \tau_2 = 50$yrs), or zero inertial terms (run $S1b$: $\tau_1 = \tau_2 = 0$)
$S2$	same as $S0$, but with temperature rate-of-change term $\dot{T}_c$ only in climate-damage costs
$S3a, b$	same as $S0$, but with abatement-cost discount time constant τ_a changed from 50 yrs to 25 yrs (S3a) and 100 yrs (S3b)
$S4a - d$	same as $S0$, but with climate-damage cost discount time constants $\tau_d = 100$ yrs (S4a), 50 yrs (S4b), 35 yrs (S4c) and 25 yrs (S4d)

for scenario SA depends on the unrealistic assumption of a continual constant emission level of 38 GtC/yr after 200 years, a major climate warming is predicted also for scenario SB.

Although it is useful to remind oneself of the drastic climatic impact of a laissez-faire climate policy, the limitations of the present linearized climate-response model for scenarios SA and SB are, in fact, irrelevant for the present optimization study. We shall need to refer to the BAU emission curve e_A only to compute the abatement costs for the determination of optimal reduced-emission scenarios, all of which – assuming a climate-protection strategy consistent with a policy of sustainable development – yield significantly smaller climate changes lying within the linear climate-response regime.

The frozen emission scenarios SF and SG yield only a relatively minor warming of the order of 1°C above 1995 temperatures within the next 100 years, but also produce significant warmings of the order of 4°C in the long term. They can, therefore, be regarded only as interim solutions that gain time for the unavoidable gradual transition to lower emission levels required for the prevention of a significant long-term climate change, as discussed in the following.

7.3.2   Sensitivity Studies

The baseline scenario

A baseline optimal reduced-emissions scenario $S0$ (Fig. 7.5) was computed for the cost-function parameters values $T_c = 1°C$, $\dot{T}_c = 0.02°C/yr$ and $\tau_1 = \tau_2 = 100$ yrs, with discount time constants $\tau_a = 50$ years and $\tau_d = \infty$. The

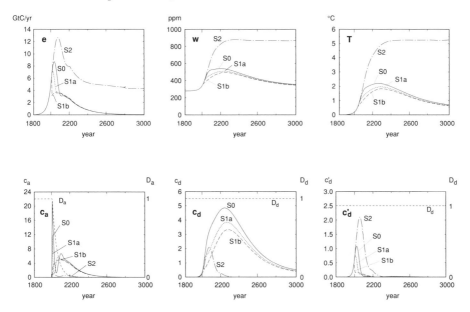

Fig. 7.5. Evolution over the period 1800-3000 of: (top, left to right) CO_2 emissions, CO_2 concentrations, global mean temperature and (bottom, left to right) specific abatement costs c_a, specific damage costs c_d and the contribution to the specific damage cost c_d' from the rate of change of temperature, for (cf. Table 7.1): the baseline reduced-emissions scenario $S0$ (full curves), the same run with reduced or zero inertial terms in the abatement-cost function (run $S1a$, dotted curves, and run $S1b$, dashed curves, respectively) and a modified baseline run in which the climate-damage costs are assumed to depend only on $\dot{T}$ (run $S2$, dash-dotted curves). Also shown in the lower panels are the exponential abatement and damage cost discount factors D_a and D_d (From HHGOS)

impact of different parameter choices is explored in the runs $S1 - S4$ (cf. Table 7.1 and Fig. 7.5- 7.7).

The critical temperature $T_c = 1°C$ and rate of change of temperature $\dot{T}_c = 0.02°C/yr$ for the climate-damage cost function are representative of typical values quoted in the literature. They lead for Scenario $S0$ to a maximum temperature increase $T_{max} = 2.2°C$ (cf. Fig. 7.5). The decrease in temperature beyond the year 2200 results from the discounting of the abatement costs, while no discounting is applied to the climate damage costs: one can more readily afford to reduce emissions in the far future to avoid climate damage costs than in the near time (discount factors are discussed further below).

Impact of economic and ecological inertia

The choice of the economic inertia coefficients τ_1 and τ_2 was found to be uncritical in a broad band of values. They ensure that no discontinuities occur in the emissions or rate of change of emissions at the start time (1995) of the integration and have a significant effect only in the initial stages of the climate evolution. Initially, the emissions are restrained to follow the BAU path (see also the more detailed discussion in Wigley et al, 1996), but the long-term impact of economic inertia remains small. This is evident in Fig. 7.5 from a comparison of the baseline scenario $S0$ with runs in which the inertial terms were reduced ($S1a$) or set equal to zero ($S1b$).

The contribution to the climate-damage costs from the rate-of-temperature change term in (7.10) was found to be small for our choice of critical parameters T_c and $\dot{T}_c$ (cf. cost curves shown in HHGOS). This is illustrated by the optimal emissions scenario $S2$, also shown in Fig. 7.5, in which the climate-damage costs were represented solely by the quadratic term in the rate of change of temperature. The maximal temperature increases to $6°C$ within 300 years and then remains at this level. (The results of Tahvonen et al (1994), who considered only the quadratic $\dot{T}$-dependent term in their climate-damage costs, should therefore be regarded only as illustrative, as pointed out by the authors. However, the relative contribution of the temperature and time rate of change of temperature to the climate damage costs deserve closer scrutiny, including a more careful discrimination between different types of ecological and economic climate damages.)

Impact of discount factors

The most critical and also most controversial terms in the cost functions are the discount factors. We have argued that the discount rates for mitigation and climate damage costs should be treated differently. We consider, therefore, their impacts first separately, returning later, however, to the question of their interrelation.

Our choice of the abatement cost discount time constant $\tau_a = 50$ years (2% per year) for the baseline scenario lies at the lower range of (inflation adjusted) discount rates proposed in greenhouse-gas abatement studies (cf. Nordhaus, 1991, 1993, Cline, 1992). Figure 7.6 illustrates the impact of decreasing the time constant τ_a to 25 years (Scenario S3a) or of doubling τ_a to 100 years (Scenario S3b). A shorter discount time scale implies that one can afford to apply mitigation measures earlier, reducing global warming, while for a larger time constant it is more economic to delay abatement measures, with a resultant increase in global warming. The value of τ_a shows a strong influence on the computed optimal emission paths. However, this applies for a fixed disount rate for the climate damage costs (which we have set to zero in our baseline scenario S0 and Scenarios S3a,b). Since the computed optimal CO_2 paths depend on the ratio of climate damage to mitigation costs,

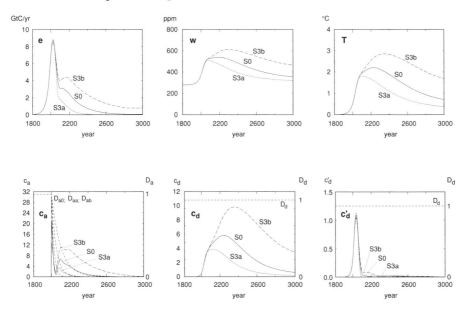

Fig. 7.6. Impact of changed abatement cost discount time constants $\tau_a = 25\,\mathrm{yrs}$ (S3a, dotted curves) and $\tau_a = 100\,\mathrm{yrs}$ (S3b, dashed curves) compared with the baseline case $\tau_a = 50\,\mathrm{yrs}$ (Scenario S0, full curves; cf. Table 7.1 and Fig. 7.5; layout as in Fig. 7.5; from HHGOS)

parallel changes in the discount rates for both types of costs tend to offset one another, as discussed further below.

The controversy over discount rates has focussed more on the question of the proper intertemporal treatment of climate damage costs. According to the traditional economic view, climate damage costs are economic costs just as any other costs, and should accordingly be discounted at the same rate as mitigation costs. This can be justified for climate damages that can be countered by appropriate economic measures, such as building higher dikes in response to rising sea levels, or a modification of agricultural practices.

However, an alternative view is that a potential deterioration of future living conditions through an irreversible change in climate represents a loss in value, which is essentially independent of the period in the future when the climate change actually takes place. The preservation of an habitable planet for future generations is regarded as a legacy which must be honored today, regardless of the time horizon over which our present actions will affect future living conditions. Thus, future sustainable development is perceived as a non-time-degradable ethical commitment to which one should assign a time-independent value.

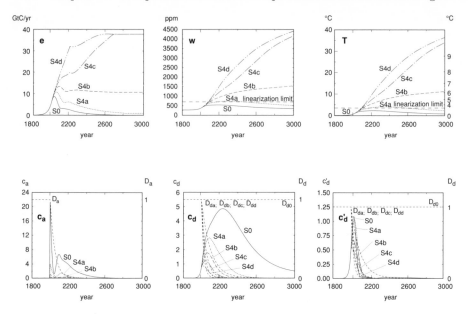

Fig. 7.7. Comparison of the baseline case $S0$ without climate damage-cost discounting (full curves) with scenarios assuming finite discount time constants $\tau_d = 100$ yrs ($S4a$, dotted curves), $\tau_d = 50$ yrs ($S4b$, dashed curves), $\tau_d = 35$ yrs ($S4c$, dashed-dotted curves) and $\tau_d = 25$ yrs ($S4d$, dashed-double-dotted curves; from HHGOS)

These alternative viewpoints are the subject of considerable debate in the current literature on integrated assessment (cf. IPCC, 1996b, HHGOS and Nordhaus, 1997). At the heart of the debate is the problem that the future climate of the earth is an asset whose present day equivalent value (judged by the present generation, acting as proxy for future generations) cannot be established objectively on the economic market. Its value is determined ultimately by ethical criteria. The standard approach for assigning monetary values to non-market assets is to apply willingness-to-pay assessments. However, very few quantitative assessments of this kind have been attempted for the climate problem. Furthermore, even if comprehensive polls were carried out, they would probably yield a wide spectrum of value assignments, and the interpretation and weighting of the results would be difficult because of the non-uniform level of information of the public on the highly complex issues involved.

The available information on the value ascribed by the public to future climate is sparse and inconclusive. Nevertheless, in order to obtain optimal emission paths which are consistent with the requirements of sustainable development, we have set the discount rate for damage costs to zero for our baseline run $S0$ and the sensitivity runs $S1a, b, S2$ and $S3a, b$. This corresponds to the value assignment of the concerned environmentalist who is willing to pay today to avert a future major climate change, independent of the time horizon of the future climate change. The application of the same or comparable discount factors to both mitigation and climate-damage costs (e.g. Nordhaus, 1991,1993) yields basically different conclusions, as discussed below. Our choice of $\tau_d = \infty$ should not be interpreted as implying that we regard a zero discount rate as the 'correct' value for climate damage costs. For political decision making – at least in an ideal democratic society – only the public and politically transmitted perception of the value of a future stable climate is relevant, which, as pointed out, is not yet well defined.

The impact of different climate damage discount rates is investigated in runs $S4a, b, c$ and d, in which the damage-costs discount time constant was set at 100, 50, 35 and 25 years, respectively. The maximal CO$_2$ concentrations and temperatures increase markedly, particularly for the last two cases. The climate changes implied by these temperature increases – noting that regional temperature changes, for example over continents, can be significantly higher than the global mean temperature increase – implies a dramatic change in the living conditions of our planet. However, this occurs only after several hundred years, when the climate-damage costs have been discounted by one or two orders of magnitude.

The character of the optimal-path solutions depends critically on the ratio of the climate-damage and abatement cost discount factors. If the discount time constant is higher for the climate damage costs than for the abatement costs, the discounted specific abatement costs become exponentially small compared with the discounted specific climate damage costs for large times, and the most cost effective path is one in which the emissions approach zero asymptotically (except for Scenario S2, in which the damage costs depended only on $\dot{T}$). Thus, the long term temperature increase for the optimal emissions path remains relatively small.

However, the form of the solution changes completely if the opposite inequality $\tau_d < \tau_a$ holds (Scenarios $S4c, d$). In this case, the climate damage costs are discounted more rapidly than the mitigation costs, and it becomes more cost effective to revert to the business as usual scenario asymptotically. Although the non-discounted specific climate damages grow with the square of the temperature, this is more than off-set by the more effective exponential discount factor for the damage costs, and $e(t) \to e_A(t)$ as $t \to \infty$. The asymptotic CO$_2$ concentrations and temperatures of Scenarios $S4c, d$ accordingly approach the BAU scenario (Fig. 7.7).

For $\tau_d = \tau_a$ (Scenario $S4b$), neither cost term is discounted more rapidly than the other. In this case, the optimal emissions path remains at a relatively high level between the BAU path and the baseline solution (cf. Fig. 7.7).

The global warming levels of the optimal path solutions of Fig. 7.7 are considerably higher than the solutions obtained assuming zero discount rates for the climate-damage costs (even for the case $S4a$ with $\tau_d = 100\,\mathrm{yrs} > \tau_a =$

50 yrs). The temperature increases exceed most estimates of the limits of global warming acceptable for sustainable development. Thus, if one subscribes to the ethical commitment of preserving a habitable planet for future generations, these solutions cannot be accepted. It follows that the social intertemporal preference relations describing the present and future costs of adapting to or mitigating climate change cannot be expressed in this case in terms of standard economic discount factors appropriate for, say, the short-term return on capital investment or intertemporal expenditure preferences for consumer goods.

We conclude from these sensitivity tests that the computed optimal emission paths depend strongly on the relative magnitudes of the discount factors assigned to climate damage and mitigation costs, and that solutions qualitatively consistent with the requirement of sustainable development are obtained only if the climate damage discount time constants are significantly larger than the discount time constants for abatement costs.

7.4   Summary and Conclusions: Single-Actor Analysis

The purpose of our single-actor analysis of optimal cooperative abatement strategies was not to provide quantitative monetary estimates of costs and benefits to aid decision makers in establishing, say, the proper level of global carbon taxes, but rather to clarify the basic input assumptions and cause-and-effect relations, which we suspect are responsible for the pronounced divergences of previously published cost-benefit analyses. This has enabled a discrimination between conclusions, which we believe represent relatively robust consequences of the basic dynamics of the climate system and predictions, which depend critically on controversial input assumptions.

The principal conclusions of our investigation can be summarized as follows:

- Since the global warming response for CO_2 emissions extends over several centuries, the costs associated with the climate impact of present and future CO_2 emissions must be optimized over time scales far beyond normal economic planning horizons.

- For all solutions yielding limited global warming, CO_2 emissions must be drawn down significantly by a factor of at least a half over a few centuries, with a continual decrease thereafter. The rate of reduction for the optimal path depends sensitively on the assumed discount rate for the mitigation costs.

- Discounting of climate damage costs at standard economic discount rates yields optimal CO_2 emission paths that are only weakly reduced relative to the Business as Usual scenario. The long-term climate warming becomes very large, of the order of 6-8°C. A necessary condition for global warming to remain below an acceptable bound for sustainable development is that the discount rate for mitigation costs is greater than the discount rate for climate damage costs. In practice, optimal CO_2 emission paths yielding limited global warming are obtained only if the discount rate for climate damages is very small or zero.

- The inclusion of economic inertia in the mitigation cost function results in optimal emission paths that continue to rise with the BAU emissions curve for a decade or two before declining. However, the omission of inertia, allowing the emissions to adjust immediately to a lower level, has negligible influence on the long-term climate response. We conclude that an effective climate mitigation strategy should focus on the long-term transition to energy technologies with zero or very low CO_2 emissions. Short term reductions through energy saving alone are insufficient and can be viewed only as a useful auxilliary measure in support of the necessary long-term technological transition process.

- The technological restructuring can be carried out without dramatic dislocations in the course of many decades or a century. This should not be interpreted to imply that there is no urgency in the implementation of policies initiating the necessary gradual transition to lower CO_2-emission levels. Any non-regulated continuation along the business-as-usual path incurs the need for larger, more costly adjustments later.

- Our optimal CO_2 emission solutions ignore the comparable global warming contributions of non-CO_2 greenhouse gases and are, therefore, too optimistic. To the extent that the abatement of non-CO_2 greenhouse gases can be achieved at a relative cost similar to that of CO_2 emissions, the impact of non-CO_2 greenhouse gases can be accounted for to first order by simply increasing the climate damage costs by an appropriate factor. This leads to somewhat lower but not drastically reduced optimal CO_2 emission paths (cf.HHGOS). As the ratio of climate damage to abatement costs is an arbitrary free parameter in our analysis, our general conclusions are not affected by this modification. However, the problem is more severe if the non-CO_2 greenhouse gases cannot be effectively abated (see discussion in HH).

- For the time scales of climate change corresponding to the optimal CO_2 emission paths, climate damages due to the rate of change of temperature are an order of magnitude smaller than the damages due to the change in temperature itself. However, these estimates are based on global critical climate damage thresholds of $T_c = 1$°C for temperature and $\dot{T_c} = 0.2°\text{C/decade}$ for the rate of change of temperature and a specific (quadratic) form of the climate damage costs. A more careful analysis of different types of climate damage is needed before definite conclusions can be drawn regarding the relative impact of the rate of change of climate and the climate change itself.

PART II: THE NON-COOPERATIVE MULTI-ACTOR CASE

7.5 The Multi-Actor Case

In the analysis so far, the world has been viewed as a single economic region controlled by a single actor. It was assumed that there existed an international agreement on the global abatement-plus-adaptation cost function, and the optimization task was then to minimize the net costs through choice of a suitable path of the globally integrated greenhouse-gas emissions.

Although providing a useful reference, this idealized picture of a single-decision-maker world economy does not represent a close approximation to the real situation. In practice, the efforts at arriving at a global climate protection strategy are better described as a multi-player game, in which each player, representing a particular political-economic region or sector, is engaged in trying to optimize his or her individual welfare function on the basis of policy decisions made under the constraints of local-interest pressures, international trade relations and various other political side-conditions.

A few authors have recently addressed the multi-actor problem using generalizations of single-actor models used in earlier studies (e.g. Nordhaus and Yang, 1996, Manne and Richels, 1995). In the following, we shall similarly address the multi-actor problem using an extension of the idealized single-actor SIAM used for our single-actor analysis. In view of the large divergences already of single-actor studies, it appears premature to embark on a multi-actor analysis using a relatively complex economic model. The goal of the following analysis is to understand the basic system-analytical implications of the multi-actor problem, and to generalize some of the conclusions of our single-actor study to the multi-actor case, using again a model designed for maximal structural transparency.

The basic structure of a multi-actor GES model was presented in Fig. 7.2, in which the sub-systems of the single-actor GES model, Fig. 7.1, were disaggregated into separate columns representing different political-economic sub-systems controlled by different actors. The disaggregation can refer to either different geographical regions or different economical sectors (for example, consumers and suppliers of fossil fuels, or private and government sectors).

The coupling between actors occurs through trade, negotiations, and the global climate, which is modified jointly by all actors, but affects each actor differently. Each actor seeks to optimize his or her individual welfare function. Whether the outcome of the interactive multi-actor optimization exercise is a non-cooperative Nash equilibrium (assuming such an equilibrium exists), an optimal cooperative solution, or some partially cooperative alliance construction, depends in detail on the type of coupling between the sub-systems, the negotiation framework (burden sharing mechanisms, possibilities of reward and retribution, etc) and the negotiation strategies of the individual actors.

We shall consider in the following the Nash equilibria for three non-cooperative scenarios: two cases of n non-trading actors who are coupled only through climate, and one case of two actors (a cartel of fossil fuel suppliers interacting with a fossil fuel user cartel) who are coupled through both climate and trade. The more complex problem of negotiations will not be addressed in this pilot analysis.

7.5.1 n Identical Non-Cooperative Actors

In general, the optimal solutions for non-cooperative or partially cooperative (cartel-type) strategies are inferior to the optimal cooperative solution of the single-actor model. We illustrate this first for the simplest example of n identical actors representing different economic regions, each of which is characterized by the same *per capita* abatement and damage costs as in the single-actor case. We use the same cost expressions as in the single-actor SIAM model. Since the normalization of the cost functions in the SIAM model may be chosen arbitrarily, we define them in the following as *per capita* costs.

In a non-cooperative game, each player seeks to maximize his or her pay-off independently, without attempting to negotiate a cooperative strategy that may yield a higher pay-off for all players. The resulting solution is normally a Nash equilibrium, namely a set of strategies such that the strategy of each player yields a maximum pay-off for the player, given the strategies of the other players (cf. Binmore, 1992). Although the Nash equilibrium generally yields a lower pay-off for the individual player than an optimal cooperative strategy, the latter is unstable: individual players can achieve higher pay-offs by not following the cooperative strategy, as illustrated by the classical example of the prisoner's dilemma. Thus, cooperative strategies are realized only if the rules of the game are changed (for example, by introducing penalties) to motivate cooperative behaviour.

The non-cooperative Nash equilibrium for the present problem is obtained by minimizing the cost function C_i for each actor i with respect to his or her own emissions e_i, keeping the emissions of the other actors fixed. From the symmetry of the problem it follows that all emission functions e_i will be equal for the Nash equilibrium solution. Thus, for any given actor i,

$$e = \sum_{j=1}^{n} e_j = ne_i. \tag{7.13}$$

The minimization problem for an individual actor i can then be reduced to the previous single-actor case, with the control path $e(t)$ replaced now by $e_i(t)$.

To determine the variations $\delta C_{ai}, \delta C_{di}$ of the abatement and climate damage costs, respectively of actor i with respect to a variation δe_i of his or her emissions, keeping the emissions of the other actors fixed, we compare the single-actor cost variations with the corresponding cost variations $\delta C_a, \delta C_d$ for the cooperative case, in which the emissions of all actors are varied by the same amount, yielding a total emission variation

$$\delta e = n\delta e_i. \tag{7.14}$$

In the case of the abatement costs, the variation in the *per capita* costs of a given actor are independent of the emission variations of the other actors, so that

$$\delta C_{ai} = \delta C_a. \tag{7.15}$$

Variations in the *per capita* climate damage costs, however, are experienced by all actors equally, regardless of the source of the emission variations. Since the temperature varies linearly with the emissions, the variations in the temperature (and rate of change of temperature) is n times larger in the cooperative case than when the emissions are varied for a single actor i only (eq.7.14):

$$\delta C_{di} = \frac{\partial C_d}{\partial T}\frac{\partial T}{\partial e}\delta e_i \tag{7.16}$$

$$\delta C_d = \frac{\partial C_d}{\partial T}\frac{\partial T}{\partial e}\delta e = n\delta C_{di} \tag{7.17}$$

Hence in computing the gradient of the *per capita* costs with respect to variations in the emissions e_i, one can use the same expression as in the cooperative case for the abatement costs, while the expression for the climate damage costs must be reduced relative to the cooperative case by a factor n^{-1}. Thus, the Nash equilibrium for the case of n identical non-cooperating actors is identical to the single-actor solution with the climate-damage costs reduced by a factor n^{-1}: each actor considers only his or her own contribution to the climate-damage costs, ignoring the contribution of the other actors.

In the limit of a large number of actors, one obtains the free-rider solution: no one carries out abatement measures (from HH).

However, for a moderate number of actors, of the order of 10 or 20, the non-cooperative Nash equilibrium emission paths do not differ as strongly from the cooperative solution as may have been expected intuitively. Figure 7.8 shows the impact of a non-cooperative strategy, compared with the cooperative single-actor solution, for the baseline reduced-emissions scenario (Fig. 7.5), for different numbers of actors.

Similar results showing the effect of a change in the ratio of climate-damage to abatement costs (but with an enhancement rather than a reduction of the ratio) were presented in HHGOS in a discussion of the role of non-CO_2 gases. The relatively minor impact can be explained by two effects:

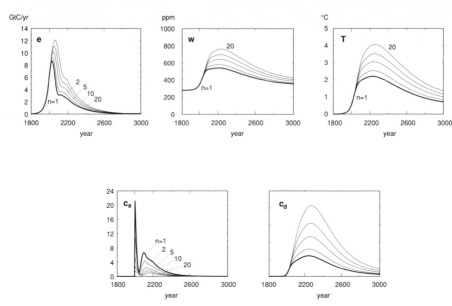

Fig. 7.8. Optimal CO$_2$ emission paths, CO$_2$ concentrations and global warming for the n-identical actor Nash equilibrium solution, computed for the parameters of the baseline reduced-emissions scenario S0, full curves (from HH)

Firstly, a decrease of the climate-damage costs by a factor n^{-1} relative to the abatement costs implies an increase of the critical climate temperature T_c (and rate of change of temperature $\dot{T}_c$) by a factor of only $\sqrt{n}$ (cf. eq.(7.10)). Thus, the climate damage costs computed by the individual actor in the non-cooperative case are raised already to the level of the true climate damage costs, as computed in the cooperative case, when the emissions are increased by a factor of only $\sqrt{n}$. However, while the computed climate-damage costs for the n-actor problem correspond to the original climate damage costs for the cooperative single-actor case, the computed abatement costs, because of the higher emission levels, are lower. But in the optimal n-actor solution, a balance is attained between the abatement and damage costs. Thus, the abatement costs will be higher and the emission levels lower than these values. Hence the increase in emission levels for the non-cooperative Nash equilibrium solution is even smaller than the factor $\sqrt{n}$.

Because of the relatively small differences between the cooperative and non-cooperative optimal emission solutions for a moderate number of actors, it could be argued that there is little incentive for attempting to reach an agreement on a more effective cooperative strategy (recall that the cooperative solution is unstable: given a cooperative strategy of the other players, it pays for an individual actor to reduce his or her abatement measures if there is no penalty imposed.) However, an essential assumption in this scenario is that all actors share the same assessment of the climate-damage and abatement costs. Once this consensus has been achieved, only a small additional step would appear to be needed to agree on a cooperative climate-protection strategy. The more difficult (and relevant) multi-actor situation arises when different actors have different climate-damage and abatement-cost perceptions, as illustrated in the following example.

7.5.2 The Single Mitigator Problem

A more realistic description of the n-actor climate-protection problem is that some actors regard the risks of adverse effects due to climate change as high, while others are more concerned with the negative effects of possible mitigation measures on the economy. To study the impact of different cost assessments, we consider a particularly simple example. Assume that one actor assigns the same values to climate-damage and abatement costs as in the SIAM model, while the remaining $(n-1)$ actors ignore all climate-damage costs, as in the business-as-usual (BAU) scenario. All actors are again assumed to represent identical economic regions, the only difference being in their assessments of the climate related costs. For the $(n-1)$ BAU actors with prescribed emissions $e_A(t)/n$, the distinction between different actors is irrelevant. We may thus, replace the set of $(n-1)$ BAU actors by a single "rest of the world" (ROW) BAU actor whose emissions are given by $e_R = [(n-1)/n]e_A(t)$. In fact, we can clearly generalize our model without change of notation by identifying also the single mitigator as a group of identical actors pursuing the same (cooperative) mitigation strategy. Thus, we may interpret n generally as the ratio of the number of BAU actors to the number of mitigators, treating both groups of otherwise identical actors as single actors. However, we shall continue to refer to the cooperative group of mitigators as the "single" mitigator.

Since e_R is prescribed, the problem reduces again formally to a single-actor problem for the first actor 1, the single mitigator. Regarding costs again as *per capita* costs, the expression for the abatement costs for actor 1 remains the same as in the global single-actor case (eq.(7.8)), with the global emission abatement factor $r = e/e_A$ replaced by the individual abatement factor $r_1 = e_1/e_{1A} = ne_1/e_A$ of actor 1. However, the climate-damage costs are now dominated by the prescribed BAU emissions of the ROW actor.

Noting that the total emissions are given by

$$e = e_R + e_1 = \frac{n-1}{n}e_A + e_1, \tag{7.18}$$

and that the climate change depends linearly on emissions, the global mean temperature T and rate of change of temperature $\dot{T}$, which determine the climate-damage costs, are given by

$$T = \frac{n-1}{n}T_A + T_1,$$

$$\dot{T} = \frac{n-1}{n}\dot{T}_A + \dot{T}_1 \qquad (7.19)$$

where T_A and T_1 represent the climate response to the emissions e_A and e_1, respectively, as defined by eq. (7.1). Substitution of (7.19) into (7.10) yields the damage costs as a function of $e_1(t)$, for given $e_A(t)$. Adding to these the abatement costs of actor 1, which are also defined as a function of e_1, the total costs of actor 1, given the BAU emissions of the remaining actors, are obtained in terms of the emissions of actor 1 and can thus, be minimized with respect to $e_1(t)$.

The optimal paths are depicted in Fig. 7.9 for various values of n. As n increases, it becomes more and more difficult for the single mitigator to influence the climate damage costs, and in the limit $n \to \infty$, actor 1 undertakes no abatement measures. However, the asymptotic solution is approached more slowly than may have been anticipated intuitively.

On the other hand, if the number of non-cooperative BAU actors is less than 10, the single mitigator compensates for the lacking abatement policy of the rest of the world by actually enhancing his or her own abatement

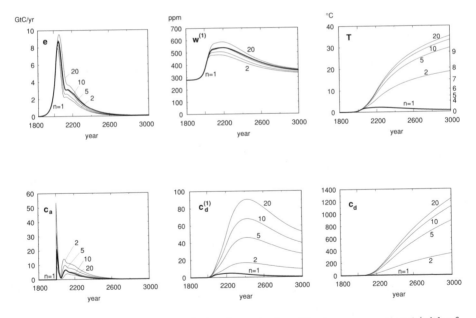

Fig. 7.9. Optimal emissions solution for a single mitigator representing $1/n$'th of the world economy while the "rest of the world" pursues a BAU emissions path. The CO_2 emissions and concentrations of actor 1 have been normalized through multiplication with a factor n, so that the curves correspond to globally scaled *per capita* CO_2 changes (i.e. the resultant CO_2 evolution if the ROW actor had adopted the same emissions as actor 1). The computations were made for the parameters of the baseline single-actor scenario, full curves (from HH)

measures. In view of the relatively small contribution of the single mitigator to the total climate damages, this result is rather surprising. It can be explained by the assumed nonlinear dependence of the climate damage costs on the climate change. The background temperature increase caused by the BAU actors amplifies the climate change impact of the single mitigator. It is shown in HH, that this amplification of the mitigator's impact more than offsets his or her smaller contribution to the total emissions for $n < 10$.

For policy makers, this simple result has important implications. Although actors who regard climate change as a potential hazard that should be mitigated through appropriate abatement measures,they will naturally strive to achieve binding international agreements on joint actions, the frequently heard argument that there is no incentive for reducing emissions on an individual basis appears to be questionable. Our model suggests that self-interest should motivate at least the larger industrial countries to undertake unilateral mitigation actions that are comparable to the measures they wish to realize ultimately in an international agreement. However, this conclusion could be modified if trading interactions are taken into account (e.g. problems of competition and leakage).

7.5.3 Trading Actors

The non-trading multi-actor model discussed in the previous section illustrates some specific features of non-cooperative optimization problems, but represents still an unrealistic simplification of the true situation. The coupling between different actors will in general be governed not only by changes in global climate, but also by trade. Moreover, in attempting to minimize his or her net costs (i.e., maximize his or her welfare), each actor will normally take into account the anticipated actions of other actors. We turn now to this more general case, but restricting ourselves as before to non-cooperative strategies. Thus, we exclude direct negotiations, but allow for foresighted interactions between individual players.

Consider a general multi-actor economic model in which the actions of each player i are described by a set of control variables $v_{i\alpha}(t)$, $\alpha = 1, 2, \ldots$, including, in addition to CO_2 emissions (or, equivalently, a carbon tax, or some other CO_2 regulatory mechanism), various other economic control factors, such as the prices of commodities, investment expenditures, tariffs, taxes, etc. In the following, it will be convenient to discretize the time variable t, writing $v_{i\alpha}(t) \equiv v_{i\alpha t}$ and denoting the set of paths of all control variables of actor i by the vector $\mathbf{v}_i = (v_{i,\alpha,t})$.

Because of the coupling through climate and trade, the individual welfare W_i of an actor i not only depends on the actor's own control variables $\mathbf{v}_i$, but on the complete set of control variables $\mathbf{V}$ of all actors. Thus, $W_i = W_i(\mathbf{V}) = W_i(\mathbf{v}_i, \bar{\mathbf{v}}_i)$, where $\bar{\mathbf{v}}_i = (\mathbf{v}_1, \mathbf{v}_2, \cdots, \mathbf{v}_{j-1}, \mathbf{v}_{j+1}, \cdots, \mathbf{v}_n)$ is the complementary set of control variables to $\mathbf{v}_i$, i.e. the set of control variables $\mathbf{v}_j$ of all actors $j \neq i$.

For given values of the control variables $\bar{\mathbf{v}}_i$ of the other actors, an actor i will strive to adjust his or her control variables $\mathbf{v}_i$ (by some strategy, discussed below) such that his or her welfare W_i is maximized. The resulting optimal control vector $\mathbf{v}_i^{opt}$ will, therefore, be a function

$$\mathbf{v}_i^{opt} = \mathbf{f}_i(\bar{\mathbf{v}}_i) \tag{7.20}$$

of the complementary set of control variables. It follows that a change $\delta\mathbf{v}_j$ in the control variable of actor j, $j \neq i$ will induce a change

$$\delta\mathbf{v}_i^{opt} = \frac{\partial\mathbf{f}_i(\bar{\mathbf{v}}_i)}{\partial\mathbf{v}_j}\delta\mathbf{v}_j \tag{7.21}$$

in the optimized control variable of actor i, or

$$\delta\mathbf{v}_i^{opt} = \mathbf{M}_{ij}\delta\mathbf{v}_j \tag{7.22}$$

where the *response matrix*

$$\mathbf{M}_{ij} = \mathbf{M}_{ij}(\bar{\mathbf{v}}_i) = \frac{\partial\mathbf{f}_i(\bar{\mathbf{v}}_i)}{\partial\mathbf{v}_j}. \tag{7.23}$$

denotes the marginal (i.e. differential) response of the optimal control variables of actor i to a marginal change of the control variables of actor j.

Depending on the actors' response strategies, one can distinguish between three forms of Nash equilibrium as outcome of the multi-actor optimization problem.

The non-interactive Nash equilibrium

In many treatments of coupled economic optimization problems (e.g. Binmore, 1992), the interactions between players are regarded as single-shot games. Each player i specifies his or her control variables $\mathbf{v}_i$ without knowledge of the action of the other player. A player's choice is based on some assumption about the control variables of the other players, but the possible 'response' of the other players to one's own choice is irrelevant, since the other players have no opportunity to respond to one's choice. The Nash equilibrium is given in this case by the solution of the simultaneous set of equations

$$\frac{\partial W_i}{\partial\mathbf{v}_i} = 0. \tag{7.24}$$

This approach was adopted in the non-interactive examples discussed in the previous section and other recent analyses of the multi-actor problem (Tahvonen, 1993, Nordhaus and Yang, 1995). However, the model is in general not very realistic for our applications. Thus, in the simple example of a two-player fossil-fuel supplier-consumer system discussed below, it will be found that only a trivial non-interactive Nash equilibrium exists. In a time dependent interactive dynamic control problem, the actors generally have an opportunity to react to the choice of control variables of the other actors, and will adjust their control variables accordingly. In setting their control variables, they will, therefore, anticipate the response of the other actors to their own strategy, and will optimize their strategy by taking the anticipated response of the other actors into account. One can then distinguish between two further Nash equilibrium solutions, depending on the manner in which the individual actors incorporate the response of the other actors in their optimization strategy.

The self-consistent interactive Nash equilibrium

If the strategic interaction between players is included, the simultaneous solutions of (7.24) no longer represent simultaneous local extrema (or turning points) of W_i with respect to the control variables $\mathbf{v}_i$. The necessary condition for a maximum of W_i, allowing for the response of the other actors, is the stationarity condition

$$\frac{dW_i}{d\mathbf{v}_i} = \frac{\partial W_i}{\partial \mathbf{v}_i} + \sum_{j \neq i} \frac{\partial W_i}{\partial \mathbf{v}_j} \frac{\partial \mathbf{f}_j(\bar{\mathbf{v}}_j)}{\partial \mathbf{v}_i} = \frac{\partial W_i}{\partial \mathbf{v}_i} + \sum_{j \neq i} \frac{\partial W_i}{\partial \mathbf{v}_j} \mathbf{M}_{ji} = 0. \tag{7.25}$$

The *self-consistent* Nash equilibrium in the interactive case is given by the simultaneous solution of the set of equations (7.25), together with the defining equations (7.23) for the response matrix. The two sets of equations are coupled and must normally be solved iteratively: to determine the solutions of (7.25), one needs to know the response matrix, but the response matrix, in turn, can be evaluated from (7.23) only when the solutions of (7.25) have been found.

The conjectured response Nash equilibrium

If the actors have no reliable information on the welfare functions of the other actors – or for some other reason are unable to solve the fully coupled optimization problem yielding the self-consistent Nash equilibrium – they cannot determine the true response of the other actors to their own actions. To optimize their welfare, they will, therefore need to make some assumption regarding the reactions of the other actors. If each actor i has some conjecture about the components of the response matrix relevant for his or her own optimization strategy, based on his or her perception of the likely strategy of the other players, one can solve the set of coupled equations (7.25) by substituting the *conjectured* response matrix $\tilde{\mathbf{M}}_{ji}$, $j \neq i$ for the true response matrix, ignoring the consistency relation (7.23). The resulting solution is termed the *conjectured* interactive Nash equilibrium.

Although the conjectured interaction equilibrium represents a mutual (pseudo-) maximum welfare solution in the sense that all actors have succeeded in modifying their control variables simultaneously such that the (conjectured) welfare gradients $dW_i/d\mathbf{v}_i$ vanish (assuming that the zero gradient values do indeed correspond to local maxima), the resultant solutions $\mathbf{f}_i(\bar{\mathbf{v}}_i)$ will not, in fact, generally represent a true joint optimum, since the conjectured response matrix will normally differ from the true response matrix, (7.23), as determined *a posteriori* from the optimized solutions. The solutions nevertheless represent a real Nash equilibrium, given the imperfect information available to the actors.

We note here a basic conceptual difference between interactive Nash equilibrium solutions, of either type, and the non-interactive Nash equilibrium. An interactive equilibrium – if it exists – is automatically realized by the actors, who systematically adjust their control variables until all actors have jointly achieved an optimal solution (or possibly one of a set of locally optimal solutions). In contrast, in the one-shot non-interactive Nash equilibrium, the solution represents only a possible outcome. Each actor is unaware of the control variables of the other actor, and the Nash equilibrium is realized only if all actors are versed in game theory and assume that the Nash equilibrium is indeed the strategy which everyone will adopt.

Numerical determination of interactive Nash equilibrium solutions

Both the self-consistent and conjectured response interactive Nash equilibrium solutions may be constructed numerically by iteration methods. The conjectured-response Nash equilibrium can be determined by a multi-component method of steepest ascent, applied simultaneously or in an iterative sequence to all welfare functions. The self-consistent interactive Nash equilibrium may be constructed from the conjectured response solution by requiring further that the derivatives $\partial \mathbf{f}_i/\partial \mathbf{v}_j$ inferred from the resulting solutions $\mathbf{v}_i = \mathbf{f}_i(\bar{\mathbf{v}}_i)$ are consistent with the conjectured response matrices, in accordance with the definition (7.23). This can be achieved by successively replacing the conjectured-response matrices by the response matrices computed from the derivatives of the conjectured response solutions. If the iterative adjustment procedure converges, one arrives at the internally consistent interactive Nash equilibrium.

Intuitively, one may expect such straightforward iterative constructions to converge for both the conjectured-response and the internally consistent Nash equilibrium solutions, provided the solutions exist. However, general conditions for the existence of Nash equilibrium solutions in these cases, and the convergence properties of appropriate numerical algorithms for their construction have not, to our knowledge, been studied.

Numerically, the derivation of internally consistent response matrices is not a simple exercise. The computation of the response matrix $\mathbf{M}_{ij}$ involves the determination of the second-derivative matrix $\partial^2 W_i/\partial \mathbf{v}_j \partial \mathbf{v}_i$, to be carried out in the high dimensional space resulting from the discretization of the time axis. The computation must, furthermore, be repeated many times to obtain internally consistent response matrices through the iterative adjustment of the conjectured response matrices to the response matrices computed from the resultant solutions.

In practice, however, the determination of the internally consistent interactive Nash solution may be an academic exercise. The construction of the solution – if it indeed exists – will be as elusive for the actors in a real situation as for the theoretical analyst. To define an iterative numerical joint optimization algorithm (which may be regarded as a mathematical proxy of the joint adjustment procedure of real life actors, which also may or may not converge), it is sufficient that each actor has a model of the response of the other actors to his or her own strategy, regardless of whether or not it corresponds to the actual response computed *a posteriori* for the resultant jointly optimized solution. The real situation may best be studied by numerical experiments, for example, by considering the sensitivity of the conjectured response Nash equilibrium with respect to the assumed marginal response matrices.

A simple example

It is instructive to illustrate the various forms of Nash equilibrium for a simple two-actor example. We choose for this purpose a prototype of the fossil fuel producer-consumer problem discussed in the following section.

Consider the interactions between a commodity supplier 1, offering a commodity at price v_1, and a consumer 2, who buys an amount v_2 of the commodity. As the welfare (utility) of the two actors we set

$$W_1 = (v_1 - \alpha)v_2, \tag{7.26}$$

$$W_2 = (\beta - v_1)v_2 - \frac{\gamma}{2}v_2^2 \tag{7.27}$$

where α is the price of production of the commodity, β is the initial utility value per unit of the commodity and γ is a coefficient representing a loss of utility of the commodity with increasing consumption. Thus, the welfare of the producer is the net earnings from the sale of the commodity, while the welfare W_2 of the consumer is the utility $\beta v_2 - \frac{\gamma}{2}v_2^2$ of the commodity minus the commodity costs $v_1 v_2$.

The utility loss term $-\frac{\gamma}{2}v_2^2$ can be regarded either as a saturation effect (there is a limit to the amount of food one can consume or cars one can drive) or, more relevant in the present context, as an environmental "degradation" factor. Interpreting the commodity as fossil fuel, the term corresponds to the climate damages. However, we make no attempt in this simple model to represent the climate damages or abatement costs explicitly. We have also suppressed the time dependence.

For positive welfare values, the control variables v_1, v_2 must be restricted to the region

$$\alpha \leq v_1 \leq \beta, \tag{7.28}$$

$$0 \leq v_2 \leq 2(\beta - v_1)/\gamma. \tag{7.29}$$

The cooperative case

As reference for the following discussion of various non-cooperative inter-action strategies, it is useful to consider first the cooperative case. Assigning the same weighting to the welfare of each actor, the cooperative goal is to maximize the total welfare

$$W = W_1 + W_2 = (\beta - \alpha)v_2 - \gamma v_2^2/2. \tag{7.30}$$

Note that the control variable v_1 no longer appears in (7.30): the price of the commodity affects only the exchange of welfare between the actors, not the total welfare.

The optimal solution is determined by

$$dW/dv_2 = \beta - \alpha - \gamma v_2 = 0, \tag{7.31}$$

yielding

$$v_2 = \frac{\beta - \alpha}{\gamma}, \tag{7.32}$$

with the total welfare

$$W = \frac{(\beta - \alpha)^2}{2\gamma}. \tag{7.33}$$

The non-interactive Nash equilibrium conditions

The necessary stationarity conditions for a Nash equilibrium in the non-interactive case are given by $\partial W_1/\partial v_1 = \partial W_2/\partial v_2 = 0$. The marginal welfare (i.e. differential welfare change)

$$\partial W_1/\partial v_1 = v_2 \tag{7.34}$$

for the producer is independent of his control variable v_1, while the marginal welfare

$$\partial W_2/\partial v_2 = \beta - v_1 - \gamma v_2 \tag{7.35}$$

of the consumer is a decreasing function of consumption, yielding decreasing welfare for $v_2 > (\beta - v_1)/\gamma$ and negative welfare when v_2 exceeds twice this value. Hence the stationarity conditions yield only the trivial solution

$$v_2 = 0, \tag{7.36}$$
$$v_1 = \beta. \tag{7.37}$$

Thus, although $v_1 = \beta$ is the largest price compatible with the inequality (7.28), no commodity is consumed, and the welfare of both actors vanishes. In fact, the stationary point corresponds to a minimum welfare for both actors rather than a maximum. None the less, given the strategy of the other actor, neither actor can increase his or her welfare through an alternative choice of his or her control variable, in accordance with the definition of a non-interactive Nash equilibrium. That the non-interactive Nash equilibrium must necessarily lead to the trivial solution (7.36),(7.37) follows immediately from eq.(7.34): for given $v_2 > 0$, the welfare gradient of the producer is positive and constant. Thus, the producer will increase his price indefinitely, if he ignores the back interaction of price on demand, until the demand falls to zero.

The conjectured response interactive Nash equilibrium

In practice, of course, the producer will not succumb to the temptation to increase his price indefinitely, as he knows that this will be countered by a decrease in consumption, at the latest, at the boundaries of the finite region defined by the inequalities (7.28), (7.29). Thus, his maximal earnings must lie somewhere within this region. To determine this maximum, he must allow for the response of the consumer to changes in price. (In the classical Cournot and Bertrand producer-oligopoly models that anticeded the Nash equilibrium concept, the response in demand to changes in price was, in fact, an essential ingredient.) We consider first the conjectured interactive equilibrium case in which each actor has no direct knowledge of the response of the other actor.

Assume that the producer conjectures that a marginal change δv_1 in price will evoke a change

$$\delta v_2 = \tilde{M}_{21}\delta v_1 \quad (\tilde{M}_{21} < 0), \tag{7.38}$$

in consumption, while the consumer conjectures, in turn, that a change δv_2 in consumption will stimulate a change

$$\delta v_1 = \tilde{M}_{12}\delta v_2 \quad (\tilde{M}_{12} > 0), \tag{7.39}$$

in price. One obtains then as stationary-point equations for the two actors (eqs.(7.25)):

$$\frac{dW_1}{dv_1} = \frac{\partial W_1}{\partial v_1} + \frac{\partial W_1}{\partial v_2}\tilde{M}_{21} = v_2 + \tilde{M}_{21}(v_1 - \alpha) = 0, \tag{7.40}$$

$$\frac{dW_2}{dv_2} = \frac{\partial W_2}{\partial v_2} + \frac{\partial W_2}{\partial v_1}\tilde{M}_{12} = \beta - v_1 - \gamma v_2 - \tilde{M}_{12}v_2 = 0, \tag{7.41}$$

which yields as functions for the optimal control variables, eq.(7.20):

$$v_1 = f_1(v_2, \tilde{M}_{21}) = \alpha - \frac{v_2}{\tilde{M}_{21}} \tag{7.42}$$

$$v_2 = f_2(v_1, \tilde{M}_{12}) = \frac{\beta - v_1}{\gamma + \tilde{M}_{12}} \tag{7.43}$$

The solution of the pair of equations (7.42),(7.43) yields as stationary point for the conjectured interaction Nash equilibrium

$$v_1 \;=\; \alpha + \frac{\beta - \alpha}{1 - \tilde{M}_{21}\left(\gamma + \tilde{M}_{12}\right)} \tag{7.44}$$

$$v_2 \;=\; \frac{\beta - \alpha}{\gamma + \tilde{M}_{12} - \tilde{M}_{21}^{-1}} \tag{7.45}$$

It can be readily verified that the stationary point does indeed correspond to a maximum for both welfare components, and that for all values of the conjectured response components, the solution lies within the region defined by the inequalities (7.28), (7.29). In fact, for non-negative $\tilde{M}_{12}$, eq.(7.43) yields, in place of the upper bound of (7.29), the more restrictive inequality

$$v_2 \le (\beta - v - 1)/\gamma \tag{7.46}$$

A reasonably realistic conjectured response model is that the consumer simply optimizes her consumption regarding the price as given, $\tilde{M}_{12} = 0$, while the producer correctly anticipates the response of the consumer to changes in price. In this case eq.(7.43) becomes

$$v_2 = \frac{\beta - v_1}{\gamma} \tag{7.47}$$

and the producer computes from this relation $M_{21} = -\gamma^{-1}$, yielding for his optimal price, from (7.42),

$$v_1 = \alpha + v_2\gamma. \tag{7.48}$$

The simultaneous solution of (7.47) and (7.48) yields

$$v_1 \;=\; (\alpha + \beta)/2 \tag{7.49}$$
$$v_2 \;=\; (\beta - \alpha)/2\gamma \tag{7.50}$$

Thus, only half as much is consumed as in the cooperative solution (eq.(7.32)), and the associated total welfare

$$W = \frac{3}{8\gamma}(\beta - \alpha)^2 \tag{7.51}$$

is reduced by a factor $3/4$ relative to the cooperative solution (7.33).

The self-consistency conditions for an interactive Nash equilibrium

The conjectured interaction Nash equilibrium will not, in general, representaself-consistent equilibrium. From the functions (7.42),(7.43) for the optimal control variables, one can compute the true response matrix:

$$M_{12} \;=\; \frac{\partial f_1(v_2, \tilde{M}_{21})}{\partial v_2} = -\tilde{M}_{21}^{-1} \tag{7.52}$$

$$M_{21} \;=\; \frac{\partial f_2(v_1, \tilde{M}_{12})}{\partial v_1} = -\left(\gamma + \tilde{M}_{12}\right)^{-1} \tag{7.53}$$

The consistency conditions $M_{12} = \tilde{M}_{12}$, $M_{21} = \tilde{M}_{21}$ then yield

$$\tilde{M}_{12} = \tilde{M}_{12} + \gamma, \tag{7.54}$$

which, for finite $\tilde{M}_{12}$, requires $\gamma = 0$. This must be rejected for a meaningful model: $\gamma = 0$ implies that consumption is not stabilized through decreasing utility for large v_2, and there exists no finite optimal solution even for the cooperative case.

Equation (7.54) can be satisfied, however, in the limit $\tilde{M}_{12} \to \infty$, which implies $\tilde{M}_{21} \to 0$ and $v_2 \to 0$; the limiting value of v_1 depends on how the limit is approached. An attempt to construct a self-consistent interactive Nash equilibrium by iteration, described in the following, converges to this solution.

Iterative interaction strategies

Our result that a conjectured response Nash equilibrium lying within the permissible control variable region (7.28), (7.29) exists for all conceivable values of the conjectured response matrix, while at the same time a self-consistent interactive Nash equilibrium exists only in the degenerate limit of infinite or zero components of the response matrix, is a little surprising. It is of interest to investigate the origin of this counter-intuitive conclusion by considering two players engaged in a sequence of conjectured response optimization exercises. After each optimization step, the players' conjectured response matrix components are adjusted, making use of the information on the players' response displayed in the previous optimization exercise. This corresponds to the iterative construction of the self-consistent Nash equilibrium outlined in Section 7.5.3.

Explicitly, we consider the following iterative scheme. Assume that at the n'th iteration step, a set of values $v_i^{(n)}$, $\tilde{M}_{jii}^{(n)}$ corresponding to a conjectured response Nash equilibrium, in accordance with eqs. (7.42), (7.43), has been determined:

$$v_i^{(n)} = f_i(v_j^{(n)}, \tilde{M}_{ji}^{(n)}), \tag{7.55}$$

where $i = 1, 2$, and the index $j = 2, 1$, $j \neq i$ denotes the complementary index.

From (7.55) we can determine the true response relations of the actors,

$$M_{ji}^{(n)} = \partial f_j(v_i^{(n)}, \tilde{M}_{ij}^{(n)})/\partial v_i^{(n)}. \tag{7.56}$$

Since there exists no non-degenerate self-consistent interactive Nash equilibrium for this problem, the inequality $M_{ji}^{(n)} \neq \tilde{M}_{ji}^{(n)}$ must hold for at least one pair of indices i, j ($i \neq j$). If the iteration procedure was intended for a system for which a finite self-consistent interactive Nash equilibrium was expected to exist, one would simply set $\tilde{M}_{ji}^{(n+1)} = M_{ji}^{(n)}$ and seek new solutions $v_i^{(n+1)}$ of eq.(7.55) using the updated conjectured response coefficients.

However, since convergence to such a solution is not possible in the present
case, a more cautious update

$$\tilde{M}_{ji}^{(n+1)} = \epsilon M_{ji}^{(n)} + (1 - \epsilon)\tilde{M}_{ji}^{(n)}, \tag{7.57}$$

of the response coefficients, with constant $\epsilon \ll 1$, is more appropriate. Depending on the value of ϵ, the relation (7.57) ensures a slow migration of
the updated control variables $v_i^{(n+1)}$, determined as the solutions of updated
conjectured-response equilibrium equations

$$v_i^{(n+1)} = f_i(v_j^{(n+1)}, \tilde{M}_{ji}^{(n+1)}), \tag{7.58}$$

along some path in the control-variable plane v_1, v_2. The path must converge
either to a limit cycle or to some point on the boundary of the region defined
by the inequalities (7.28), (7.29)/(7.46).

Figure 7.10 shows the evolution of the control variables v_1, v_2 (left panel)
and the conjectured response matrix elements $\tilde{M}_{12}, \tilde{M}_{21}$ (right panel) for the
parameters $\alpha = \gamma = 2$, $\beta = 4$ and the iteration parameter $\epsilon = 0.1$, starting
from a distribution of initial values in the permitted region of the v_1, v_2
plane. The iteration is seen to converge for all starting values to the point
$v_1 = 3$, $v_2 = 0$ on the edge of the permissible region, while the response
coefficients tend to the limiting values $\tilde{M}_{12} \to \infty$, $\tilde{M}_{21} \to 0$. Different values
of the iteration parameter ϵ have little influence of the curves as long as the
parameter is small. The iteration (7.57) also converges to the same endpoints
(although no longer as a smooth curve) for larger values of ϵ, even for the
maximum value $\epsilon = 1$, corresponding to an update without memory.

For small ϵ, the iteration (7.57) may be expressed as the discretization
of a pair of differential path equations for the response coefficients. Writing
(7.57) in the form

$$\left(\tilde{M}_{ji}^{(n+1)} - \tilde{M}_{ji}^{(n)}\right)/\epsilon = M_{ji}^{(n)} - \tilde{M}_{ji}^{(n)}, \tag{7.59}$$

and substituting the relations (7.52), (7.53), the iteration eqs.(7.59) can be
recognized as the discretized first order, forward difference equations of the
pair of coupled path equations

$$\frac{dx}{d\tau} = -x - \frac{1}{\gamma + y}, \tag{7.60}$$

$$\frac{dy}{d\tau} = -\frac{1}{x} - y, \tag{7.61}$$

where $x = \tilde{M}_{21}$, $y = \tilde{M}_{12}$ and the discretization of the path parameter τ is
given by $\tau^{(n)} = n\epsilon$.

For $\tau \to \infty$, the asymptotic path is given by $x(< 0) \to 0$, $y \to \infty$. Since
$\frac{dx}{d\tau}$ also approaches zero for large τ, it follows from (7.60) that $x(\gamma + y) \to 1$.
This yields for the associated control variables, from eqs (7.44), (7.45),

$$v_1 \to (\alpha + \beta)/2, \tag{7.62}$$

$$v_2 \to 0, \tag{7.63}$$

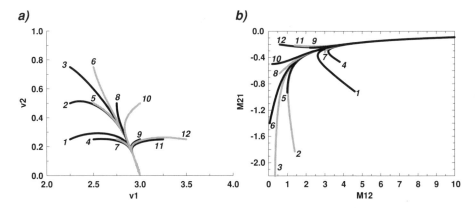

Fig. 7.10. Evolution of control variables (left panel) and conjectured response coefficients (right panel) for an iterative interaction strategy given by eq.(7.58) for different starting values and the iteration parameter $\epsilon = 0.1$

for $\tau \to \infty$, in accordance with our numerical iteration result.

Since $v_2 \to 0$ for $\tau \to \infty$, it follows, as in the case of the non-interactive Nash solution (7.36), (7.37), that the welfare vanishes for both actors in the degenerate limiting case of a self-consistent interactive Nash equilibrium. Thus, it appears that complete foresight in the non-cooperative attempt of each actor to optimize his or her own welfare yields a solution of the 'prisoner's dilemma' type: the welfare of each actor is zero, but cannot be increased, given the iterative strategy of the other actor.

We conclude that for this elementary problem there exists neither a non-trivial non-interactive Nash equilibrium nor a finite self-consistent interactive Nash equilibrium. However, conjectured response equilibrium solutions with finite welfare for both actors exist for arbitrary choices of the conjectured response matrix.

7.5.4 The Two-Actor Fossil Fuel Producer-Consumer Problem

In the following, we generalize the simple supplier-consumer example of the previous section to a somewhat more realistic fossil fuel producer-consumer model, including now both the time dependence and explicit expressions for the climate damages and mitigation costs. The world fuel producers are assumed to be organized in a cartel, operating as a single actor, while the consumers have similarly agreed on a joint strategy, also acting in unison.

In contrast to the cooperative single actor case, the abatement measures of the fuel consumers can be counteracted in this scenario by the fuel suppliers, who can respond to reduced consumption measures by stimulating consumption through reduced fuel prices (cf. Blank and Ströbele, 1994, Richels *et al,*

1996).

Generalizing eq.(7.26) to the time dependent case, the welfare expression for the fuel producer is given by

$$W_1 = \int \alpha p(t)e(t))D_a(t))dtt, \tag{7.64}$$

where $p(t)$ (the control variable v_1) is the price of a unit of fossil fuel associated with a unit of emissions $e(t)$ (control variable v_2), and we have neglected here for simplicity the price of extraction. The notation is the same as in Section 7.2, the earnings being discounted with the same discount factor $D_a(t)$ (eq. (7.9)) as used for the abatement cost function in the single-actor case. A normalization factor α of dimension [(emissions × price]$^{-1}$ has been introduced to relate W_1 to the dimensionless climate damage and abatement costs of Section 7.2.

The welfare of the fossil fuel consumer, actor 2, is given by the difference

$$W_2 = W - W_1 \tag{7.65}$$

between the welfare W (negative costs), as defined previously for the single-actor case, and the payments W_1 to the producer. In the cooperative case, the dependence on price of the total welfare $W = W_1 + W_2$ vanishes, as pointed out: the supplier's gain is the consumer's loss.

In the non-cooperative case, each actor tries independently to maximize his individual welfare expression: the fuel supplier by selecting a price path $p(t)$ that maximizes his earnings, the fuel consumer by choosing an emission path $e(t)$ that minimizes the sum of her abatement, climate damage and fuel costs. As in the previous example, it can immediately be seen that there exists no non-trivial non-interactive Nash equilibrium solution: for given $e(t) > 0$, the earnings of the fuel supplier increase monotonically with increasing $p(t)$.

A meaningful Nash equilibrium must, therefore, again be sought as an interactive equilibrium. We have not investigated whether a self-consistent interactive Nash equilibrium exists for this problem, but suspect that no such solution exists: the same instability to continual mutual adjustments of price and consumption found in the previous simple model may be expected to apply also to this more complex case. We have accordingly investigated a conjectured response strategy, for which a Nash equilibrium can be found.

In the previous simple model, we considered as special case a conjectured response scenario in which the consumer ignores the response of the producer to consumption changes, while the producer correctly predicts the consumer's response to price changes. This corresponds to the normal market situation, in which the producer monitors the response of the consumer to price changes, while the individual consumer, realizing that she has negligible impact on total consumption, ignores this back-interaction. The net result is that the consumers as a group also ignore their impact on price, although the net impact is, of course, no longer negligible. We investigate in the following the complementary situation, more appropriate for a set of fossil fuel consumers pursuing a cooperative abatement strategy, in which the consumer correctly anticipates the response strategy of the producer, while the producer has no detailed information on the consumer's response. The producer assumes instead a linear response relation

$$\delta p(t) \rightarrow \delta e(t) = -\eta \, \delta p(t). \tag{7.66}$$

with some (possibly time dependent) marginal response coefficient η.

As maximization condition for the fuel supplier's earnings, eq.(7.64), we obtain

$$
\begin{aligned}
\delta W_1 &= \int \alpha \left\{ \delta p(t)e(t) + \delta e(t)p(t) \right\} D_a(t)dt \\
&= \int \alpha \delta p(t)(e(t) - \eta p(t))D_a(t)dt = 0,
\end{aligned}
\tag{7.67}
$$

or

$$e(t) - \eta p(t) = 0, \tag{7.68}$$

so that

$$p(t) = e(t)/\eta. \tag{7.69}$$

The earnings of the fossil-fuel supplier for this optimal price is given by (cf. eqs (7.64),(7.69))

$$W_1 = \int \alpha \eta^{-1} e^2 D_a dt, \tag{7.70}$$

or, expressed in terms of non-dimensional variables,

$$W_1 = \int \beta r^2 D_a dt, \tag{7.71}$$

where $r = e/e_A$ and

$$\beta = \alpha \eta^{-1} e_A^2. \tag{7.72}$$

The coefficients α and η occur in (7.70) and in all following dimensionless relations only in the non-dimensional combination $\alpha \eta^{-1} e_A^2 = \beta$. In the numerical examples presented below we shall, therefore, assume that the dimensional parameters η and α scale with the BAU emissions (the only relevant externally prescribed dimensional variable) such that the non-dimensional parameter β remains constant.

Since the fossil fuel consumer is aware of the fuel supplier's strategy, she can calculate the marginal response of the fuel supplier to a marginal change in emissions from eq.(7.69):

$$\delta e(t) \rightarrow \delta p(t) = \delta e(t)/\eta. \tag{7.73}$$

Substituting (7.73) into (7.64),(7.65), we obtain then for the gradient of the cost $C_2 = -W_2$ for the fuel consumer

$$g_2(t) = -\frac{\delta W_2}{\delta e} = g(t) - g_1(t) \tag{7.74}$$

where $g(t) = -\delta W/\delta e$ is the gradient of the total costs (sum of climate damage and abatement costs), or negative total welfare, as defined for the single-actor case, and

$$g_1(t) = -\frac{\delta W_1}{\delta e} = -\alpha\left(p(t) + \eta^{-1}e(t)\right)D_a(t) \tag{7.75}$$

is the additional cost gradient term arising from the price interaction with the fuel supplier. The term proportional to $p(t)$ in eq.(7.75) represents the incremental costs transferred to the fuel supplier for an incremental increase in consumption (emission), while the term proportional to $\eta^{-1}e(t)$ represents the feedback term resulting from an increase (decrease) in fuel price by the fuel supplier in response to an increase (decrease) in consumption.

Applying (7.69) to eliminate $p(t)$, eq. (7.75) may be written more simply

$$g_1(t) = -2\alpha\eta^{-1}e(t)D_a(t), \tag{7.76}$$

or, in terms of non-dimensional parameters,

$$g_1(t) = -2\beta e_A^{-1}rD_a, \tag{7.77}$$

Since both gradient contributions g and g_1 in the expression (7.74) for the fuel consumer cost gradient g_2 are given as functions of the emissions $e(t)$, the minimal cost $C_2 = min$ with respect to the emissions $e(t)$ can be determined using a method of steepest descent.

However, before carrying out the computations we must modify our definition of the abatement cost function. In the cooperative single-actor case, the abatement costs were defined as the additional costs incurred through a deviation from the BAU emission path $e_A(t)$, where $e_A(t)$ maximized the total welfare W in the absence of climate damage costs at the value W_A. Thus, the functional derivative $\delta C_a/\delta e$ with respect to the emissions path $e(t)$ vanishes for the BAU path $e = e_A$. However, in the fossil fuel supplier-consumer problem, the emissions are controlled by only one of the two actors, the fossil fuel consumer, who will chose the BAU emission path e_A to maximize its own welfare $W_2 = W - W_1$, rather than the total welfare W. If the expression (7.8) for the abatement costs in the single-actor case is retained also for the two-actor case, W_2 will not be maximized for $e = e_A$, since for this path $\delta W_2/\delta e = -\delta W_1/\delta e = -2\beta e_A^{-1}D_a\Delta t \neq 0$ (cf. eq. 7.71).

The reason for this is that the optimal BAU emissions path for the fuel consumer is depressed relative to the single-actor BAU path $e_A(t)$ through the negative fuel cost term W_1 in the welfare expression W_2. If we wish to recover the original single-actor BAU path also for the two-actor case, the depression factor must be appropriately balanced. This can be achieved, for example, by rescaling the ratio $r = e/e_A$ in the abatement costs expression (7.8): $r \to \tilde{r} = \lambda r$. The rescaling factor $\lambda \, (< 1)$ can be chosen such that one recovers $r = 1$ as the optimal solution for the BAU case.

With this modification, the net costs for actor 2 in the unregulated case, i.e. without climate damage costs, are given by (cf. eqs. (7.8),(7.71))

$$
\begin{aligned}
C_2 &= C_a(\lambda r) + W_1 \\
&= \int \left\{ \left(\tfrac{1}{\lambda r} - \lambda r \right)^2 + \tau_1^2 (\lambda \dot{r})^2 + \tau_2^4 (\lambda \ddot{r})^2 + \beta r^2 \right\} D_a dt
\end{aligned}
\tag{7.78}
$$

The necessary condition for the minimization of C_2 is accordingly (noting that the first and second time derivative terms of r do not contribute to the gradient if λ is chosen such that $r \equiv 1$, i.e. $\dot{r} = \ddot{r} = 0$, for the optimal solution)

$$
\frac{\delta C_2}{\delta r} = 2\lambda \left(\lambda r - \frac{1}{\lambda r} \right) \left(1 + \left(\frac{1}{\lambda r} \right)^2 \right) + 2\beta r = 0,
\tag{7.79}
$$

or

$$
(\lambda r)^4 - 1 + \lambda^2 r^4 \beta = 0,
\tag{7.80}
$$

which yields

$$
r = \left(\lambda^4 + \beta \lambda^2 \right)^{-1/4}.
\tag{7.81}
$$

Thus, to recover the BAU solution $r = 1$ we must set

$$
\lambda^4 + \beta \lambda^2 = 1,
\tag{7.82}
$$

or (taking the relevant positive real root)

$$
\lambda = \left\{ -\frac{\beta}{2} + \left(\frac{\beta^2}{4} + 1 \right)^{1/2} \right\}^{1/2}
\tag{7.83}
$$

With this modification, the welfare of the fossil fuel consumer is given by

$$
W_2 = W_A - \tilde{C}_a - C_d - W_1,
\tag{7.84}
$$

where W_A is the total welfare in the BAU case, the abatement costs are given by $\tilde{C}_a(r) = C_a(\tilde{r})$ (eqs.(7.8), (7.83)), the climate damage costs C_d have the same form (7.10) as in the single-actor case, and the fuel costs W_1 are defined by (7.71). The expression (7.84) may be written as

$$
W_2 = W_{2A} - C_{2a} - \Delta W_1 - C_d,
\tag{7.85}
$$

where

$$
W_{2A} = W_A - \tilde{C}_a(1) - W_1(1)
\tag{7.86}
$$

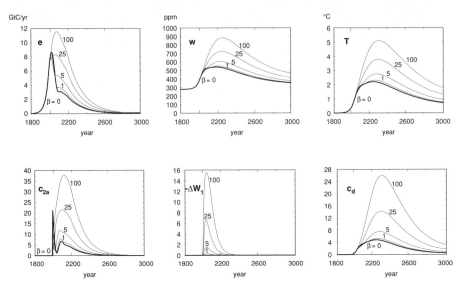

Fig. 7.11.   Impact of the price-consumption response parameter β on the conjectured-response optimal emissions solution for the non-cooperative two-actor consumer-supplier problem. The solutions are computed for the parameters of the single-actor baseline scenario $S0$ shown in Fig. 7.5, reproduced here as the thick curves $\beta = 0$. (From HH.)

denotes the welfare value of actor 2 in the BAU case, $r = 1$, without consideration of the climate damage costs,

$$C_{2a} = \tilde{C}_a(r) - \tilde{C}_a(1) \tag{7.87}$$

represents the direct abatement costs for actor 2 arising from a deviation from the BAU path, and

$$\Delta W_1 = W_1(r) - W_1(1) \tag{7.88}$$

denotes the change in fuel costs arising from the change in emissions relative to the BAU path. All cost terms are defined as functions of the emissions path $e(t)$, so that the welfare W_2 can be maximized with respect to $e(t)$.

Figure 7.11 shows the computed optimal solutions for different constant values of the non-dimensional price-consumption response parameter β. The smaller the marginal response coefficient η, i.e. the larger the parameter β, the larger the impact on the optimal emissions path. Small values of η imply that fossil fuel suppliers respond to reductions in fossil fuel use with large decreases in the fossil fuel price. This stimulates consumption, thereby counteracting abatement measures.

The impact of the fuel costs on the computed optimal emission paths is seen to increase at a less than linear rate with β. This is an artifact of our side condition that, in the absence of climate damage costs, the BAU optimal emission curve $e_A(t)$ should remain the same, independent of the response parameter β. Our adjustment of the abatement cost expression to satisfy this condition (eqs. (7.78) - (7.88)) has the effect that, as β is increased, the abatement costs grow more rapidly with decreasing emission reduction factor $r = e/e_A$: the economic system becomes 'stiffer', increasing the costs of counteracting the negative effects of climate change by reducing emissions. This cross-coupling of two opposing effects could presumably be avoided in a more realistic economic model, including a specific description not only of the costs incurred through a deviation from the BAU path, as in the present analysis, but also of the basic economics determining the reference BAU path itself.

Specific implications for policy can clearly be derived from interactive optimization analyses of the type presented here only if they are based on more realistic economic models than we have used. Nevertheless, we anticipate that the general methodological approach and some of the qualitative conclusions will carry over to more detailed quantitative models.

7.6 Summary and Conclusions: Multi-Actor Case

The implementation of a global climate protection strategy with optimized emissions of CO_2 and other greenhouse gases in the real world of many interacting, interdependent decision makers with diverse interests and different assessments of climate change impact is a complex multi-actor problem. The optimization problem has been investigated with quantitative models so far only in two limiting cases, both of which are rather far from reality: full cooperation, assuming an agreement has been reached on joint mitigation goals, for which the problem reduces to the single-actor case; and the non-cooperative n-actor problem, that ignores all negotiatory aspects, including the various options of forming alliances. Despite these shortcomings, the two limiting cases are useful in identifying basic structural features of the problem and defining a space of possibilities which may span some of the key conclusions of more realistic multi-actor models.

Augmenting the principal conclusions of our single-actor optimization study summarized in Section 7.4, the main conclusions of our n-actor investigation may be summarized as follows:

- The reduction in abatement measures for non-cooperative strategies relative to the optimal cooperative solution is not as pronounced as may have been anticipated intuitively on the basis of free-rider considerations.

- Thus, in the case of n identical non-trading actors sharing the same assessment of climate change damages and mitigation costs, non-cooperation has only a minor impact for n less than about 10, and the limiting solution of zero abatement is approached only gradually for n greater than about 100.

- A similar picture emerges in the case of a single mitigator representing $1/n$'th of the world economy opposed by the rest of the world pursuing a business-as-usual emissions path. Although the contribution of the single mitigator to global emissions is small, because climate damages grow quadratically with climate change, his incremental mitigation efforts are amplified by the climate change induced by the rest of the world. For $n < 10$ it is even cost effective for the single mitigator to increase his abatement efforts above the normal cooperative level. As n increases, the single mitigator's impact on the climate damage costs decline, and his abatement efforts are reduced accordingly. However, the fall-off is relatively mild, the limiting zero-abatement solution for $n \to \infty$ being approached at a similar slow rate as in the case of n identical non-cooperative mitigators.

- In the general multi-actor case in which several players interact through trade, the non-cooperative Nash equilibrium solution, if it exists, depends on the marginal response matrix characterizing the trade coupling between actors. Three forms of Nash equilibrium may be distinguished: the non-interactive Nash equilibrium, in which the response matrix is set equal to zero; the conjectured-response equilibrium, in which the response of other players to one's own actions is unknown and is therefore conjectured; and the self-consistent interaction equilibrium, for which the assumed response matrix is consistent with the true response inferred from the resulting Nash equilibrium solution. The concepts were illustrated for a simple two-actor producer-consumer problem, for which a non-trivial Nash equilibrium existed only for the conjectured response case.

- In analyzing the impact of trade interactions, one needs to consider not only different fossil fuel users with different climate damage and mitigation cost assessments, but also fossil fuel suppliers. A simple two-actor supplier-user model illustrates that for a suitable (conjectured) consumption-price marginal response relation, reductions of the fossil fuel price by the suppliers can effectively counteract efforts of fossil fuel users to mitigate climate change by curtailing consumption.

7.7 Outlook

The purpose of the present exploratory paper was primarily to clarify some of the basic concepts and problems posed by single- and multi-actor optimization problems in the context of global warming, rather than to offer concrete solutions for particular practical situations. A number of basic improvements in the climate-socio-economic model and other extensions of the analysis are needed before useful quantitative predictions can be made.

In addition to a more realistic multi-sectoral, multi-regional description of the economic system and a regional multi-component representation of climate change, an improved Global Environment and Society (GES) model would also need to simulate the inherent internal variability of the coupled system. This is an essential dynamical feature of both climate and the socio-economic system. It has been shown (Hasselmann, 1976) that long-term fluctuations in the climate system can be generated by the stochastic forcing exerted by short-term random weather fluctuations acting on the slow components of the system (the oceans, biosphere and cryosphere), in analogy with the Brownian motion of heavy molecules excited by random collisions with lighter molecules. Stochastic forcing may be expected to produce also slow fluctuations in the socio-economic system, which similarly contains both slow elements, for example in the form of energy technology or the cultural values of a society, and more rapidly fluctuating components, such as business cycles, consumer fashions, speculation shocks, and other short-term adjustment processes. A realistic representation of the interactions governing the different spectral frequency bands of the natural variability spectrum is an important test of our understanding of the dynamics of the GES system, and our ability to properly represent the response of the system to external anthropogenic forcing.

A further aspect, which would need to be included in more realistic integrated assessment studies, is the problem of decision making under uncertainty. This would need to include the probabilistic assessment of risk and the impact of an anticipated future reduction of uncertainty on the timing of decisions.

Finally, the present analysis needs to be extended to include negotiations between actors, in order to bridge the present gap between investigations of purely cooperative and purely non-cooperative optimization strategies.

Despite the limitations of the present study, we believe that several general features of the optimal emission-path solutions we have presented here will survive later improved insights and more quantitative treatments. These concern, in particular, the implications of the long time scales of the climate response, the general time history and order of magnitude of the reduction in CO_2 emissions required to avert a major global warming, the need to assess future climate damages using only very weakly discounted intertemporal cost relations, if one wishes, to recover optimal emission solutions compatible with a commitment to long term sustainable development, and the central role of conjectured response matrices in determining optimal control strategies in the general multi-actor case.

7.8   Acknowledgements

This work was supported in part by the Office of Naval Research, grant no. N00014-94-1-0541 and the Bundesministerium für Bildung, Wissenschaft, Forschung und Technologie, project ICLIPS, 96/3. We are grateful to Hans von Storch for constructive comments on a first draft of this paper.

Chapter 8

"Mastering" the Global Commons

by Nico Stehr

Abstract

The question of "mastering" the global commons will increasingly become a central socio-political issue, if it has not already attained this status. For example, the dilemmas brought about by anthropogenic climate change are in many ways unprecedented. They call for massive efforts to plan global climate change. In this context, knowledge about the physical nature of global climate changes is adequate in order to move from a comprehension to a solution of the problem. The record shows that past generations, too, have been fascinated with and concerned about the impact of climate on society, as well as, anthropogenic climate change. But these efforts have, for the most part, been informed by the doctrine of climate determinism. In much the same vein, the concept of climate policies as an "optimal control problem" is inadequate. Impact research has to be cognizant of the social construct of climate, as well as, fundamental secular societal changes that profoundly alter modern societies and the value orientations of its citizens. Climate policies as a form of large-scale and deliberate climate change, therefore, have to draw extensively on social science expertise.

8.1 Prologue

The analysis of climate and its possible change prompted by human intervention, as well as, the development of practical knowledge about ways of interfering with nature to mitigate the initial intervention, cannot be separated from an understanding of the evolution of the intellectual and material relationships of society to nature established especially over the course of the last two or three centuries. We are living on a constantly evolving planet. In this process, we are actors and not merely spectators. But what makes us actors and may allow us to be actors in the future is the equally persistent evolution of society. In the context of this analysis, dramatic transformations of the modern economy and, in its wake, possible novel value-orientations of the citizens of modern societies are of particular relevance.

There is little agreement on most of the **practical** issues that follow the present outpouring of scientific efforts informing us what to be concerned about although the question of "what to be apprehensive about" is one of the central troubling topics of our age.

Nor is there agreement on some of the fundamental **theoretical** issues: For example, there is no consensual theory pertaining to the linkages between nature and society, to the fragility or robustness of nature, the relations between resources and economic well-being, or between environment and development (cf. Kates, 1988:9). But there ought to be agreement, that long-term advancement depends on theoretical progress and not merely on solid empirical information about what to fear – be it climate change or any other risk we are supposed to be concerned about. A lack of theoretical consensus does not signal a lack of theory. There is no scarcity of theories. I will focus critically on some of the relevant theories hoping to advance our understanding of the dynamics of the intersections between society and nature under contemporary circumstances.

8.2 Overview

I would like to offer a summary of the topics I plan to address in this presentation, but I also want to highlight some of the issues and perspectives I am unable to deal with in any detail.

It is not only the lack of space that prevent me from discussing a host of interesting and controversial issues that emerge in the context of responding to the possibility of climatic change. A number of these issues have been thoroughly aired, perhaps not directly with climate change in mind but then in relation to environmental changes and policies in general and these discussions are readily available. In some respect, one can even observe an emerging consensus irrespective of ideological position. For instance, it is widely accepted that environmental degradation - once detected and defined as such -

Table 8.1. Paradigmatic differences in theoretical approaches to the nature/society divide (Adopted from Redclift (1988))

Paradigm one	Paradigm two
Control of Dominance over Nature	Harmony with nature
Environment as a resource	Values in nature/biosphere
Economic growth	Non-material goals/sustainability
Ample reserves or substitutes	Finite natural resources
High technology/science solutions	Appropriate/soft technical solutions
Consumer culture	Basic needs/recycling
Central/large scale	Decentralized/small scale
Coercive social/political structures	Participatory social/political structures
One-dimensional causality	Interactive causality

arises from market deficiencies created by the free-rider problem.[1]

Among the more contentious questions I am not discussing are, first, the history, as well as, ongoing debate in the scientific community and elsewhere which disciplinary perspective best captures environmental problems in the social and political arena. Economists of course have taken a lead and have argued that it is possible to consider the environment within governing economic paradigms. Economic perspectives are perhaps even capable of integrating the analysis of natural and economic systems (cf. Pearce, 1991). For example, an extended cost-benefit analysis is said to be of considerable value as a decision-making tool for practical knowledge. Others have argued skeptically that economic models are ill-suited for a marriage with environmental matters (cf. Norgaard, 1985).

Assuming that the economic perspective is applicable at all, the debate among economists and political scientists, as well as, other interested parties then shifts toward the question of whether a free-enterprise orientation toward environmental policy problems is warranted or if the pursuit of environmental ends requires policies that are not only based on free-market prices, private property rights, and a legal system that delineates and defends such conditions and rights but on government regulation in the form of administrative instruments and economic policies.[2]

---

[1] Yet, this broad assertion ignores contextual factors and tends to reduce environmental dilemmas to property rights. It is important to be reminded, for example that all "good" things in life are stratified and the way in which they tend to be stratified varies enormously.

[2] An affirmative answer to this question may be found in a collection edited by Block (1989); a much more skeptical response is for example Weale's (1993) assessment of the positions defended by the authors of the Block anthology. From the perspective of classical liberal, neoclassical, the Austrian marginal utility or public choice economic theories, the answer is always quite straightforward, a market driven response to environmental degradation is superior in the sense of being more effective than are government designed and administered instruments.

Nor am I able to offer a comprehensive account of pertaining to legal or geographical discourse or, for that matter, the question of international relations and organizations as they become pertinent for any response to climate change.

The boundaries I am drawing, signal that I will approach my topic from the perspective of the cultural sciences and not those offered by political science, economics or law.

Finally, I will not refer to immense variety and diversity of political positions that tend to be associated with modern environmental/ecological concerns and environmental movements (cf. Paehlke, 1989:194; Harvey, 1993:20).

The issues I will discuss in some detail are:

Do we need to re-examine the location of nature in social science discourse? One of the more difficult issues that arises in any effort to mobilize social science expertise in climate research activities concerns the essentially contested concept of nature and natural within recent traditions of social science discourse.

Is climate impact research a new scientific discipline? I will show that it is not new but a forgotten "science" that has fascinated countless generations in many societies. However, past climate impact research was mainly of the "climatic determinism" genre, a paradigm which disappeared for good reason from scientific discourse, for example because of its intimate relation to racial theories. But even if it disappeared from the scientific agenda, climate determinism is still a most vivid concept among the public in contemporary society and that includes decision makers and politicians.

Is it sensible to consider the social consequences of global warming pre-eminently as an "optimal control problem" requiring the construction of "climate policies" that balances expected abatement costs against expected climate change damage costs? I will assert that framing the issue of climate change within the context of such an understanding of economic theory and development is flawed, because it disregards the dynamics of social value attribution, the importance of social and cultural inequalities, as well as, differences in economic and political power and, last but not least, profound changes in the social structure of modern societies. In addition, the "optimal control" model severely overestimates, as I will argue, the ability of the state and multi-national organizations to implement policy choices in modern societies because the ability of the state and states to govern, that is, to impose their will, is increasingly reduced.

Can one reconcile ecological, economic and modern life-styles imperatives? Discussing the ambivalent status of nature in social science discourse can easily leave the impression as if the primary route or approach to the issue of mastering the global commons consciously is a mere matter of cognitive or intellectual change. Undoubtedly, developments beyond the current intellectual division of labor in science are important but also equally significant are changes in modern society itself that, in the end, may dramatically alter what are now often seen as contradictory trajectories of social development and/or entrapments of one sort or the other, that are almost inescapable and iron clad limits of social, political and economic action. I want to show that significant societal developments allow for the possibility that traditional linkages between nature, society and the economy become uncoupled.

8.3 Introduction

It may be said that the issue of climate change and appropriate political, social, economic and legal strategies to cope with changes in the global climate to which climate sciences have alerted us (that is, the variety of national, international and regional social institutions that characterize the complex make-up of modern societies) do not represent a novel problematic - although in other respects it clearly constitutes a new way of comprehending our natural environment. What is not new and what, therefore, has numerous precedents in the history of science and in politics is the demand that natural phenomena be mastered for the benefit of humankind. As a result I have called my essay "'Mastering' the Global Commons." The demand or need to intervene in nature in order to arrest, slow or reverse climate processes in operation amounts to the demand to master nature in a certain fashion.

My essay is also concerned with the transition of authoritative scientific analysis of "climate change" issues into the arena of public and political discourse. The intention of communicating these findings to the public by climate scientists is not merely to enlighten but to encourage work designed to understand the "damage" climate change imposes on social and economic systems; but is also intended to advocate interventions to mitigate the kind of interference in physical processes that have produced the observed effects, in the first place. We are dealing, therefore, with research and policy tasks on an unprecedented global scale. We are talking about intentional or planned efforts to "manage the global commons".[3] The ultimate aim are societal

---

[3] I am employing a variant of a metaphor from Nordhaus (1994:35) who describes the "task of understanding and controlling interventions on a global scale" as the task of "managing the global commons". The metaphor resonates of course with the term of "the tragedy of the commons" as introduced by the biologist and ecologist Garrett Harding (1968) and the for centuries widely discussed dilemma that common property tends to get less care than individually owned property. Harding's term also refers to contradictions between individual and system interests, in particular, those posed by essentially free and unregulated access to scarce and commonly "owned" resources (for a critique of the model see McCay and Acheson, 1986). My own use of the former term signals that any climate policy that may be devised *itself* amounts to deliberate climate change or, to an effort to effect the self-reorganization and self-regeneration of the natural ecosystem. In distinction to the forces that bring about the changes that prompt our discussion, in the first place, I am of course concerned not with unplanned or unanticipated change but only with planned, managed climate change. Planned climate change requires policies and strategies to implement and realize policies. Whether policies, treaties, education and other measures designed to effect climate can be expected, under present and future socio-

adaptations to changing environmental conditions and/or radical changes in the environment itself. Policies aimed at healing or limiting dangers that may follow from climate change amount to *planned climate change.* [4]

The speed and the success with which climate issues – once discovered after a delay of perhaps a decade by the media (Mazur and Lee, 1993) – have become part of political discourse is unusual.[5] But this is not to say that there are no historical precedents pertaining to efforts to deal in practice with large-scale environmental transformations [6] nor is this to maintain that it is mainly authoritative scientific discourse that structures and frames public and political discourse on the environment. Historical precedents are of value but they cannot substitute for an analysis of the present context within which sometimes vigorous or at other times more lethargic and disinterested communication (cf. Mazur, 1996), as well as, action and inertia about climate issues takes place.

In the following, I take a skeptical stance towards the exclusive relevance of (natural) scientific information about climate and climate change for society and draw attention to the ways in which such information in fact enters into highly contested debates about the ways in which we should respond. The practical political answer to authoritative scientific claims about climate change appear to be quite straightforward, at least from within its own perspective, namely to embark upon an unprecedented form of planned and regulated climate change.

Science may attempt to influence but it cannot choose the ways in which it is interpreted in the public arena nor can it claim with conviction that planned climate change is without its own pains and unanticipated consequences.

---

political conditions, to be effective in reaching its goals depends first and foremost on the ability of all political institutional levels to "impose their will" on societies. There is good reason to surmise that success on this score will be difficult to come by (cf. Stehr and Storch, 1997).

[4]It is perhaps not so much the unprecedented nature and range of environmental matters confronting contemporary societies that is unparalleled but the *discovery of the problem* by science - and not political discourse - that makes many of these concerns unique political issues. Of course, the origins of the (possibility of the) greenhouse effect can also be traced to scientific and technological developments. After all the evolution of industrial civilization is last but not least closely linked to technical developments made possible by science in the last century.

[5]The practical political relevance and the extent to which environmental issues have become an ideological tradition associated with social movements and (new) political parties on par with other modern ideological cleavages varies to a significant degree from society to society. Modern environmentalism is still evolving and it remains to be seen whether one can for example link its success or failure to the degree to which welfare capitalism has developed (e.g. Eder, 1966) in specific societies, the nature of different political systems, the role of mass media and/or other social and political processes.

[6]As a matter of fact, the evolution of industrial societies has from the beginning been accompanied in different countries by the intense public discussion of environmental concerns, for example, from deforestation, soil erosion, resource depletion to the effects of the commercial use of nuclear power.

My observations about "mastering" the global commons should not be interpreted as an attempt to deny the reality of anthropogenic climate change. Instead I am persuaded that anthropogenic climate change due to increasing atmospheric greenhouse gas loading, caused by ongoing anthropogenic emissions, is indeed happening and can be identified in the near surface temperature record (Hegerl et al., 1996), and it is found to be consistent with the best projections by quasi-realistic climate models (e.g. Cubasch et al., 1992, 1995). Figure 6.12 shows the globally distributed warming trend over the last 30 years and indicates that there could well be a signal embedded in the noise of natural climate variability.

Having underlined that I consider the ongoing climate changes as anthropogenic changes, one needs to emphasize that the *statistical significance* of the present and likely future warming trends does not automatically imply that this warming is in any uncomplicated and direct sense significant for society. What matters more for the public debate and public policy, especially for policies designed to deal with the consequences of climate change, is the framing of discourse within the scientific community that specifically deals, on the one hand, with the impacts of climate change in the form of a balance sheet of damages and benefits for social and economic systems and, on the other hand, with the ways in which heretofore unplanned climate change can be transformed into deliberate or managed change. At the same time, of utmost importance in this context are major secular changes in modern society, especially its economic and political realities that occur quite independently of changes in the environment but likely have a tremendous effect on any public discussion let alone societal response to climate change. I plan to deal with this important set of issues in the last section of this paper.

Therefore, any authoritative analysis and more particularly, confirmation of the global warming hypothesis is quickly and closely bound to research and discussions about climate impact and one's assessment of damages and benefits.[7] This implies, in turn, that any examination of the consequences of climate change and the discussion of managed climate change has to be placed firmly within the context of social science discourse. Given the nature of contemporary social science this is by no means an easy task free of major controversies.

8.4   The Location of Nature

It is easy and perhaps even self-evident to call for a more up front and central role of social science expertise in climate research (cf. von Storch and Stehr, 1996). But the modern social sciences have quite an ambivalent relation to

---

[7]Independent of the contemporary global warming hypothesis and attendant demands for adaptation, there has been a long tradition within science aligned with an intensive curiosity about the effects of climate on human conduct. And, there is of course a prior common sense image of the impact of climate in everyday life (cf. Stehr, 1996a).

anything defined as "nature" or "natural". As a matter of fact, genuine advances in climate impact research likely require a shift in some of the dominant meanings associated with nature in social science discourse.

The term nature in the social sciences has a variety of usages: For example, nature often refers to those elements of social conduct that are beyond the control of all of us and, therefore, act as a firm and fixed limit to social conduct.[8] Nature is regular and unchanging. It is permanently frozen in form, outline, and proportion. In social science discourse, this usage of the term nature was a widely accepted assumption in much of Western thought during the nineteenth century and much of our century. Contemporary social science discourse continues to echo this assumption.

But what is natural is at times also seen as that which is authentic and pre-cultural. In that sense, nature reminds us what we were and perhaps could be. Nature as a kind of organic (or aesthetic) home to which we yearn to return. Yet, within the same meaning complex, natural also can indicate that something is lacking, namely that which is cultural or civilizational which sets us rigidly apart from the merely natural.

While these meanings of nature in social science are also linked to contradictory perspectives and projects, for example, in the context of more liberal of conservative political philosophies, these platforms share a basic similarity, namely to excise or set aside nature from social science discourse and natural science discourse.[9]

Even if social science discourse does not attempt to define nature as outside of the boundaries of its interests and the procedures that give it its peculiar identity, much of social science certainly subscribes in most of its paradigms to a strong distancing from nature.

Such distancing is based on at least two good reasons. One related to theoretical reasons, the other to practical social relations: (1) The proven dangers that a mixture of social and natural attributes in discourse almost invariably leads to reductionist claims, for example, discourse exclusively governed by biological considerations. (2) In the course of social and economic development of the last two centuries, modern societies have succeeded in largely isolating humans from nature. For example, humans flourish in all climates of the globe.[10] Thus, if the latter is thrown into question because of

---

[8]The meaning of nature as limit has some affinity to the more recent conception of nature as an "agent" that can be firmly trusted. Nature not only knows best but in the human context becomes a process that is superior to the market and should replace it (cf. White, 1996:140).

[9]See also Grundmann and Stehr (1997). Other usages of "natural" within social sciences discourse, for instance one that equates it with habitual asserting that a routine belief or some form of conduct such as a greeting becomes a kind of "second nature", I will not consider in this context.

[10]As Nordhaus (1994:50) puts it in his examination of the modern impact of climate on economic activity: "For the bulk of economic activity, variables like wages, unionization, labor-force skills, and political factors swamp climatic considerations...moreover, the process of economic development and technological change tend progressively to reduce climate-sensitivity..."

profound climate changes then the usage of nature in social science discourse has to be reconsidered and reformulated in some way,[11] avoiding of course the reductionist trap. In any event, the analysis of the relation between climate and society can never merely be a (natural) scientific and technological issue. One has to be sensitive to the essentially contested ways of using the term nature in the social sciences and we have to be cognizant that specific views of nature constrain and influence reflections about the impact of climate on society.

8.5 Climate Impact Research

If we can conquer climate, the whole world will become stronger and nobler.
Huntington, [1915] 1924:411

Speculation about the considerable effect of climate on humans is evident from the earliest time of civilization. For example, in classical Greece, Hippocrates already suggests in his treatise on "Air, water and places" that knowledge about climate ought to be used to explain the psychology and physiology of humans. The Greeks were convinced that observed differences in habits of life and character between East and West were determined by differences in climate. During the enlightenment period, the educated part of the population of France, Germany and England spent enormous intellectual energy to argue about the climatic determinants of the civilizational peculiarities of entire nations. Philosophers such as Montesquieu in his influential "Esprit des Lois" or Herder in his "Ideen zur Philosophie der Geschichte der Menschheit" advanced widely discussed ideas about the significant constraint that climate represents.

But even in our century, climatic explanations of history and the theory of significant climatic influences on individuals and societies have flourished. While earlier speculation about the impact of climate was largely derived from casual observation, the American geographer Ellsworth Huntington pioneered the quantitative method. In his "Civilization and Climate" (1915), and for more than three decades of vigorous publications (cf. Huntington, 1945), he advanced and defended - despite early criticism (cf. Sorokin, 1928) - the hypothesis that the formation of a civilization would be possible only in areas where favorable climatic conditions prevail. Huntington's conclusions were based for example on a statistical analysis of work records of factory workers and marks of college students. Huntington claimed to be able to show

---

[11]The result of such a reformulation of the role which the concept of nature may play in social science could for example be a clearer understanding of what kinds of climate change dimensions are particularly relevant to impact research. Thus, is information about gradual, especially global small average temperature changes significant or is there a greater need for information about changes in regional climatic extremes? A reconsideration of the role of nature could assist in gaining insights into the ways in which discourse about relation between climate and society is best cast under contemporary circumstances.

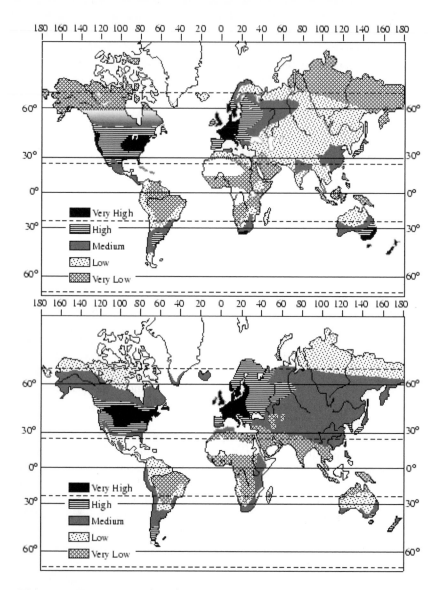

Fig. 8.1. Huntington's (1925) key arguments for his "climate hypothesis of civilization": the global distribution on climatically determined "health and energy" and of "civilization", the latter based on a survey of "experts".
Top: Huntington's "Distribution of Civilization"
Bottom: Huntington's "Distribution of Human Health and Energy on the Basis of Climate"

that humans are most energetic and productive at a temperature of about sixty to seventy degrees Fahrenheit, as well, as a moderate annual range of temperature and the presence of short-term variability. The latter was thought to be stimulating both in terms of mental and physical energy output and health. Such climatic conditions prevail in modern times in western and central Europe, most of North America, to some extent in Japan, and in Australia and some parts of southern South America.

Conversely, Huntington claimed that both physical and mental activity decline with extremes of either heat or cold. As a verification of his hypothesis, Huntington showed two maps (redrawn as Fig. 8.1) displaying the distribution of "health and energy" as derived from climatic conditions, and the distribution of "civilization", as determined by a survey among "experts". Similar ideas were en vogue among scientists in many countries; the German social psychologist Willy Hellpach for instance maintained in an essay entitled "Culture and Climate" (Wolterek, 1938) : *"Prevalent in the North ... are the character traits of sobriety, harshness, restraint, imperturbability, readiness of exertion, patience, stamina, rigidity, and the resolute employment of reason and determination. The prevalent traits of the South are liveliness, excitability, impulsiveness, engagement with the spheres of feelings and imagination, a phlegmatic going-with-the-flow or momentary flare-ups. Within a nation, the northerners are more practical, reliable, but inaccessible, and the southerners devoted to fine arts, accessible (sociable, likable, talkative), but unreliable."* [12] After the fall of racial science, their intellectual siblings geographic and climatic determinism were also discredited. The incorporation of environmentally determined impacts on human behavior is today almost considered a taboo within most of social science discourse. [13] In the natural sciences, the concept survived to some extent. Such a scientific perspective is pursued by "biometeorologists", investigating for instance the effect of heat waves on domestic violence or mortality rates, and by psychological research on "seasonal disorder".

In the contemporary social and natural sciences, the relation between climate, social conduct, value-orientations and abilities is examined, if at all, in a most cautious and circumspect manner while the general public continues to accepts variants of climate determinism. A survey among 2,900 college students from 26 countries conducted by Pennebaker et al. (1995), finds support for the persistent resonance among the young and educated segment of the population for Hellpach's notion and stereotypical image of different Northern and Southern personality types.

Thus, one should not underestimate the relevance of a widespread belief in "climatic determinism", as well as, other relevant cultural climate-

---

[12] For a more detailed description analysis of Hellpach's views see Stehr, 1996a.

[13] Compare for instance the repeated controversies surrounding the thesis of the essential heritability of intelligence, most recently as the result of the publication of the Bell Curve (Herrnstein and Murray, 1994) or, the highly contested debates about the impact of "nature" on criminal behavior.

related doctrines, for example, the notion that climate is constant and that climatic extremes only confirm its essentially normality (see Stehr, 1997), in any decision-making processes in modern society about climate matters and in possible conflict with the "hard" information provided by natural sciences.

Another course of inquiry of earlier climate research and climate impact research has concerned variations, or changes, of climate. Attentive observers detected already in the 18th century that climate is not constant, and researchers speculated about the reasons for such changes. As a result, the dichotomy of natural and anthropogenic climate change was introduced. In 1770, the American physician Williamson, for instance, described a change in climatic conditions in the North American colonies, and linked this favorable change to the ongoing settlement that produced increased drainage and deforestation. Similarly, the saying *"The rain follows the plough"* describes the idea of a beneficial climate change caused by the transformation of the North American prairies into agriculturally managed farm land.

In the 19th century, widespread discussions took place in Europe, in Australia and in North America about climate change due to de- and, sometimes, reforestation (cf. Brückner, 1890). The debate was not confined to the scientific community of the day but found a considerable echo in the media and in politics. Moreover, the discussion was rather similar to the present one about the interpretation of the current warming trend, that is: Are we faced with just another long-term swing in the course of natural variability or is it becoming warmer because of anthropogenic modifications of the environment?

The conviction that the changes were anthropogenic lead in several countries to the establishment of governmental and parliamentarian committees with the purpose to design proper "response strategies". The AAAS demanded a reforestation program to mitigate adverse climatic changes.

The other point of view that climate change would be a matter of natural processes was advocated by other researchers. Eduard Brückner documented that climate would vary for natural reasons on decadal time scales and continental spatial scales. Specifically he proposed a 35-year cycle as the dominant signal of this global variability. Interestingly, after his analysis of the climatological data, he turned his interest to the impact of these climatic variations on health, transportation, international trade, migration pattern etc. [14]

As an example for this type of analysis by Brückner, I reproduce his statistical analysis about the simultaneous variations of rainfall amount in Europe and the arrival of emigrants in the United States in Fig. 8.2.

On the basis of the historical record, it can be concluded:

That climate impact research is by no means a new line of research as is sometimes supposed. It has been pursued for centuries. However, present-day scientists are mostly unaware of these earlier discussions, hypotheses and theories.

---

[14]For a review of Brückner's work see Stehr et al. (1996).

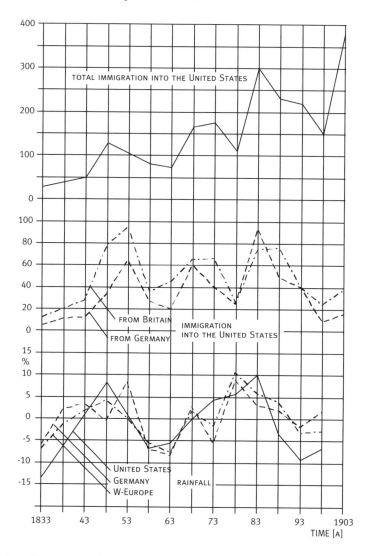

Fig. 8.2. Emigration from Britain and Germany to the United States in the 19th century and rainfall in the US, Germany and Western Europe (cf. Brückner, 1915)

Historical climate impact research maneuvered itself into a blind alley by trying to attribute most or even all social and economic facts, such as health conditions, as well as, an endless variety of patterns of social conduct to climatic (and other geographical) factors. At the same time, there has not been a critical public discussion leading to a questioning of the doctrine of "climatic determinism", or perhaps this discussion did not have much of an impact and has been forgotten.

The significance of these conclusions is that it seems that present day climate impact research to a large extent has tacitly returned to the old concepts, and there is a real danger that it eventually will end up in the same blind alley as their predecessors, who certainly were no less intelligent, educated and careful than contemporary researchers.

8.6 Climate Policies as an Optimal Control Problem?

So far I have dealt with climate as a factor that effects society in ways that are not anticipated, often poorly understood if at all and certainly not planned. As a result, adaptation is passive. Individuals and society respond to climate and its variation in an ad-hoc fashion. However, from time to time people have contemplated to actively influence and change the climate, either to reverse adverse developments or to directly "improve" their climate (e.g. the Soviet plans of rerouting rivers in Siberia). In that sense, there is a history of "managed", and even "planned" climate change. In the following section I will discuss the problem of managing climate change in greater detail.

From a macro-economic perspective, the climate change problem may be conceptualized as a situation in which the creation of an optimum of welfare has the secondary effect of causing damages to the environment. In the case of anthropogenic climate change, the harmful side effect is the result of the emission of carbon dioxide into the atmosphere. These potential damages such as rising sea levels or health risks create the need for adaptation measures, for example, the construction of dikes or a relocation of human settlements. These measures require economic resources that could alternatively be used for the creation of welfare.

The problem is related to the "tragedy of the commons" (Harding, 1968): The collectivity of actors exhausts a resource without individual property rights, in this instance the atmosphere as a dump for gaseous by-products of energy generation. By doing so, individual profits are attained. The effect for the common good, however, results in adverse effects for all concerned, independent of the amount of emissions injected into the atmosphere by individuals or aggregate actors (e.g. corporations).

Assuming no action of all pertaining to emissions, economists refer to a monotonic increase of greenhouse gas emissions, the so-called "business as usual" effect. An alternative would be a treaty signed of the world's governments agreeing on joint policies aimed at limiting damages by way of regulating emissions.

One option in the context of such a scenario is to aim for the "social optimum". The social optimum is an emission plan for the entire world balancing the costs associated with the reduction of emissions with the expected damage costs in the foreseeable future. In more economic terms, a time-dependent emission-path is aimed for, so that the marginal abatement costs equal the marginal adaptation costs. [15]

Hasselmann has condensed this approach into the "Global Environment and Climate"-model (GES), in which two dynamical entities, namely the climate system and the economic system interact (Fig. 7.1). The economic system affects the climate system by wastes such as carbon dioxide, and the climate systems responds with a change of, say, sea level or precipitation pattern. Any waste reduction is associated with costs. Climate changes incur costs as well. The role of public policy is to minimize the total costs, the exact measure of which is left to society. Assuming that the various countries cannot agree on a joint policy, then the problem may be cast into the format of differential game theory (Hasselmann and Hasselmann, 1996).

This "optimal control"-approach is perhaps not only intellectually tempting but may also appeal to policy makers. Undoubtedly, it represents a rewarding and informative perspective for discussing the problem at hand. On the other hand, it functions only on the basis of various assumptions, some of which are implicit assumptions. Some simplifications, such as the absence of natural climate variability, could easily be accounted for by modifying the involved dynamical models. Other static assumptions are more difficult to justify. For example, a major assumption of the model is that future generations will accept our values and our concept of a healthy environment. Indeed, future generations will have difficulty entering the market place. The macro-economic models assume that the assignment of value is mostly constant, perhaps with an exponential discounting, but without a significant change in the relative designation of values for, say, healthy forests and religious prescriptions. But we know of course that societal values undergo complex and hardly predictable transformations (see also the next section). What is of utmost relevance for significant segments of the public at large today, may be totally irrelevant even a few years or decades later. In other words: Models like GES lack a module describing not only the dynamics of social value assignment, for example, as the result of learning processes, but also ongoing significant transformations of the social structure of modern societies.

To illustrate the general point, I refer to an example from medieval times (Stehr and von Storch, 1995): In the early part (1315-19) of the 14th century, parts of Europe suffered from severe weather induced shortages of food; the problem was severe in England, among other countries. A persistent anomalous cyclonic circulation over Central Europe brought unusual cold and rainy conditions for the summer with disastrous consequences for the harvests.

---

[15] The idea was pioneered in economics by Nordhaus (1991) and in climate research by Hasselmann (1990). Cast in these terms, the climate problem reduces itself to an optimal control problem, with the emission path as control variable and climatic conditions as state variables. The case of a transient evolution of a highly idealized system has been worked out by Tahvonen et al. (1994).

The hostile climatic conditions were interpreted by the contemporary authorities, i.e., the church, as an optimal control problem. The adverse climate was seen and understood as being brought upon society by God in response for sinful conduct. In a sense, society was confronted with anthropogenic climate change. Any "business-as-usual" response would be associated with unbearably high "damage costs" (famine, epidemics, high mortality apart of unfavorable perspectives such as the purgatory). Thus, the damage, or adaptation costs, were assessed as being infinite.

Abatement measures considered were a general Christian betterment of people's life style. (Analogous to the present situation, such an abatement policy was considered as generally benevolent apart from the immediate harvest problems it might remedy.) Also, the costs of such a course of action were perceived as considerably smaller, if not negative, than the expected damages. Consistent with such a perspective, the authorities advised their flock *"to atone for their sins and appease the wrath of God by prayer, fasting, alms giving, and other charities"* (Kershaw, 1973).

After a few years, climate conditions returned to normal. One can easily imagine that these developments then counted as strong evidence for the public and the authorities that their climate policy was successful.

Within the context of our contemporary knowledge about climate dynamics, the 1315-case appears rather absurd. But we cannot really be confident that our comprehension of many present environmental crises, and their management by society and policy will not appear to future generations equally incongruous. Indeed, what can be learned from this case, is that the GES model (Fig. 7.1) is overly simplistic because it implicitly assumes that the costs refer to actual processes. What happens in reality is, however, that the costs are estimated for the perceived processes, and that this understanding is subject to its own dynamics largely independent from the real processes. Therefore, the GES model should be modified to the PES ("Perceived Environment and Society") model, as sketched in Fig. 8.3. The effects of human activity on the environment, as well as, the changes of the environment are explained to the public by authorities, in the case of the present-day environmental crises, the scientific advisory committees such as IPCC, and any subsequent policy is effected but not determined by this interpretation.

The authoritative scientific interpretation does not settle the nature and the execution of climate policy. Instead stake-holders confront the received interpretations with their own cognitive models and doctrines, that is, their understanding of many processes and interests that may or may not be related to the problem at hand. This dynamic mixture can be called the "social construct" of climate (or of the environment generally). It is this complex "social construct" which ultimately affects the design of and the compliance with any climate policy. Thus, the mapping of the social construct, in different times and societies, is of utmost importance for a successful solution to the "climate problem".

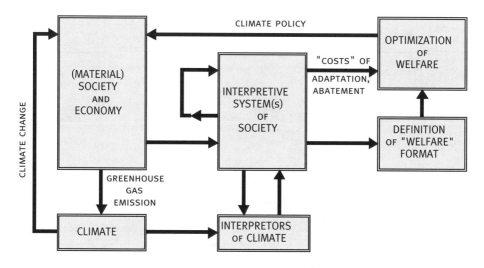

Fig. 8.3. The "Perceived Environment and Society" (PES) model, which deviates from the GES model in Fig. 8.3 by two additional boxes representing societal processes

I conclude, therefore, that models such as GES are informative and useful to discuss the general format of the problem, but lack crucial modules, namely the modules that would assist in describing the evolution of social preferences and relevant socio-structural transformations of society. For a few years, in times of accelerating social change, value assignments may be taken to be constant but beyond that time scale this process is likely to exhibit significant variations created by social, economic, political and cultural processes.

8.7 Linking Nature, the Economy and Society

One of the more serious drawbacks of any model that attempts to capture societal responses to the major environmental challenges that face modern society, be it in the form of integrated assessment models or any other mode of representation, is the lack of any general conception of the fundamental ways in which theses societies are bound to evolve in the next decades.[16] For

---

[16]The chapter that deals with the role of integrated assessment models in the most recent assessment report of the Intergovernmental Panel on Climate Change (IPCC) makes reference to various purposes these models are supposed to live up to. Among these goals is "the coordinated exploration of possible future trajectories of human and natural systems" (cf. Bruce et al., 1996:371). However, the remainder of the chapter relegates the discussion of future societal trends to insignificance; in fact the explication of the variety of models that are being developed indicates that the relative intellectual energy that is invested in tracing such trajectories is very limited indeed.

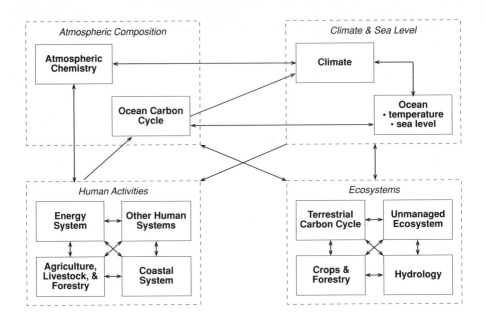

Fig. 8.4. Key elements of full scale integrated assessment models

the most part, such discussions are relegated to the margins of the reflections if they occur at all.

On the other hand, broadly speaking, the links between progress and nature are highly contentious issues. We have moved from the widely accepted conception that progress equals the conquest of nature to the equally universal assertion that progress constitutes the end of nature.[17] The discovery, as Christopher Lasch (1991:529) has pointed out for example, that the global ecology will not sustain an indefinite expansion of the productive forces, represents, therefore, for some observers "the final blow to the belief in progress". The negative by-products of progress, especially economic progress, now loom as large or larger than the ostensible benefits. The bitterness with which defenders of the economic growth from time to time attack environmentalists and the virulence with which they in turn are abused by defenders of progress unmistakably signals the profound contradiction in the

---

[17]Richard White (1996) has attempted to interpret the history of ideas, at least in the case of the United States, in a much broader sense. Such a reading does not allow for the simple reduction of the notion of progress to coincide with the conquest of nature. For example, less than centuries ago, to many observers and not only Robert Malthus it appeared that nature was about to dispatch with progress. In the United States, contrary to the current pessimism, a harmonious linkage between nature and culture became a very common idea and the "conquest of nature was not only a recipe for progress, but also a corrective to the dangers of progress" (White, 1996:125).

minds of many between economic development, particularly in the form of excessive consumption and environmental sustainability. The apparent difficulties in achieving what are widely seen as contradictory goals are reflected in contrasting interests pursued by political parties, social movements, business and other social institutions.

The contradictions between economy, society and ecology also find their expression in the scientific community: The main theoretical perspectives advanced by say economics, ecology and sociology do not seem to converge into a common framework. On the contrary, each discipline advocates different frameworks: The economists emphasize the preeminent role of the market, the ecologists or environmentalists stress the overriding significance of the biosphere while sociologists consider political or normative issues to be of utmost importance (cf. Mulberg, 1996). We cannot expect that a common framework for the pressing issues that link the economy, society, and the environment will emerge from opposing disciplinary concerns.[18]

Moreover, and turning attention to the existing economic, social and ecological conditions, prevailing conceptions often either assume that established economic, political and social trends can be extended into the future, creating a kind of business-as-usual scenario for the realm of social, economic and political developments or, the same trends are seen to come abruptly to a catastrophic end. But even such a projection is but a variation of the business-as-usual conception except that the outcome is viewed much more pessimistically.

What correct is about this scenario, is that there is every indication that economic growth - in terms of per capita income and wealth, as well as, in terms of the quality of goods and the amount of leisure - is accelerating and will continue to speed up. However, the foundations for the persistence of economic growth will not be what they were in the past, namely driven by an exploitation of the traditional factors of production labour and property. Growth increasingly is driven by knowledge or, as some may prefer, more narrowly, by technological progress that feeds upon itself. Any new idea makes the development of subsequent knowledge easier. I have called the trend, I have in mind, the trajectory toward the emergence of modern societies as knowledge societies.

---

[18]Robinson and Tinker (1997) in their paper on reconciling ecological, economic and social imperatives make the point that it is possible to consider the "biosphere, the market and human society as three interacting 'prime systems', sharing many common characteristics, and each co-equal to the others in that each has an equivalent primacy and importance." In contrast to Robinson and Tinker, I emphasize the degree to which the now contradictory imperatives converge or, are moving toward reconciliation as the result of changes in the trajectory of modern society; for example, values change as does the economy enabling novel links between society and the economy or the environment. Robinson and Tinker stress intentional changes that are required, for example, in terms of policies and strategies designed to achieve reconciliation. My emphasis is on unintended outcomes.

In other words, if there is any meaningful, broad-based assertion one is able to offer about societal trends, it would be the assertion that the tempo of social and economic change is accelerating. Nor can there be any doubt that these changes in turn will have a significant impact on how, if at all, it will be possible to "master" the global commons.

These changes, particularly, those observed in the area of economic production and economic relations, perhaps afford a way of reconciling conflicting imperatives as the unintended outcome of these very changes.

Central to my analysis is the thesis that the origin, the social structure and the development of knowledge societies is linked foremost to a radical transformation in the **structure of the economy** including a set of novel and largely unintended consequences, for example in the area of terms of trade, inflation, productivity, competitiveness and employment (cf. Stehr, 1994).

Productive processes in **industrial society** are governed by a number of factors all of which appear to be on the decline in their relative significance as conditions for the possibility of a changing, particularly **growing** economy: The dynamics of the supply and demand for primary products or raw materials; the dependence of employment on production; the importance of the manufacturing sector which processes primary products; the role of labor (in the sense of manual labor); the close relation between physical distance and cost and the social organization of work; the role of international trade in goods and services; and the nature of the limits to economic growth. The most common denominator of the changes in the structure of the economy seems to be a shift from an economy driven and governed, in large measure, by "material" inputs into the productive process and its organization to an economy in which transformations in productive and distributive processes are determined much more by "symbolic" or knowledge based inputs and outputs. However, social science discourse and official data collection still tend to think of economic activities primarily in terms of the production of commodities.

The economy of industrial society is initially and primarily a material economy and then changes gradually to a monetary economy. Keynes' economic theory, particularly as outlined in his General Theory (1936), reflects this transformation of the economy of industrial society into an economy affected to a considerable extent by monetary matters; it becomes, as evident recently, a symbolic economy. The changes in the structure of the economy and its dynamics are increasingly a reflection of the fact that knowledge becomes the leading dimension in the productive process, the primary condition for its expansion and for a change in the limits to economic growth in the developed world. In short, the point is that for the production of goods and services, with the exception of the most standardized commodities and services, factors other than "the amount of labor time or the amount of physical capital become increasingly central" (Block, 1985:95) to the economy of advanced societies.

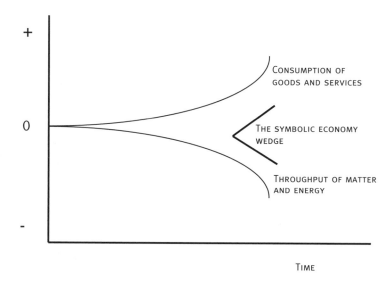

Fig. 8.5. The Symbolic Economy Wedge. Adopted from Robinson and Tinker, 1997

The striking and relevant change here is the "uncoupling" of the raw material economy from the industrial economy. The uncoupling has been accompanied in recent decades, perhaps slowed, by a secular decline in the price of commodities when compared to the price of manufactured goods. The decline of commodity prices has been uneven. It has been particularly strong in the case of metals (cf. Grilli and Yang, 1988; for recent trends see IMF, 1992:86-87). In general these developments imply that the recent "collapse in the raw materials economy seems to have had almost no impact on the world of industrial economy" (Drucker, 1986:770). The traditional assumption of economists has of course been that changes in the price structure, most surely dramatic changes, ought to have a profound impact on the cycle of economies. However, the significant decline in the price of most raw materials has not brought about an economic slump, except perhaps in those countries which rely to a large extent on trade with raw materials. On the contrary, production has grown. The results of these developments produce what is called the "symbolic economy wedge".

A noteworthy and relevant second development that affects dominant value-orientations and life-style choices in modern society relates to the growing affluence (cf. Stehr, 1996b). I am referring to the relative decline in the immediate and unmediated importance of the economy for individuals and households. I mean a decline in the direct material subordination of individuals and households on activities centered in the market economy, in particular, their occupational roles and, therefore, their dependence on what is still for many their basic role as economic actors. What diminishes is the tightness of the linkage in the material dependence of many actors on their occupational status only and what increases is the relative material emancipation from the labor market in the form of personal and household wealth. The decreasing material subordination to one's occupational position of course not only affects those who work but applies with even greater force, paradoxically perhaps, to the rising segment of the population who is out of work and who, therefore, involuntarily is cut off from the labor market.

Compelling evidence illustrating the extent and the relative significance of this transformation are difficult to obtain because considerations of the distribution of personal wealth, households assets, various entitlements etc. still have mainly been driven, for ideological reasons, by an interest in the concentration of wealth, especially the proportion of the wealth controlled by the upper percentiles of the wealthy or the focus has exclusively been on attempts to measure poverty. Enduring wealth inequalities, which at times defy comprehension or the real prospect of an increasingly divided society, should not lead one to simply ignore the substantial rise in the general level of wealth and prosperity and, of course, to ask what consequences this may have in highly developed nations.

One of the consequences in the general rise of affluence is a change in value-orientations, life-styles choices and consumption patterns away from purely materialistic to what Ronald Inglehart (e.g. 1977, 1987) has described as modification in beliefs toward a post-materialistic outlook.

The emergence of a post materialist world view commences with those generations in the postwar era who spent their formative years in conditions of relative economic and physical security. The trend toward post materialist values implies new political priorities, especially with regard to communal values and life-style issues and leads to a gradual neutralization of political polarization based on traditional class-based loyalties.

To sum up, there is an urgent need for a better understanding of these transformations of modern society and the implications these developments may have for any progress towards the goal of "sustainable development" (cf. also Grove-White, 1996). There is the requirement to better comprehend the economic, cultural and moral changes that are now taking place in modern society in order to gauge whether and to what extent these changes constitute transformations that radically alter any discussion about the linkages between nature, society and the economy.

8.8 Summary

Climate research has thrived within the scientific community for the past decades. To date climate research has mainly dealt with questions raised by the scientific community and to a lesser extent the political system about the physical dynamics of climate understood as a natural phenomenon. Accurate

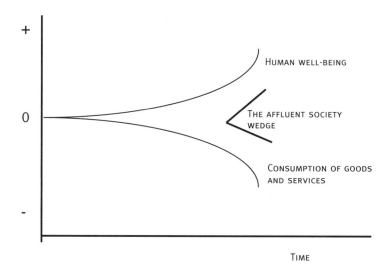

Fig. 8.6. The Affluent Society Wedge. Adopted from Robinson and Tinker, 1997

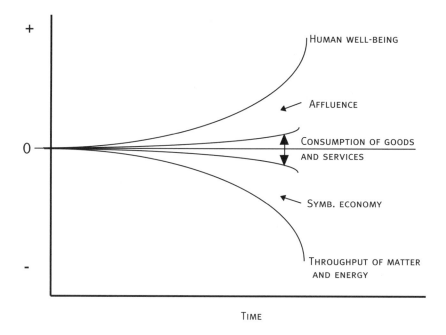

Fig. 8.7. Combining the Wedges. Adopted from Robinson and Tinker, 1997

numerical and system-analytical answers were considered sufficient answers while the translation of such knowledge into practical decisions in the societal and political realm were largely taken-for-granted.

But the success of climate research so far did not lead to the institution of practical policies by balancing expected damages and abatement costs to mitigate, or even avoid, the detrimental consequences of expected anthropogenic climate change. Instead, the (often misinterpreted) information provided by climate research is responsible for the creation of alarm ("climate catastrophe") in the public and for political inactivity. In everyday life, the magic terms - greenhouse effect, global warming - are now widely known; but equally widespread is confusion about the nature of these concepts. Political actions are mostly limited to verbal announcements and more or less generous funding of climate research.

In that sense, present-day natural scientists continue to be as naive and well-meaning as Svante Arrhenius - who is responsible for the theory of the warming effect of atmospheric carbon dioxide loading (Arrhenius, 1896) - was in his times, when he believed that scientific knowledge alone would improve the world. Such a view is wishful thinking.

In order to mitigate such reasoning, it is imperative that social science expertise is brought into the center of climate research. Specifically, the following aspects should be studied:

What are the main trajectories of social, political and economic change of modern society and how do these changes impinge upon the relationships between environment, the economy, the political system and prevailing value-orientations in and among societies?

What has happened to the doctrine of climate determinism and what climatic events do influence societies and under what conditions? For as some might see it, extreme climates do influence social performances, say in arctic or arid regions. But how far has and can society emancipate itself from climatic conditions? What are the fundamental errors made by Huntington and others? Only a few examples of a critical assessment of the doctrine of climate determinism, such as Sorokin (1928) and Nordhaus (1994), are now available.

Do the discussions from the last century about natural and/or anthropogenic climate change represent a useful analogue for the understanding of the present debate and the present decision process on national and international levels? How can we incorporate the dynamics of social value assignment let alone the major changes in the social structure and culture of modern society to transform a GES model into a more realistic PES model?

How do we beneficially re-combine the boundaries of social and natural science discourse in the area of climate research?

Chapter 9

Climate Science and the Transfer of Knowledge to Public and Political Realms

by Dennis Bray and Hans von Storch

9.1 Collegial Comments

"The questions [in this survey] are very much flawed as subjective, and the replies will constitute nothing more than an opinion poll. Hence, I put very little credence in these results as having much bearing on affecting our work." (sic) case usa072: US physicist with more than 20 years in the field.

"This survey contributes to the problem. The questions are worded very precisely. There is the potential for misusing the results of this survey." case usa079: US physicist with 0-5 years experience.

"Thank you for the opportunity to respond in this survey." case usa096: US meteorologist with more than 20 years experience

"There should be more surveys like this" case usa098: US meteorologist with 11-15 years experience.

"This is a good survey and I hope you repeat it in about 10 years." case usa127: US meteorologist with more than 20 years experience.

Abstract

This paper presents the results of a survey of the perspectives of climate scientists on the topic of global warming. It addresses both internal and external elements of the science. A total of 412 responses from climate scientists in Canada, USA and Germany are analyzed. Differences among those groups with higher levels of involvement with policy makers, with the media, and the less vocal members of the scientific community are the focus of this paper. Statistically significant differences were found among these three groups on a number of pertinent issues. These differences were more often among those areas which were beyond the areas of the scientists' areas of expertise. More precisely differences were found in: The assessment that global warming is a process already underway, the nature of the impacts of climate change, the knowledge transfer process, and the conduct of the climate sciences. These perspectives are of considerable importance for they relate to the transfer of scientific knowledge to the public and political realms. In short, this paper contributes to the discussion of the socio-scientific construction of the climate change issue.

The paper describes the data collection method and provides a description of the sample. In the results, a series of t-tests are used to identify differences that exist among three identified groups; 1. those scientists with a high level of contact with the media; 2. those scientists with a high level of contact with policy makers, and; 3. those scientists with a low level of external contact outside of the scientific community. Greater differences were found when considering the extension of knowledge to matters outside of the scientists' areas of expertise.

This analysis raises the question of how scientific knowledge is transformed into high levels of public and political significance. This transition could not, as of yet, be attributed to the human experience since the experience of any expression of climate change, with the exception of extreme events (a highly contested relationship), is typically well below the thresholds of human climatic perception. However, it could not be denied that the *issue of climate change* has had, and creates the potential, for significant social impacts.

What we discuss in this paper represents only one aspect of the science-politico-society triad. More specifically, we address the role of the human element in the interpretation of scientific "fact" or, even more specifically, the *scientific construction* of the climate change issue. Not only do we suggest, in light of the now *globalness* of many contemporary issues, the requirement to make assessments of all of the triadic interactions, but also to address the process by which multiple interpretations stem from a single scientific artifact.

9.2 Introduction

In recent years global warming has been among the most publicized of environmental issues, raising both public and political debate. Climate change however, at least in terms of the thresholds of human sensory experience, remains a future event, and consequently it remains, as of yet, a resident (mostly) of the lay imagination. (Science, of course, does more to substantiate its claims.) Certainly some climatic events have sparked this imagination at times, of both the general public and the scientific community. At the times of the occurrence of these events, the public imagination now well versed in the climatic terminology, is further excited by the sometimes ominous commentary of some of the climate experts, at least as it is reported via knowledge brokers. Public comments to the contrary, that is, positive interpretations of the events referring to the potential for positive benefits due to climate change are rather rare since, perhaps because they do not make good copy, or, at least they do not arouse a similar response as a report couched in fear.

The impact and content of what reaches the public ear and how it is interpreted by the public has become a timely topic represented in the body of literature addressing the social construction of climate, climate change, and numerous other environmental issues (for examples see Kempton et al., 1995, or Dunlap et al., 1993). While the public's interpretation of the global warming issue and the social construction of climate change and other environmental issues has been a well addressed topic, little attention has been given to the construction of the issue within the scientific community. To that extent this paper discusses the *socio-scientific construction* of the climate change issue.

The conception of risks associated with climate change, and indeed the phenomenon itself, are assumed herein, in light of them being a *future* event, to be socio-scientific constructions which are open to multiple interpretation. This is a significant consideration since the issue of climate change has arisen, coincidentally or otherwise, with the rise of the new environmental ideology or environmental ethic and environmental organizations intent on addressing global environmental issues. This gives rise to the opportunity for sentiments to begin to feed back into science. This influence is obvious in public declarations of environmental organizations, as for example the declaration of the WWF: "This is what makes WWF's European Policy Office so important. Started in 1989, it acts as the eyes and ears of the worldwide WWF Network. Based in Brussels, its job is to influence the political developments in the "capital" of Europe which have far-reaching environmental effects for the rest of Europe and worldwide."(see www.panda.org) Assuming that the political ear is attentive, we can also assume that it has some ramifications for the conduct of science. Furthermore, being human, scientists cannot fully escape the influence of contemporary ideologies, and as the discussion will demonstrate, the threat to scientific objectivity is well noted by scientists.

Moreover, while within the scientific community the debate continues regarding the numerous aspects of global climate change, in the public disclosure of science the debate often seems settled. According to the WWF, a well founded and publicized organization, "99 % of scientists agree, global warming is real, it's happening now, and it's getting worse."
(www.panda.org/climate/impact.html). In fact, according to Auer et al. (1996:145) "Public opinion and mass media have taken over the topic to such an extent that in the meantime it begins to repenetrate and influence scientific discussion." This line of reasoning is discussed in detail in the body of this paper. If this is indeed the case, then climate change aside, we are at the risk of a mutual relationship between science and ideology, a relationship that has not always proved favorable. In the following, the inner workings of the climate science community are explored using the results of a survey questionnaire distributed to climate scientists in the USA, Canada and Germany.

9.3   The Survey

A series of in-depth interviews were conducted with scientists in major institutions in the USA, Canada and Germany. A list of pertinent themes were drawn from the interviews and used to construct a survey questionnaire. The questionnaire, consisting of 74 questions was pre-tested in a German institution and after revisions, distributed to 1,000 scientists in North America and Germany.

Most questions were designed on a seven point rating scale. A set of statements was presented to which the respondent was asked to indicate his or her level of agreement or disagreement, for example, 1 = strongly agree, 7 = strongly disagree. The value of 4 can be considered as an expression of ambivalence or impartiality depending on the nature of the question posed. In spite of the pretesting, comments made on the survey indicated that some of the respondents were critical of some of the questions. Additional space was left on the booklet for respondents to make comment, some of which were presented at the beginning of this paper.

9.4   The Sample

The sample chosen for this study was largely a result of available funding. The initial intent was to limit the study to the perceptions and interpretations of the German climate science community. Upon the suggestion of those who endorsed the project, the project was redesigned to allow for a comparative study of the German and North American climate science communities. (For a more detailed discussion of the differences based on host societies, see Bray and von Storch 1996.)

An anonymous, self-administered questionnaire was distributed by post to samples of US climate scientists, Canadian climate scientists and German climate scientists. The sample for the North American component was drawn from the EarthQuest mailing list. Due to the fact that the mailing list is more extensive than the discipline of climate science, a true random sampling technique was not employed. Rather, subjects were selected according to institutional and disciplinary affiliations. This resulted in a final sample of 460 US scientists and 40 Canadian scientists. The sampling of German scientists, due to reasons of confidentiality, was beyond full control. A random sample of German scientists was drawn from the mailing list of the Deutsche Meteorlogische Gesellschaft by its administration, resulting in the distribution of 450 survey questionnaires. A further 50 questionnaires were distributed to members of the Max-Planck-Institut für Meteorologie, Hamburg, and members of the University of Hamburg. Returns of the German sample extended beyond Germany and included 13 respondents reporting to be other than German. In the analysis, this group are included in the 'German' category since they originated from a German mailing list. A description of the sample is presented in Tables 9.1 through 9.3.

The mail-out occurred only once and no follow-up letters of reminder were distributed. The number of completed returns were as follows: USA 149, Canada 35, and Germany 228, a response rate of approximately 40 %. Additional questionnaires were returned due to noncurrent addresses. A response rate of 40 % and the total number of respondents can be considered as quite good when compared to other similar surveys. Stewart et al. (1992) for example in a SCIENCEnet electronic survey received 118 responses from "a computer-based network ... which has over 4000 subscribers" (p.2); the National Defense University Study (NDU, 1978) based its conclusions of the responses from 21 experts; the Slade Survey (1989) based conclusions on responses from 21 respondents; the Global Environmental Change Report Survey (1990) had a response rate of approximately 20 % from a sample of 1500; the Science and Environmental Policy project (Singer, 1991) received a 32 % response rate from a sample of 102, and later a 58 % response rate from another sample of 24; the Greenpeace International Survey received 113 responses from a sample of 400, and; Auer et al (1996) report that "about 250 questionnaire were distributed [by method of personal contact at conferences] and 101 were sent back".

As Tables 9.2 and 9.3 indicate, the sample is well diversified. While categories in Table 9.3 could possibly be collapsed, they are presented in their entirety to demonstrate self constructed divisions within the climate sciences, in short, a preliminary basis for the social construction of the science, each label possibly representing different perspectives and different vested interests.

Table 9.1. The institution in which I work is located in:

Country	Total
USA	149
Canada	35
Germany	128
Grand Total	412

Table 9.2. The number of years I have worked in the climate sciences are:

Years	USA	Canada	Germany	Grand Total
0-5	15	3	104	122
6-10	16	12	46	74
11-15	20	4	23	47
16-20	24	7	10	41
> 20	72	9	39	120
missing	2	0	6	8
Grand Total	149	35	228	412

9.5   Results

This analysis attempts to determine, through a series of t-tests, if there are statistically significant differences in the perspectives of the vocal and non-vocal members of the scientific community. (All tests of statistical significance are at the level of .05) "Vocal" is further subdivided into two categories, those scientists with a high level of involvement with policy makers and those scientists with a high level of involvement with the media.

Should we fail to reject the null hypothesis, we can assume that the *voice* of science is representative of the scientific consensus. This is not to say, however, that the voice is representative of reality. On the contrary, it might indicate a discipline well inoculated with ideology, sharing a common imagination.

Before beginning the analyses it is necessary to present the data in terms of media and policy involvement so as to distinguish these sectors of the climate science community. This is indicated in Fig. 9.1, 9.2 and 9.4 below. Responses of less than three have been chosen to represent the 'high' contact category, in other words, those scientists who responded with a value of less than 3 are for the sake of this analysis considered to be the vocal members of the climate science community. Figures 9.1 and 9.2 indicate that only a small proportion of the climate science community represents the voice of the scientific community to the more public consumers of knowledge. What this voice might convey is the topic of the following discussion.

Table 9.3. The area in which I conduct most of my research is (U: USA, C: Canada, G: Germany, GT: Grand Total):

Country	U	C	G	GT		U	C	G	GT
impact assess.	8	5	1	14	science policy	1	0	0	1
geosc. instrum.	0	1	0	1	biochemistry	1	0	0	1
oceanography	2	1	3	6	physical chem.	1	0	0	1
observations	22	9	34	65	chemistry	2	0	3	5
bio-geo cycles	0	1	1	2	atmos.processes	2	0	9	11
climate sc.	0	1	1	2	climate theory	1	0	2	3
modeling	38	8	48	94	air-sea int.	1	0	2	3
measurement	4	1	1	6	diagnostics	3	0	0	3
nutrient cyc.	0	1	0	1	convection	1	0	0	1
administration	5	1	2	8	turbulence	1	0	0	1
fluid dynamics	13	2	1	16	engineering	1	0	0	1
monitoring	0	1	0	1	cloud physics	2	0	5	7
boundary layers	0	1	0	1	strato. dyn.	1	0	1	2
ecology	0	1	2	3	solar influ.	1	0	1	2
ecosystems	0	1	0	1	snow ice	1	0	0	1
physical proc.	17	0	23	40	public forecast	0	0	2	2
radiation	1	0	1	2	agrometeorol.	0	0	1	1
nonlinear dyn.	1	0	1	2	regional clim.	0	0	6	6
computer apps.	1	0	0	1	thermodynamics	0	0	1	1
ocean modeling	1	0	0	1	flight meteor.	0	0	2	2
environ. chg.	1	0	0	1	economic geogr.	0	0	2	2
physics	1	0	1	2	stochastic proc.	0	0	2	2
remote sensing	2	0	0	2	forecasting	0	0	3	3
global policy	1	0	0	1	data systems	0	0	2	2
experimentation	2	0	11	13	synoptics	0	0	2	2
atmospheric rad.	1	0	0	1	climate change	0	0	14	14
interseasnl. cl.	1	0	0	1	meteorology	0	0	2	2
biometeorology	1	0	0	1	meso climat.	0	0	1	1
palaeoclimatology	1	0	0	1	hydrodyn.	0	0	4	4
fluid dynamics	1	0	0	1	missing	4	0	24	28
					other	0	0	6	6

The response rate indicates that 48 people (12 % of the sample) reported a high level of contact with policy makers, and 45 respondents (11 % of the sample) reported a high level of contact with the media. In Fig. 9.4, the main diagonal indicates frequencies that are above those expected by chance. Seventeen of the scientists reported themselves as being at a level of high involvement with both media and policy makers, whereas 48 scientists reported as having a high level of contact with policy makers and 45 reported as having a high level of contact with the media. The number of scientists claiming almost no contact with either the media or policy makers was 150. Regarding only media contact, 197 claimed very little or no contact, and re-

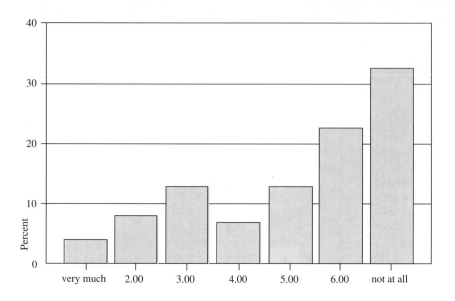

Fig. 9.1. How much have you been involved with those people who make climate related policy decisions?

garding contact with policy makers, 224 scientists claimed very little or no contact. Twenty eight percent of the scientists, often contacted by the media, claimed to have almost no contact with policy makers and, twenty four percent of scientists claiming a high level of involvement with the media reported to have almost no contact with policy makers. As Fig. 9.4 indicates, those scientists with a high level of contact with the media are not necessarily the same scientists with a high level of contact with policy makers. Furthermore, Fig. 9.4 indicates considerable differences between the observed frequencies and frequencies expected of the two variables were independent.

Logically, the best place to begin is the consensus regarding the phenomenon of global climate change. Figure 9.3 refers to the scientist's response to his or her level of certainty that global warming is underway, the 99 % claim of the WWF. Here, only 10 % of the respondents express *no* doubt that global warming is a process already underway. With a marginal expression of doubt, those scientists responding with a value of 2 or 3, the percentage of the sample of scientists that would hedge towards agreement that global warming is underway includes an additional 55 % of the respondents. When asked if global warming would definitely occur in the future if human behavior did not change (Fig. 9.4) there is a large shift in the perspectives of the respondents. Here 29 % strongly agree that without change in human behavior, global warming will definitely occur sometime in the future. Again, taking

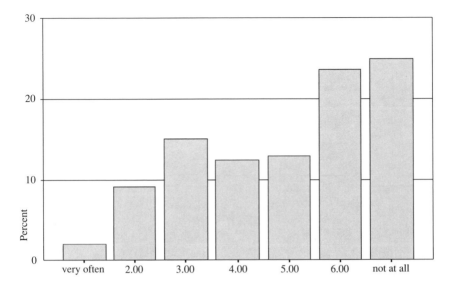

Fig. 9.2. How often are you contacted by the media for information pertaining to climate change?

those respondents more inclined to agree (values 2 and 3) with this prospect the percentage increases greatly to include another 50 % of the respondents. In short, 65 % of the respondents express some level of agreement that global warming is a process underway while 79 % of the respondents express a level of agreement that without change in human behavior, global warming will occur sometime in the future.

A further analysis of the data in Fig. 9.3 and 9.4 indicates that scientists more involved with those people who make climate related policy decisions (Fig. 9.5) demonstrate a statistically significant difference from the other categories in the their level of certainty that global warming is a process already underway, being more inclined to agree and express that we are, at present, experiencing global warming. ("high policy involvement" mean: 2.82; "others" mean 3.63). If there would be no link between high level of contact with policy makers and a high level of agreement that global warming is a process already underway, then we would expect the observed frequencies to closely approximate the expected frequencies in the four upper left cells of Fig. 9.5. This however, is not the case with the observed frequency being 26 and the expected frequency being 17. In short, there are more scientists than would be expected by chance that have a high level of contact with policy makers and strongly agree that global warming is a process already underway. There is no statistically significant difference between that group

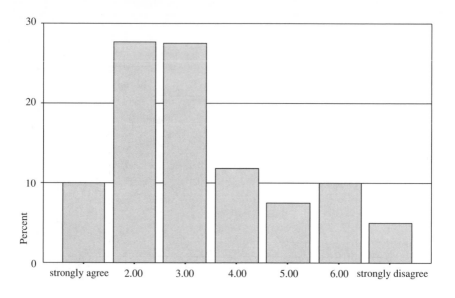

Fig. 9.3. We can say for certain global warming is a process already underway

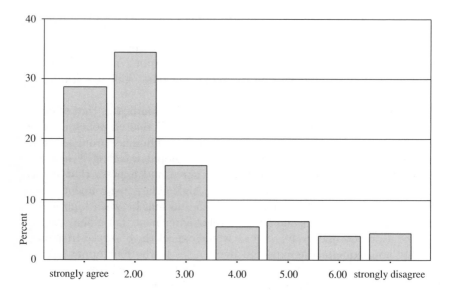

Fig. 9.4. We can say for certain that without change in human behavior global warming will definitely occur sometime in the future

Table 9.4. Crosstabs of Policy and Media Involvement
Normal size: observed frequency, small: expected frequency, **bold:** below expected
frequency, *italic:* above expected frequency

Media	Policy							
	1.00 often	2.00	3.00	4.00	5.00	6.00	7.00 never	row total
1.00 often	*2*	*3*	1	**0**	**0**	*2*	**0**	8
	.3	*.6*	1	*.6*	**1**	*1.8*	**2.6**	2.0%
2.00	*2*	*10*	*5*	*3*	*6*	**6**	**5**	37
	1.5	*2.9*	*4.8*	*2.6*	*4.8*	**8.2**	**12.2**	9.1%
3.00	*4*	*7*	*16*	*9*	**7**	**11**	**7**	61
	2.4	*4.8*	*7.8*	*4.4*	**7.8**	**13.6**	**20.2**	15.1%
4.00	*3*	4	*13*	*7*	*10*	**9**	**4**	50
	2.0	4.0	*6.4*	*3.6*	*6.4*	**11.1**	**16.5**	12.3%
5.00	**0**	**3**	**5**	*5*	*9*	*19*	**11**	52
	2.1	**4.1**	**6.7**	*3.7*	*6.7*	*11.6*	**17.2**	12.8%
6.00	**3**	**3**	**8**	**2**	*15*	*32*	*33*	96
	3.8	**7.6**	**12.3**	**6.9**	*12.3*	*21.3*	*31.8*	23.7%
7.00 never	**2**	**2**	**4**	**3**	**5**	**11**	*74*	101
	4.0	**8.0**	**13.0**	**7.2**	**13.0**	**22.4**	*33.4*	24.9%
Column total	16	32	52	29	52	90	134	405
	4.0%	7.9%	12.8%	7.2%	12.8%	22.2%	33.1%	100.0%

which frequently speaks to the media and that group which does not (Fig. 9.6)
in regards to a greater expression that global warming is already underway,
with the observed frequency being 21 and the expected frequency 17. It
would seem then that those scientists who have higher contact with policy
makers are more inclined to agree that global warming is indeed a process
already underway while those scientists with high levels of media contact
perhaps draw from a broader range of perspectives and that the media might
be more likely to seek out opposing views. Nonetheless, the advice given to
the policy arena is that global warming is a process already underway while
commentary directed to the public (via the media) is typically less committed
to such a position. Consequently, while perhaps not being able to spark the
public imagination and maintain a degree of significance in light of absence
of dramatic changes, it would seem the issue of climate change is presented
so as to maintain a high priority in the political realm.

The data in Fig. 9.3 and 9.4 might, as stated, also be an indication, since
we are dealing with statistical artifacts, that the press might be inclined to
seek out opposing extremes, thereby negating statistical differences. This is
evident in a comparison of Fig. 9.5 and 9.6. Under the categories of strongly
disagreeing that climate change is underway, of the 47 scientists with a higher
level of contact with policy makers (i.e. values 1 and 2) there is evidence of
only four individuals from the sample claiming a strong level of disagreement

Table 9.5. Crosstabs of Level of Policy Involvement and Certainty of Global Warming Underway
Normal size: observed frequency, small: expected frequency, **bold:** below expected frequency, *italic:* above expected frequency

Policy	Certainty							row total
	1.00 agree	2.00	3.00	4.00	5.00	6.00	7.00 disagree	
1.00 often	*2* 1.5	*10* 4.2	1 **4.1**	0 **1.8**	0 **1.1**	1 **1.5**	*1* .8	15 3.7%
2.00	*4* 3.2	*10* 8.9	8 **8.7**	*5* 3.9	*3* 2.4	2 **3.2**	0 **1.7**	32 7.9%
3.00	*7* 5.3	12 **14.7**	*15* 14.5	5 **6.4**	4 4.0	7 **5.3**	*3* 2.7	53 13.0%
4.00	1 **2.9**	*10* 8.1	4 **7.9**	*5* 3.5	*6* 2.2	2 **2.9**	1 **1.5**	29 7.1%
5.00	*6* 5.2	12 **14.4**	*18* 14.2	5 **6.3**	3 **4.0**	4 **5.2**	*4* 2.7	52 12.8%
6.00	9 **9.4**	24 **25.8**	23 **25.4**	10 **11.2**	*10* 7.1	*12* 9.4	5 **4.8**	93 22.9%
7.00 never	12 **13.4**	35 **36.9**	*42* 36.3	*19* 16.0	5 **10.1**	13 **13.4**	7 **6.9**	133 32.7%
column total	41 10.1%	113 27.8%	111 27.3%	49 12.0%	31 7.6%	41 10.1%	21 5.2%	407 100%

(values 6 and 7) while when considering the 45 scientists claiming higher levels of media contact there is evidence of 8 individuals claiming that global warming is *not* underway (values 6 and 7).

In an effort to account for any bias in the above line of questioning, scientists were asked to comment on the *future* possibility of global warming, that is, if "We can say for certain that without change in human behavior, global warming will definitely occur sometime in the future." In essence, this gives credence to the theory of global warming in lack of the manifestation of the event (Fig. 9.4).

The analysis of the data indicates that there is little difference in the acceptance of the *theory* of global warming between those who speak to policy makers and those who do not. The question of differences remains in the manifestation of the event, evident in the higher level of acceptance of the theory (means: high policy involvement group 2.23, high media involvement group 2.8) than of the manifestation of the event (means: policy involvement 2.8, media involvement 3.6). This is also the case when responses were given regarding reporting to the media, that is, those who spoke to the media also had a higher level of consensus regarding the *theory* of global warming than they did the event of global warming.

Table 9.6. Crosstabs of Level of Media Involvement and Certainty of Global Warming Underway
Normal size: observed frequency, small: expected frequency, **bold:** below expected frequency, *italic:* above expected frequency

Media	Certainty 1.00 often	2.00	3.00	4.00	5.00	6.00	7.00 never	row total
1.00 often	*2* .8	*4* 2.2	**2** **2.2**	**0** **1.0**	**0** **.6**	**0** **.8**	**0** **.4**	8 2.0%
2.00	*4* 3.7	*11* 10.2	**7** **10.1**	**4** **4.5**	*3* 2.9	*5* 3.7	*3* 1.8	37 9.2%
3.00	**3** **6.0**	**15** **16.6**	**16** **16.5**	*10* 7.3	*6* 4.6	*7* 6.0	*3* 3.0	60 15.0%
4.00	*8* 5.0	*18* 13.8	*14* 13.7	**3** **6.1**	**3** **3.9**	**3** **5.0**	**1** **2.5**	50 12.5%
5.00	**4** **5.2**	**14** **14.4**	**9** **14.3**	*8* 6.4	*8* 4.0	*6* 5.2	*3* 2.6	52 13.0%
6.00	**9** **9.5**	**25** **26.3**	*30* 26.1	*14* 11.6	**5** **7.3**	**9** **9.5**	**3** **4.7**	95 23.7%
7.00 never	*10* 9.9	**24** **27.4**	*32* 27.2	**10** **12.1**	**6** **7.7**	*10* 9.9	*7* 4.9	99 24.7%
column total	40 10.0%	111 27.7%	110 27.4%	49 12.2%	31 7.7%	40 10.0%	20 5.0%	401 100.0%

Overall, the theory of global warming seems to be less contentious than the actual event, although even the theory is far from a state of unanimous consensus. However, it would seem that those scientists from the sample with a higher contact to policy makers are more convinced that the process of global warming is already underway and those scientists with higher levels of media contact are more likely to present both extremes. Given that the theory of global climate change has moved beyond the scientific context (in spite of less than consensual acceptance of the theory) it is necessary to ask what global climate change might mean in terms beyond academic significance, since it is at the social and political levels of interpretation that the phenomenon becomes interpreted into action and policy. To this end scientists were asked to make some initial assessments regarding impacts. These are presented in Fig. 9.5 through 9.7.

Figure 9.5 summarizes the responses to a question regarding the potential for a rapid onset of climate change and a lack of preparation to result in devastation of some areas of the world. Results indicate there are no statistically significant differences between the general scientific community and those speaking to the media and/or policy makers. The distribution in the bar chart indicates panic is not the general status of the scientific community.

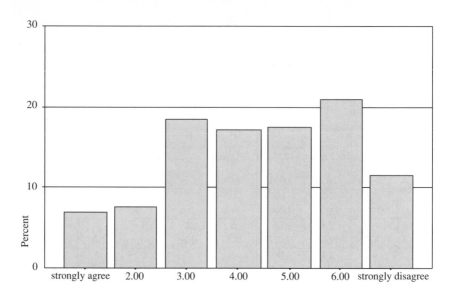

Fig. 9.5. Assuming climate change will occur, it will occur so suddenly that a lack of preparation could result in the devastation of some parts of the world

Figure 9.6 represents the responses when scientists were asked if the potentially detrimental impacts can yet be identified. As the data indicates, there is a high level of uncertainty within the scientific community regarding this assessment. Figure 9.6 implies that there might be a split in the scientific consensus as to the level of achieved scientific knowledge regarding the impacts of climate change. However there are no statistically significant differences between those groups with high levels of media and policy maker contact and the general scientific community. We can assume that both perspectives are represented in the public and political forums.

Pursuing further the general nature of the possible impacts of climate change, scientists were asked to respond to similar lines of questioning, one pertaining to the world in general, the other to his or her more familiar host society. The results are presented in Fig. 9.7 and 9.8.

When asked about the respective host society (Fig. 9.7) regarding the explicit nature of the impacts of climate change, there appears to be somewhat more of a consensus that climate change is perceived of as having a greater potential for negative consequences. There were no statistically significant differences between the groups. It seems that the people who have a high degree of contact with policy makers share a similar perspective as the general scientific community as do those who have a high degree of contact with the media. In short, there is a considerable degree of consensus that climate

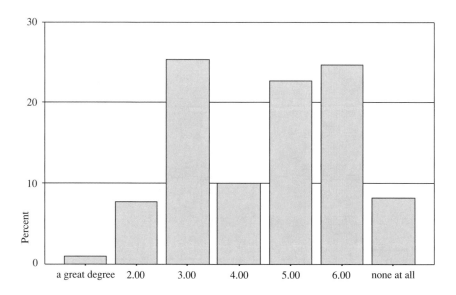

Fig. 9.6. To what degree can we explicitly state the *detrimental* effects that climate change will have on society?

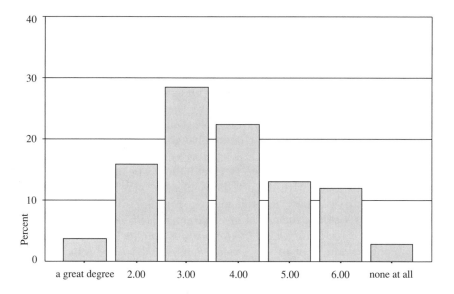

Fig. 9.7. To what degree do you think that climate change will have a detrimental effect for the society in which you live?

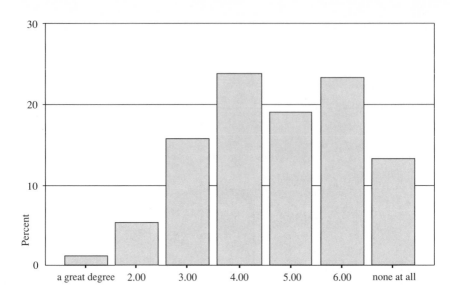

Fig. 9.8. To what degree do you think that climate change might have some positive effects for the society in which you live?

change is prone to produce negative consequences in the scientist's host society while at the same time, the scientific community is far from making the claim of being able to state the characteristics of the impact. This has implications for the construction of science-policy dialogue procedure, particularly in light of the fact that scientists with a higher degree of policy contact are more inclined to perceive global warming as underway but are not necessarily able to state the specific nature impacts, only that they will be detrimental, an attitude which would be conveyed to policy and public attention.

To determine if the impact of climate change was limited to a negative connotation, scientists were asked if they perceived climate change as possibly resulting in *benefits* for the host society (Fig. 9.8). It appears in general that climate change as a negative impact outweighs the perspective of climate change as a positive impact (overall means: Fig. 9.7: 3.9, Fig. 9.8: 4.2) however, *scientists with a high level of involvement with policy makers were more inclined to perceive the possibility for positive benefits for the host society* which, among a variety of other reasons, might add to the reluctance of political bodies to engage in political action in terms of climate policy and to raise differing perspectives in international negotiations. When asked to what degree climate change might have some positive effect there was a statistically significant difference between those scientists with a high level of contact with policy makers and the remainder of the sample. The mean response of those

Table 9.7. Crosstabs of Level of Policy Involvement and Positive Benefits
Normal size: observed frequency, small: expected frequency, **bold:** below expected
frequency, *italic:* above expected frequency

Policy	Positive benefits							
	1.00 great	2.00	3.00	4.00	5.00	6.00	7.00 none	row total
1.00 often	*1* .2	*2* .8	*4* 2.5	*5* 3.8	**1** **3.0**	**0** **3.7**	*3* 2.1	16 3.9%
2.00	*1* .3	*3* 1.6	*6* 4.8	*11* 7.3	**4** **5.9**	**4** **7.1**	**2** **4.0**	31 7.6%
3.00	*1* .5	**1** **2.7**	*13* 8.2	*15* 12.5	*15* 10.0	**6** **12.2**	**2** **6.9**	53 13.0%
4.00	**0** **.3**	*3* 1.4	*6* 4.2	**6** **6.4**	**3** **5.1**	*8* 6.2	**1** **3.5**	27 6.6%
5.00	**0** **.5**	*3* 2.7	**8** **8.2**	**7** **12.5**	*15* 10.0	*15* 12.2	**5** **6.9**	53 13.0%
6.00	*1* .9	*5* 4.8	**11** **14.4**	**19** **21.9**	**16** **17.6**	*26* 21.4	*15* 12.1	93 22.8%
7.00 never	**0** **1.3**	**4** **6.9**	**15** **20.8**	*33* 31.8	**23** **25.5**	*35* 31.1	*25* 17.5	135 33.1%
column total	4 1.0%	21 5.1%	63 15.4%	96 23.5%	77 18.9%	94 23.0%	53 13.0%	408 100.0%

with high contact to the policy makers was 4.0 as compared to 4.8 for the
remainder of the sample with a value of "1" representing "a great" potential
for positive impacts and a value of "7" representing "no" possibility at all.

In brief, while the scientists with a high level of contact to policy makers
might be more inclined to believe that climate change is underway, they might
also be more inclined overall to feel that climate change has the potential to
spawn some positive benefits for society, as well as, negative impacts. This
potential was worthy of further analysis and is elaborated in Fig. 9.7 and
9.8. Again, using the values of 1 and 2 to indicate high levels of involvement
with policy makers and with the media, the data indicates that from among
the scientists who responded to this question and also claimed a high level of
policy contact, 22 % saw the potential for positive outcomes stemming from
climate change (values 1-2) and 29 % saw little potential for positive outcomes
(values 6-7). Of those claiming a high level of contact with the media who
responded to this question, however, only 11 % saw the potential for positive
benefits while 44 % saw little or no potential for a positive outcome from
climate change. Consequently we could assume the likelihood that more
often than not the perception of the climate change issue being presented to
the public is one of mostly negative impacts.

Scientists were also asked to what degree they felt that global climate
change was one of the leading problems facing society. As Fig. 9.9 indicates,

Table 9.8. Crosstabs of Level of Media Involvement and Positive Benefits
Normal size: observed frequency, small: expected frequency, **bold:** below expected
frequency, *italic:* above expected frequency

media	1.00 great	2.00	3.00	4.00	5.00	6.00	7.00 none	row total
			Positive benefits					
1.00 often	0 .1	*1* *.4*	1 **1.3**	0 **1.9**	*3* *1.5*	*2* *1.8*	1 **1.1**	8 2.0%
2.00	*1* *.4*	*2* *1.9*	*6* *5.6*	8 **8.4**	6 **6.8**	*9* *8.1*	4 **4.7**	36 9.0%
3.00	*1* *.6*	*5* *3.2*	*13* *9.6*	*17* *14.3*	8 **11.5**	11 **13.8**	6 **8.0**	61 15.2%
4.00	*1* *.5*	*4* *2.6*	*8* *7.7*	10 **11.5**	*13* *9.3*	10 **11.1**	3 **6.5**	49 12.2%
5.00	*1* *.5*	2 **2.7**	6 **8.0**	*11* *11.9*	*13* *9.6*	*13* *11.5*	5 **6.7**	51 12.7%
6.00	0 **1.0**	4 **5.0**	*20* *15.0*	*24* *22.4*	13 **18.1**	*21* *21.7*	14 **12.7**	96 23.9%
7.00 never	0 **1.0**	3 **5.3**	9 **15.8**	*24* *23.6*	*20* *19.1*	*25* *22.9*	*20* *13.3*	101 25.1%
column total	4 1.0%	21 5.2%	63 15.7%	94 23.4%	76 18.9%	91 22.6%	53 13.2%	402 100.0%

the status of global climate change, in spite of areas of contention, is generally assigned a high global priority from among the members of the climate science community, and opinions would suggest this is drawn from the perception of negative consequences assigned to the event, in spite of the inability to explicitly state these consequences. Among the groups being considered, there were no statistically significant differences.

On this basis one would assume that the theory, at least, would have a high degree of consensus regarding its derivation. To this end, scientists were posed questions pertaining to the tools of their trade. In regard to the internal assessment of the science, scientists were asked to assess the inner workings of their science, the results of which (as indicated in the above discussion) have led to multiple interpretations reaching far beyond some scientists' intentions.

Table 9.9 presents the responses in rank order. The most problematic area indicated by the data is the ability of atmospheric models to deal with the influence of clouds while the least problematic area is the ability of atmospheric models to deal with hydrodynamics.

Figures 9.10 through 9.14 demonstrate graphically the responses of scientists to questions pertaining to some of the components of modeling, the test of climate theory. There were no statistically significant differences among the public and political voices and the more silent majority. The ambiva-

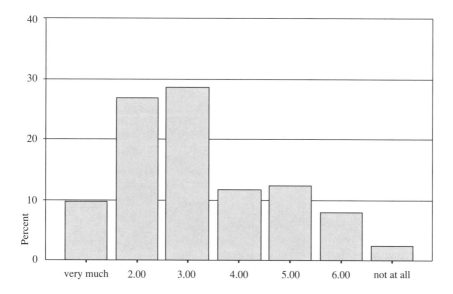

Fig. 9.9. How much do you think global climate change is one of the leading problems facing humanity?

lence indicated by the mean in Fig. 9.10 (overall means: 4.6, Std. Dev. 1.5) however, suggests less than unanimous faith in the output of climate models. This tendency is repeated when specific aspects of climate science are addressed in Fig. 9.11 through 9.15.

When questioned about the ability to deal with precipitation (Fig. 9.11), particularly in light of some of the more public claims of the impacts of global warming, scientists, vocal and otherwise, demonstrated limited faith in the ability for precipitation to be accounted for in climate modeling. Scientists were then asked to comment on the ability of models to incorporate the influence of clouds (Fig. 9.12) and this produced similar results. Scientists were then asked to comment on the perceived abilities of ocean models (Fig. 9.13). The question of the ability to deal with convection in ocean models resulted in responses similar to the questioning of the ability to deal with the components of atmospheric models. When asked about the coupling of atmospheric and ocean models (Fig. 9.14) similar responses were forthcoming.

While other similar questions were posed it is redundant at this point to discuss each in detail as similar responses were presented throughout the entire line of questioning. As Fig. 9.10 through 9.14 indicate, the scientific community makes no claim of perfection. Nonetheless, much rests on the products of these tools, namely the future direction of climate policy. It is the output and interpretations of these models that are employed by

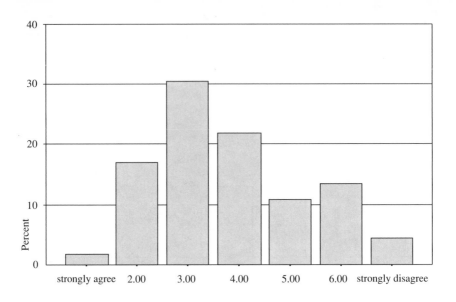

Fig. 9.10. Climate models accurately verify the climatic conditions for which they are calibrated

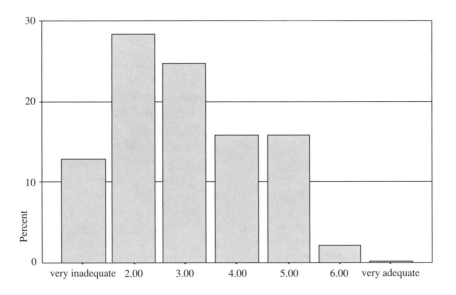

Fig. 9.11. How well do atmospheric models deal with precipitation?

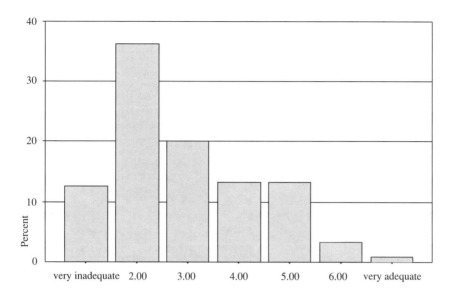

Fig. 9.12. How well do atmospheric models deal with clouds?

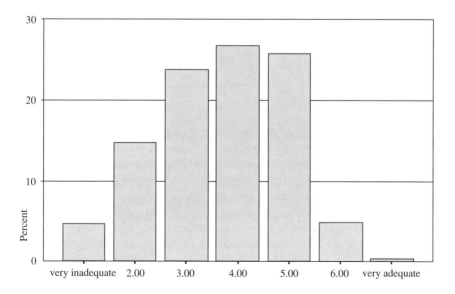

Fig. 9.13. How well do ocean models deal with convection?

Table 9.9. Assessment of Components:

Component	Mean	Std.Dev.	N
Climate models	1 = strongly agree 7 = strongly disagree		
accurately verify the conditions for which they are calibrated	3.8	1.5	407
accurately predict climatic conditions of the future	4.6	1.5	407
How well do atmospheric climate models deal with	1 = very inadequate 7 = very adequate		
the influence of clouds	2.9	1.4	405
precipitation	3.0	1.4	405
atmospheric convection	3.5	1.3	404
atmospheric vapor	3.5	1.4	406
radiation	4.6	1.4	406
hydrodynamics	4.7	1.4	406
How well do ocean models deal with	1 = very inadequate 7 = very adequate		
coupling of ocean and atmospheric models	3.3	1.3	401
oceanic convection	3.7	1.3	396
heat transport in the ocean	4.5	1.2	397
hydrodynamics	4.7	1.3	396
The state of knowledge allows for reasonable assessments of	1 = strongly agree 7 = strongly disagree		
turbulence	3.6	1.4	397
land surface processes	3.6	1.4	399
sea-ice	3.9	1.3	402
green-house gases	4.5	1.5	404
surface albedo	4.6	1.3	403

policy makers. In short, it is the predictive powers of the science that are anxiously awaited and often, by what ever means, extrapolated for policy considerations. In the process of employing the current state of knowledge, it is sometimes the case that the expert opinion of uncertainty demonstrated in the above is sometimes transformed into, or misinterpreted as, an expert *prediction* of the future.

As the following responses suggest, most scientists would tend to agree that this ability is well beyond the state of the science. Considering some of the lay perceptions of the issue of global climate change, it is interesting to note that the limited faith in the ability of climate models was prevalent among both the vocal and non-vocal sectors of the climate science community, suggesting the public applies its own interpretation.

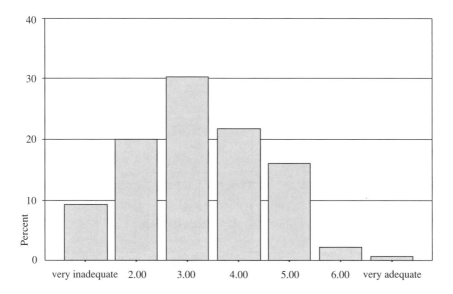

Fig. 9.14. How adequate is the coupling of atmospheric models and ocean models?

The phenomenon of the extension of knowledge beyond its capabilities is further expressed when scientists were asked to comment on the predictive ability of the current state of the science (Fig. 9.15). This raises the question of how, in the transformation of knowledge, a major degree of uncertainty is removed.

In an attempt to further explore the prediction potential of climate science, questions were posed incorporating the specification of time spans (Fig. 9.16 through 9.18).

Figure 9.16 indicates the responses to a short time span perspective. Data indicates there is far from a consensus regarding the ability to predict inter annual variability. In Fig. 9.17, the time frame is extended to a period of 10 years. And, again, as would be expected, the scientific community expressed little faith in the ability of the models to predict future climatic conditions. The time span was further extended to a period of 100 years (Fig. 9.18). It was suspected that perhaps a more general trend could be better endorsed in terms of predictions. However, as Fig. 9.18 indicates, this was not the case. The inability to state with any level of certainty the future characteristics of climatic conditions again points to the role of the imagination in the socio-scientific construction of climate change. On the matter of predictive capabilities, no statistically significant differences between or among any of the groups were evident.

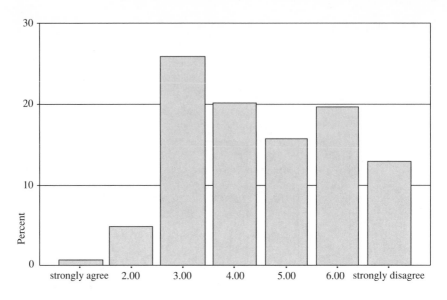

Fig. 9.15. Climate models can accurately predict climatic conditions of the future

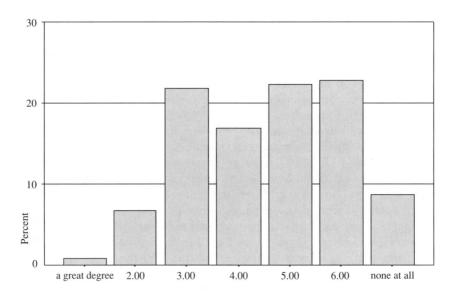

Fig. 9.16. To what degree can climate models provide *reasonable* predictions of inter annual variability?

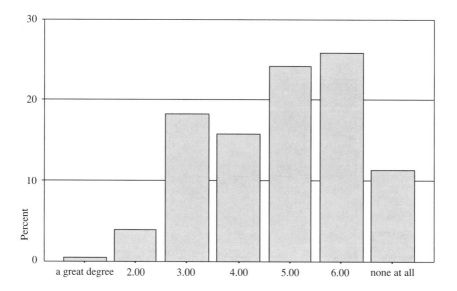

Fig. 9.17. To what degree can climate models provide *reasonable* predictions of climate variability of a time scale of 10 years?

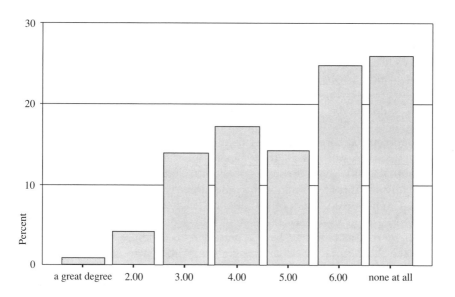

Fig. 9.18. To what degree can climate models provide *reasonable* predictions of climate variability of time scales of 100 years?

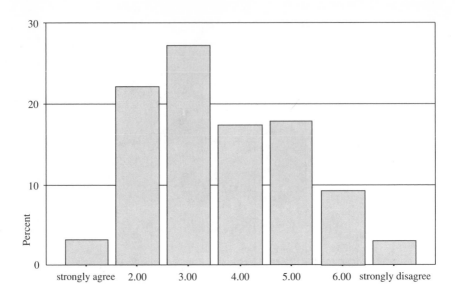

Fig. 9.19. In general, those scientists producing GCMs are knowledgeable about what data are needed by those scientists that endeavor to study the impacts of climate change

In short, Fig. 9.15 through 9.18 indicate that the scientific community is far from being convinced of its predictive powers. Nonetheless, the products of the science are in many circumstances interpreted precisely as being predictions. Undoubtedly, if this interpretation was without the possibility of climate change having a significant impact sometime in the future, climate change would not be perceived as a major issue. Furthermore, the nature of the impact is also a concern. Impacts must be presented to coincide with human perceptual thresholds in terms of both time and magnitude. To this extent, scientists were asked to comment on the transformation from the physical to socially relevant aspects of climate change. It is at this point that statistically significant differences become more readily apparent between the vocal and non-vocal sectors of the climate science community. Figure 9.19 pertains to the intra-science transfer of knowledge. As the data indicates, this is perceived as being far from a perfect relationship.

Notable here is the statistically significant differences among those vocal members of the scientific community and those less vocal members, particularly between those who have a high level of interaction with policy makers (mean 4.5) and those who do not (mean 3.5). Those who deal with policy makers expressed a lower level of confidence in the typical level of expressed awareness of the knowledge needs of the more external parts of the climate

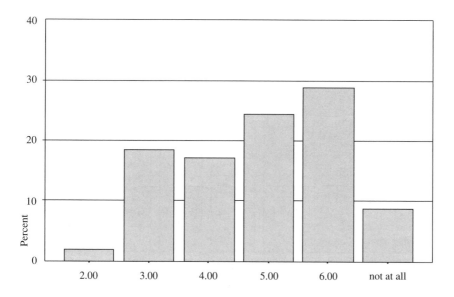

Fig. 9.20. To what degree, through the process of downscaling, is it now possible to determine local climate impacts?

science community. This is perhaps due to their increased exposure to the needs of other sectors in the debate, as derived from multiple levels of involvement.

When asked of the ability to determine local climate impacts, through the process of downscaling (Fig. 9.20) the mean response indicated that this ability is perceived also to be in need of great development and again, a greater awareness of this problem was expressed by those with a higher level of contact with policy makers, with a statistically significant difference between the means of those with a high level of contact with policy makers (mean 5.2) and the remainder of the sample (mean 4.8). Among those with high media contact and the remainder of the sample there was no statistically significant difference. In short, what gets reported to the media does so with a greater faith in the ability of science, and the voice to policy makers is more likely to heed on the side of caution.

When asked if the science is developed to the degree that information could be provided for local social impact assessments (Fig. 9.21), the need for further research was again made very evident, but no statistically significant differences were found among the groups. This is noteworthy in light of the numerous local impact assessments that can be found in the literature and have become the basis of policy discussion. We can assume that policy debates often proceed on the findings of "what-if" as opposed to "what-is" analyses.

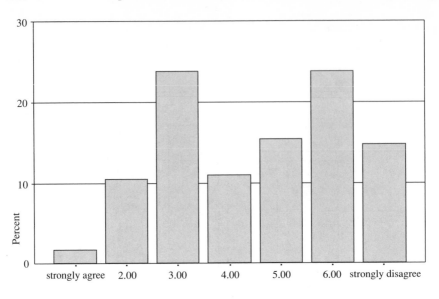

Fig. 9.21. The climate sciences are developed well enough to provide information for local social impact assessment

In spite of the lack of consensus as to the ability of the climate sciences to provide information for local impact assessments, a minority of scientists felt the issue of global warming should extend beyond a discussion limited to the physical world and begin to incorporate the social scientists in the discussion. These tendencies are evident in Fig. 9.21. Here, there is the indication of a bipolar distribution when it comes to the readiness of the science to point to social relevance. While there is a tendency for scientists to claim knowledge of what is required, the ability to generate such knowledge is still a contested area. Regarding the data summarized in Fig. 9.22, no statistically significant differences were found among the groups under consideration.

Regarding the broader scope of the implications of climate change, and in light of the perception that climate change will have (mostly negative) social impacts, climate scientists were asked of their perceptions of the integration of the social and physical worlds (Fig. 9.23). Here there is a statistically significant difference between those with higher levels of involvement with policy makers (mean 4.8) and those with lower levels of involvement (mean 3.7). Those with less involvement with the political realm tended to perceive that scientists have a higher level of understanding of the sensitivity of the human social system than did those with higher levels of involvement with the political realm. Perhaps higher levels of policy involvement create a greater awareness of the complexities of the social world, while those with

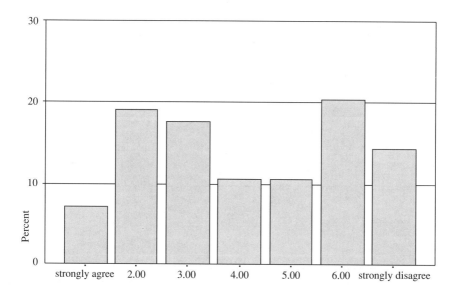

Fig. 9.22. Natural scientists have established enough physical evidence to turn the issue of global climate change over to social scientists for matters of policy discussion

a lesser degree of integrated involvement, that is limited activity in the political sphere, maintain a somewhat naive interpretation of the social world. There were no statistically significant differences when concerning the group with higher levels of media contact, perhaps suggesting a contribution to the reason why somewhat naive statements often find their way to the press, and ultimately, why media might begin to feedback into science.

That scientific claims are reaching the public and political ear could not be contested. The next section of this paper explicitly asks climate scientists how they perceive the relationship between the scientific community and the external users and reporters of scientific knowledge. This is a descriptive account of the *transfer and transformation* of knowledge. In light of the data presented above, this is an issue of particular relevance, for once outside of the scientific community, hypotheses and propositions are often ascribed the status of scientific fact. In addition to the science community, three key players enter into the debate of climate change, namely those who design climate related social and economic policies, the media and the general public.

The interaction of the scientific community and policy makers is by no means a new phenomenon. What is historically (somewhat) unique in this case however is the fact that climate policy has global implications. The following provides a descriptive account of some elements of the perceived

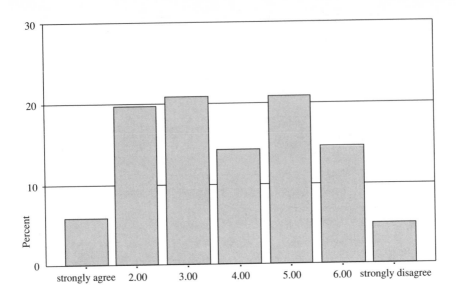

Fig. 9.23. Climate scientists are well attuned to the sensitivity of human social systems to climate impacts

relationships that exist among science, policy makers, the media, and the general public. It is assumed that ultimately these relationships all constitute a system of feedbacks. The importance of such considerations is further emphasized when one considers that social-feedback loops are most noticeably absent in all but a few impact scenarios.

The science-policy relationship is addressed in Fig. 9.24 and presents the results of the assessment of the relationship made by climate scientists. The analysis of the data produced a statistically significant difference between the means of those scientists who have a high level of contact with policy makers (mean 4.2) and those who do not (mean 4.7). As would be expected, it was those scientists who claimed a high level of contact with policy makers that perceived the relationship with policy to be more satisfactory. There is no evidence of a statistically significant difference when considering that group that claimed a high level of involvement with the media and the rest of the climate science community in regards to the science-policy relationship.

Figure 9.25 indicates scientists' perceptions of the impact of their scientific endeavors on political matters. When asked about the impact of scientific knowledge on the political sphere an analysis of the data resulted in a statistically significant difference between those claiming a higher level of involvement with the policy sphere (mean 3.4) and those scientists claiming a lesser degree of involvement (4.0). Those scientists with a higher level of

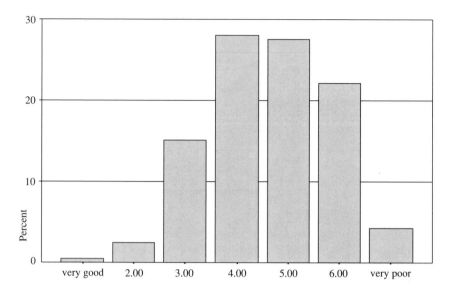

Fig. 9.24. How would you describe the working relationship between climate scientists and policy makers?

contact with policy makers perceived the efforts of science to be slightly more effective than those less involved with policy. This difference might simply indicate that those scientists involved with the policy arena might, as would be natural, have the tendency to slightly overstate their own impact. More noteworthy, however, is the apparent overall dissatisfaction with the process of the transfer of knowledge.

In short, the data indicates that while some scientists are inclined to believe that their efforts are instrumental in shaping policy, less perceive the working relationship that exists between policy makers and scientists to be satisfactory.

In an effort to determine if scientists felt there were a two way casual relationship between science and politics, that is, if as well as, a process of the scientification of politics there was a process of the *politicization of science*, they were asked if they felt that politics might influence the direction of scientific research. These results are summarized in Fig. 9.26 (Fig. 9.26 overall mean 4.4). The data indicates this to be a somewhat contentious area within the scientific community, with a wide variation of perspectives. In this instance, an absence of a statistically significant difference between those who have a high level of involvement with policy makers and those who do not is noteworthy. This line of reasoning was continued as represented in

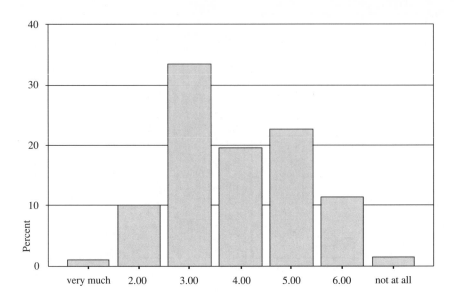

Fig. 9.25. To what degree do you think that the results of scientific inquiry are instrumental in causing policy makers to redefine their perceptions of climate related issues?

Fig. 9.27, where the influence of politics on the direction of science is given more prominence (overall mean 3.2).

When comparing the data of Fig. 9.26 and 9.27 it appears more unanimous that while politics are not perceived of as having a great influence on the individual perspectives of scientists, politics are perceived of as having a strong influence on the collective scientific community. This suggests the possibility that scientists are undertaking scientific research against that which would be of a natural inclination and raises the question of how influential public demand might be on politics and, in turn, on science. Again, it is noteworthy to mention the lack of a statistically significant difference between the mean responses of those with a higher level of interaction with policy makers and those who do not.

To determine how this political influence might be put into effect, scientists were asked to comment on the necessity to justify research in terms of policy relevance. As Fig. 9.28 indicates, scientists felt there is a great demand to justify research in terms of policy relevance.

Due to the political and social implications of the climate change issue, the fact that there was the inclination to perceive that research now has an added requirement of justification in terms of policy relevance is no surprise. Neither should it be surprising to note that those scientists with a higher

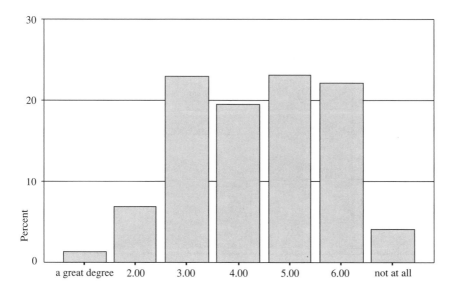

Fig. 9.26. To what degree are policy makers influential in causing scientists to redefine their perceptions of the climate issue?

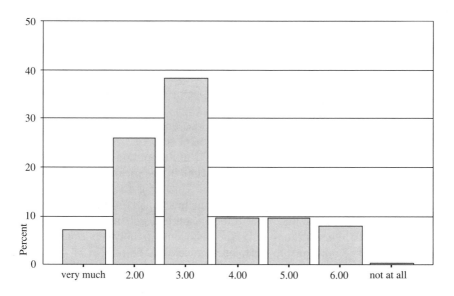

Fig. 9.27. How much do you think the direction of research in climate science has been influenced by external politics?

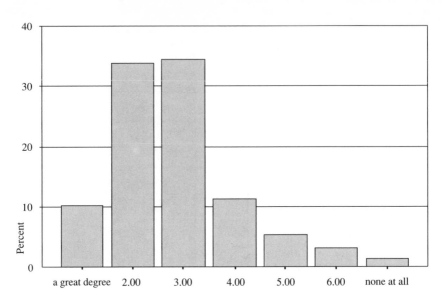

Fig. 9.28. To what degree do you think there is growing pressure for climate research to be justified in terms of policy relevance?

degree of contact with policy makers are slightly more aware of the situation (as represented by a statistically significant difference from the group with a lower level of contact (mean high policy contact 2.6, mean remainder of sample, 2.9). However, if, as Fig. 9.26 tends to indicate, there is a perception that policy makers are at least sometimes perceived as being influential in causing climate scientists to redefine their perspectives of the climate issue, not simply address policy relevant research, then there is a risk of the development of *science-for-politics*, a somewhat less than favorable situation and in contradiction to the notion of a value-neutral science.

If, as Auer et al. suggested (1996:145) "Public opinion and mass media have taken over the topic to such an extent that in the meantime it begins to repenetrate and influence scientific discussion.", there is the potential for the public to have an influence on science, then one has to ask from where does the public gain its information and what information is it getting? Consequently it is necessary to consider not only the relationship between science and policy, but also the relationship between science and the public.

The interaction between science and the public is mediated by other sources, one example being the media. To address this relationship, scientists were asked to comment on what they perceived to be the role of *scientists* in this process of knowledge transfer (Fig. 9.29). While generally the sample of the scientific community in this study was inclined to perceive

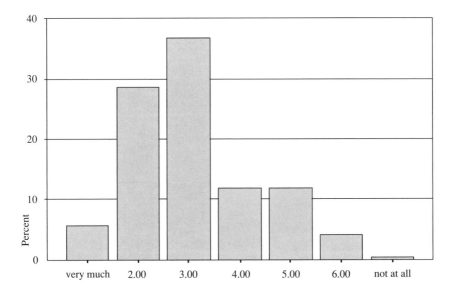

Fig. 9.29. How much have climate scientists played a role in transforming the climate issue from a scientific issue into a public issue?

itself as somewhat instrumental in the transformation of the scientific issue into a public issue, those who reported a higher level of involvement with the media attributed more to the role of the scientist in this area of concern (statistically significant different means: high media involvement 2.7, others 3.2) than those with less contact with the media.

As to the perceptions of acceptable practice in conveying scientific information to the public, scientists were asked to comment on the practice of presenting extreme statements and worst case scenarios (Fig. 9.30). There are a number of statistically significant differences. Those scientists with a high level of policy contact (mean 4.7) were less inclined to agree with the practice of presenting the extremes to the public than were the remainder of the sample (mean 4.1) although overall neither group condoned the practice. When concerning the group with the high level of media contact and the remainder of the sample there is no statistically significant difference. It is perhaps possible that those with a high level of policy contact, witness the public pressure as it is transformed back into the political spectrum and subsequently back into science. However, this remains a proposition beyond the limits of this data.

Nonetheless, as Fig. 9.31 indicates, the scientific community was quite explicit in stating the perception that the general public seldom get presented

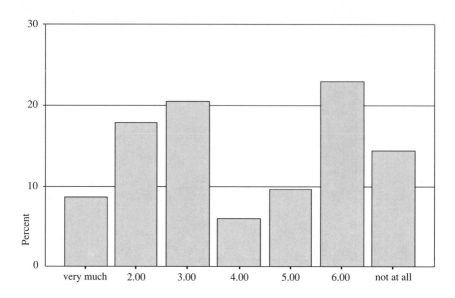

Fig. 9.30. Some scientists present the extremes of the climate debate in a popular format with the claim that it is his or her task to alert the public. How much do you agree with this practice?

with the full picture of the global warming issue. Here there are no statistically significant differences among the group.

In summary, Fig. 9.29 through 9.31 indicate that while scientists feel they have been somewhat instrumental in transforming a scientific debate into a public issue, the mechanism by which the transfer takes place is perceived as less than ideal.

If it is assumed that public wishes can be transformed into political actions, then the potential exists for an indirect public influence into the workings of science. Consequently, the content and context of information presented to the public might be an issue of greater significance than it is typically attributed. To reach a broad and diversified public audience requires a broker with the ability to transform scientific information into a format palatable for the general public. Here lies the role of the media. Consequently, scientists were asked to comment on both his or her personal experience with the media and on his or her perception of the general pattern of interaction between science and the media. First, scientists were asked if they felt a high level of media contact could influence the individual (Fig. 9.32). The results indicate mixed perceptions although some members of the scientific community express a high degree of concern. There were no statistically significant differences among the group of scientists with a high level of media contact,

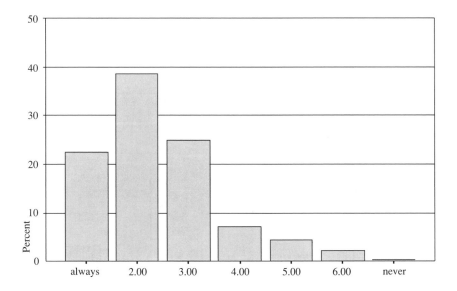

Fig. 9.31. How often are the general public presented only part of the picture?

a high level of political contact or the less vocal members of the scientific community.

Scientists were then asked if they felt the contact with the media could play a role in influencing the direction of future research (Fig. 9.33). The results of the analysis of the data presented in Fig. 9.33 indicate some concern that publicity via the media has the potential to influence the direction of future research. Here the data indicates statistically significant differences among all three groupings of scientists (means: high policy involvement 4.2, others 3.7, high media involvement 4.4). When compared to the rest of the sample, the scientists with a higher level of contact with the political realm saw less potential for publicity to influence the direction of research. When compared to the rest of the sample, scientists with higher levels of contact with the media perceived the same, that is, a lessor potential for publicity to influence the shaping of research. Nonetheless, the overall means (Fig. 9.32: 3.8, Fig. 9.33: 3.9) suggest concern within the scientific community regarding the science-media-public interface.

Assuming the potential influence of politics and the potential influence of media have come to play a role in the shaping of climate science, scientists were asked to assess this impact on the conduct of science. Scientists were questioned regarding one of the basic tenets of science, that is, to what degree he or she felt climate science has remained value-neutral (Fig. 9.34). Statistically significant differences evident between the group with

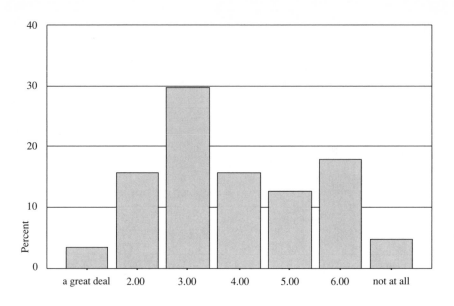

Fig. 9.32. To what degree does exposure to the media have the potential to change the attitudes of a scientist?

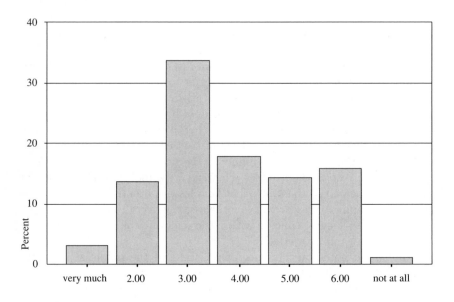

Fig. 9.33. How much do you think a scientist's exposure to publicity, influences the direction of his or her future research?

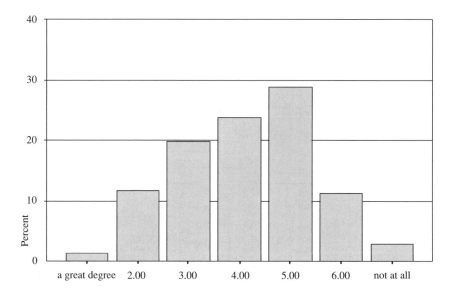

Fig. 9.34. To what degree do you think climate science has remained a value neutral science?

high levels of contact with policy makers (mean 4.5) and the remainder of the group.(mean 4.1) indicate that those with a higher degree of involvement with policy makers are more inclined to notice the shift away from value-neutrality, perhaps due to the greater insight of the external impositions on science. There was no statistical significant difference among the sample and the group with higher media contact. This perhaps suggests that those scientists that are privilege to the interactive process at the policy level are more aware of the process and outcomes. Nonetheless, the expression of the entire scientific community is less than favorable, depicting considerable concern that climate sciences are questionable in terms of value-neutrality, perhaps again pointing to the infusion of public and political concerns into an area of science with high levels of external vested interests, and again to the uniqueness of the global phenomenon at hand.

Furthermore, and unfortunately, scientific research is often forced into the position to enter into competition for limited funds. This too might, in the long run, have a negative impact on the progress of science since the continuation of research may require a reformulating of the problem so as to fit into the current vogue (Fig. 9.35).

There were no statistically significant differences among the groups when asked "How often do you think experts frame problems so that the solution fits his or her area of expertise?" The overall mean of the responses was

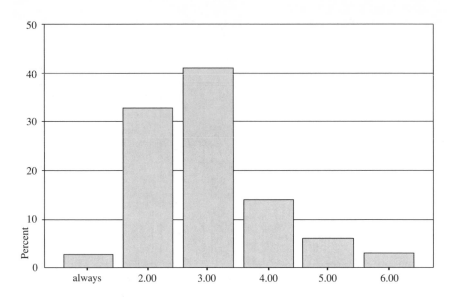

Fig. 9.35. How often do you think experts frame problems so that the solution fits his or her area of expertise?

2.97. This is highly indicative of a perception of such practices. While this practice might not necessarily be condoned by the scientist it may be a necessary means in times of limited and competitive funding. Unfortunately, this tendency may act to perpetuate disciplinary isolation at a time when interdisciplinarity and transdisciplinarity are demanded by the magnitude of the phenomenon at hand.

9.6 Conclusion

Climate science, according to the responses of the sample of this survey, has not developed the level of ability or accuracy of prediction with which it has been popularly attributed. Consequently, on the acceptance of *theory*, limited scientific knowledge has been transformed into a social and political issue. In spite of the inability to make certain, robust and reliable predictions, and to explicitly state the potential impacts, the impacts have been assigned a mostly negative connotation. One could assume this interpretation acts to maintain the issue as publicly and politically significant.

The reasonable consensus among the sample regarding the imperfection of the components of the science and the reasonable consensus among the sample regarding the inability of the science to provide a detailed picture of the future do not necessarily correspond to the differing perspectives that climate scientists have regarding the impacts of climate change that seem to reach the general public or the political body. One can only assume that larger forces than scientific objectivity are at play and that the cognitive interpretative mechanisms employed are shaped by the influence of personal persuasions, as well as, by scientific fact. However, those scientists with a greater level of outside contact tended to be more cautious in their claims than those scientists with less contacts, perhaps suggesting the potential that the levels of prominence of the issue is assigned *outside* of the scientific community, as much, if not more, than from within science.

Briefly, in spite of the discrepancies in the scientific interpretation of the general nature of global warming, there does seem to be a reasonable level of agreement that global climate change does indeed pose a major problem to be confronted. Considering the apparent lack of agreement as to the ability to *state* the detrimental effects (Fig. 9.8) it is difficult to determine on what basis scientists made their judgments in Fig. 9.7 through 9.9. This may be a case of sentiment, public, political or popular scientific, feeding back into the body of science.

Furthermore, the statistical mean 3.3 associated with the ability to say for certain that global warming is a process already underway (Fig. 9.5) suggests that this possibility still remains highly contested within the scientific community. The *potential* for the event, that is, the *theory* of global warming, however is much more readily accepted (Fig. 9.6) with a statistical mean of 2.6, although as the means suggests, even the acceptance of the theory is far from a unanimous consensus. Here, however, "theory" does not necessarily apply to the green house theory as put forward by Arrhenius, but to the notion that anthropogenic greenhouse effect will have a significant effect on, for example, mean global temperature.[1]

As to what it might mean when global warming arrives, the climate science community is much more reluctant to make a committed estimate, with responses falling mostly within the range of ambivalence. For example, when asked if a lack of preparation for climate change would lead to devastation (Fig. 9.7) the statistical mean of the response was 4.4, when asked about the ability to explicitly state the detrimental effects of climate change (Fig. 9.8) the mean of the responses was 4.5. Only when questioned about the detrimental effects for the *host* society (Fig. 9.7) does the mean (3.7) shift to an extremely weak position of commitment.

Yet, in spite of this expression of ambivalence and uncertainty, scientists in general were slightly more inclined to agree that global warming is a leading problem facing society (mean 3.3). Nonetheless, a mean of 3.3 is not representative of a collective call for alarm coming from the scientific community. It does, however, suggest that in spite of the lack of any specifics,

---

[1] These findings do no imply that the authors necessarily share the views put forward as responses to the survey. Indeed, Hans von Storch is co-author of an earlier study claiming the detection of climate change in the present global mean temperature record (see Hergel et al, 1995).

global warming appears to have been assigned, through whatever process, a negative connotation.

This analysis raises the question of how scientific uncertainty is transformed into high levels of public and political significance. This transition could not, as of yet, be attributed to the human experience since the experience of any expression of climate change, with the exception of extreme events (a highly contested relationship), is typically well below the thresholds of human climatic perception. However, it could not be denied that the *issue of climate change* has had, and creates the potential, for significant social impacts.

What we have discussed in this paper represents only one aspect of the science-politico-society triad. More specifically, we have addressed the role of the human element in the interpretation of scientific "fact" or, even more specifically, the *scientific construction* of the climate change issue. Not only does this suggest that, in light of the now *globalness* of many contemporary issues, the requirement to make assessments of all of the triadic interactions, but also to address the process by which multiple interpretations stem from a single scientific artifact.

9.7   Acknowledgements

We would like to thank the Thyssen Stiftung whose generous funding made this project possible, the Max-Planck-Institut für Meteorologie, Hamburg, and GKSS, Geesthacht for ongoing support, and all of those scientists who made time to participate in interviews and time to respond to the survey questionnaire.

References

The numbers given in parantheses at the end of the items in this bibliography refer to the sections in which the publication is quoted.

Abdulla, F.A., D.P. Lettenmaier, E.F. Wood and J.A. Smith, 1996: Application of a Macroscale Hydrological Model to Estimate the Water Balance of the Arkansas-Red River Basin, J. Geophys. Res. 101 (D3), 7449-7459 (3.3)

Ackerman, A.S., O.B. Toon, and P.V. Hobbs, 1993: Dissipation of marine stratiform clouds and collapse of the marine boundary layer due to the depletion of cloud condensation nuclei by clouds. Science 262, 226-229 (5.6)

Ackerman, A.S., O.B. Toon, and P.V. Hobbs, 1994: Reassessing the dependence of cloud condensation nucleus concentration on formation rate Nature 367, 445-447 (5.6)

Adler, R. F., G. J. Huffman and P.R. Keehn, 1994: Global rain estimates from microwave-adjusted geosynchronous IR data. Remote Sens. Rev., 11:125-152 (3.4)

Agbu, P., B. Vollmer and M. James, 1993: Pathfinder AVHRR land data set, NASA Goddard Space Flight Center, Greenbelt, MD (3.1)

Albrecht, B.A., 1989: Aerosols, cloud microphysics, and fractional cloudiness. Science, 245, 1227-1230 (5.6)

Arola, A., D. P. Lettenmaier and E. F. Wood, 1994: Some preliminary results of GCIP modeling activities in the Red-Arkansas River basin, First International Scientific Conference on the Global Energy and Water Cycle, Royal Society, London (3.4)

Arrhenius, S.A., 1896: On the influence of carbonic acid in the air upon the temperature of the ground. Philosophical Magazine and Journal of Science 41, 237-276 (1.1)

Arrhenius, S.A., 1908: Das Werden der Welten. Leipzig: Akademische Verlagsanstalt. (8.8)

Auer, I., R. Böhm and R. Steinacker, 1996: An opinion poll among climatologists about climate change topics. Meteorologische Zeitschrift, N.F.5, 145-155 (9.2)

Bakan, S., A. Chlond, U. Cubasch, J. Feichter, H. Graf, H. Grassl, K. Hasselmann, I. Kirchner, M. Latif, E. Roeckner, R. Sausen, U. Schlese, D. Schriever, I. Schult, U. Schumann, F. Sielmann and

W. Welke, 1991: Climate response to smoke from the burning oil wells in Kuwait. Nature 351, 367-371 (1.5)

Bárdossy, A., and E. J. Plate, 1992: Space-time model for daily rainfall using atmospheric circulation patterns. Water Resour. Res. 28, 1247-1259 (1.4)

Barnett, T.P. 1983: Interaction of the Monsoon and Pacific trade wind system at interannual time scale. Part I. Mon. Wea. Rev. 111, 756-773 (4.1)

Barnett, T.P. 1985: Variations in near-global sea level pressure. J. Atmos. Sci., 42, 478-501 (4.1)

Barnett, T.P. and M.E. Schlesinger, 1987: Detecting changes in global climate induced by greenhouse gases. J. Geophys. Res. 92, 14772-14780 (6.6)

Barnett T.P., M.E. Schlesinger and X. Jiang, 1991: On greenhouse gas detection strategies. In: M.E. Schlesinger (ed.): Greenhouse-Gas-Induced-Climatic Change: A Critical Appraisal of Simulations and Observations, Elsevier Science Publishers, Amsterdam, 537-558. (6.4)

Beljaars, A.C.M., P. Viterbo, M. Miller, and A. K. Betts., 1995: The anomalous rainfall over the USA during July 1993: sensitivity to land surface parameterization and soil moisture anomalies. Mon. Wea. Rev. 144:362-383 (3.3)

Bell, T.L., 1982: Optimal weighting of data to detect climatic change: Application to the Carbon Dioxide problem. J Geophys Res 87:11161-11170 (6.5)

Bell, T.L., 1986: Theory of optimal weighting of data to detect climatic change. J Atmos Sci 43:1694-1710 (6.5)

Bello, M., 1991: Greenhouse warming threat justifies immediate action. National Research Council News Report XLI, No. 4, 2-5 (5.6)

Ben-Zvi, A., 1997: Comments on "A new look at the Israeli cloud seeding experiments." J. Appl. Met. 36, 255-256 (5.2)

Bigg, E.K., 1997: An independent evaluation of a South African hydroscopic cloud seeding experiment, 1991-1995. Atmos. Res. 43, 111-127 (5.4)

Binmore, K., 1992: Fun and games. A text on game theory, D.C. Heath and co., Lexington, Toronto, 642 pp. (7.5)

Blank, J.E. and W.J. Ströbele, 1994: The economics of the CO_2 problem. What about the supply side? Rep., Univ. Oldenburg (7.5)

Block, F., 1985: Postindustrial development and the obsolescence of economic categories', Politics and Society 14:416-441 (8.7)

Block, W., 1989: Economic and the Environment: A Reconciliation. Vancouver: Fraser Institute. (8.2)

Boer, G.J., N.A. McFarlane, M. Lazare, 1992: Greenhouse gas induced climate change simulated with the CCC second-generation general circulation model. J Clim 5:1045-1077 (6.7)

Boer, G.J., G.M. Flato, M.C. Reader, D. Ramsden, 1998: Transient climate change simulation with historical and projected greenhouse gas and aerosol forcing. In manuscript. (6.4)

Bradley, R.S. and P.D. Jones, 1993: 'Little Ice Age' summer temperature variations: Their nature and relevance to recent global warming trends. The Holocene 3, 367-376 (6.4)

Bray, D. and H. von Storch, 1996: The climate change issue, perspectives and interpretations. Proceedings of the 14th International Congress of Biometeorology, Part 2 Vol. 3 pp. 439-450 (9.4)

Brinkop, S. and E. Roeckner, 1995: Sensitivity of a general circulation model to paramterizations of cloud-turbulence interactions in the atmospheric boundary layer. Tellus 47A, 197-220 (1.5)

Brocker, W.S., Peteet, D.M., and Rind, D., 1985: Does the ocean-atmosphere system have more than one stable mode of operation? Nature, 315, 21-26.(4.1)

Brooks, W.T., 1989: The global warming panic. Forbes, December, 97-102 (5.6)

Browning, K.A., R.J. Alam, S.P. Ballard, R.T. Barnes, D.E. Bennetts, R.H. Maryon, P.J. Mason, D. McKenna, J.F.B. Mitchell, C.A. Senior, A. Slingo and F.B. Smith, 1991: Environmental effects from burning oil wells in Kuwait. Nature 351, 363-367 (1.5)

Bruce, J.P., H. Lee and E.F. Haites (eds.), 1996: Climate Change 1995. Economic and Social Dimensions of Climate Change. Contribution of Working Group III to the Second Assessment Report of the Intergovernmental Panel of Climate Change. Cambridge: Cambridge University Press. (8.7)

Brückner, E., 1890: Klimaschwankungen seit 1700 nebst Bemerkungen über die Klimaschwankungen der Diluvialzeit. Geographische Abhandlungen, herausgegeben von Albrecht Penck in Wien; Wien and Olmütz, E.D. Hölzel. (8.5)

Brückner, E., 1915: The settlement of the United States as controlled by climate and climatic oscillations. In: Memorial Volume of the Transatlantic Excursion of 1912 of the American Geographical Society, 125-129 (8.5)

Brutsaert, W., 1975: On a derivable formula for long-wave radiation from clear skies, Water Resour. Res., 11, 742-744 (3.4)

Bryan, K., 1969: A numerical method for the study of the circulation of the world ocean. J. Comput. Phys., 4:347–376 (2.5)

Bryan, K., S. Manabe and R.L. Pacanowski, 1975: A global ocean-atmosphere climate model. Part II: The oceanic circulation. J. Phys. Oceanogr., 5:30–46 (2.5)

Bryson, R.A., 1989: Environmental opportunities and limits for development. Environ. Conserv., 16, 299-305 (5.7)

Busuioc, A. and H. von Storch, 1996: Changes in the winter precipitation in Romania and its relation to the large scale circulation. Tellus 48A, 538-552 (1.4)

Cahalan, R., 1992: Kuwait Oil Fires as seen by Landsat. J. Geophys. Res., 97, 14,565-14,57 (1.5)

Chappell, C.F., L.O. Grant, P.W. Mielke, 1971: Cloud seeding effects on precipitation intensity and duration of wintertime orographic clouds. J. Appl. Meteor. 10, 1006-1010 (5.2)

Chen, T.H. et al., 1997: Cabauw Experimental Results from the Project for Intercomparison of Land-Surface Parameterization Schemes, accepted J. Clim., 10(6), 1194-1215 (3.3)

Cline,W.R.(1992: The economics of global warming, Inst.Internat.Econ., 399 pp. (7.3)

Conway, D., R.L. Wilby, and P.D. Jones, 1996: Precipitation and air flow indices over the British Isles. Clim. Res. 7:169-183 (1.4)

Cooper, W.A., and R.P. Lawson, 1984: Physical interpretation of results from the HIPLEX-1 experiment. J. Climate Appl. Meteor. 23, 523-540 (5.2)

Cooper, W.A., R.T. Bruintjes, and G.K. Mather, 1997: Some calculations pertaining to hygroscopic seeding with flares. J. Appl. Met., In press (5.4)

Copeland, J.H., R.A. Pielke, T.G.F. Kittel, 1996: Potential climatic impacts of vegetation change: A regional modeling study. J. Geophys. Res. 101, D3, 7409-7418 (1.5)

Cotton, W.R., 1972a: Numerical simulation of precipitation development in supercooled cumuli, Part I. Mon. Wea. Rev., 100, 757-763 (5.3)

Cotton, W.R., 1972b: Numerical simulation of precipitation development in supercooled cumuli, Part II. Mon. Wea. Rev., 100, 764-784 (5.3)

Cotton, W.R. and R.A. Pielke, 1992: Human Impacts on Weather and Climate. ASTeR Press Ft. Collins, (ISBN 0-9625986-1-5), 288pp. (5.1)

Cotton, W.R., and R.A. Pielke, 1995: Human Impacts on Weather and Climate. Cambridge Univ. Press, 288 pp. (5.1, 5.2, 5.4, 5.5)

Crowley, T.J. and G. R. North, 199: Paleoclimatology. Oxford University Press, New York, 330 pp.(1.2)

Cubasch U, K. Hasselmann, H. Höck, E. Maier-Reimer, U. Mikolajewicz, B.D. Santer, R. Sausen, 1992: Time-dependent greenhouse gas warming computations with a coupled ocean-atmosphere model. Clim. Dyn. 8, 55-69 (6.4, 8.3)

Cubasch, U., G. Hegerl, A. Hellbach, H. Höck, U. Mikolajewicz, B. D. Santer and R. Voss, 1995: A climate change simulation starting at 1935. Climate Dynamics 11, 71-84 (8.3, 6.4)

Cubasch, U., G.C. Hegerl, R. Voss, J. Waszkewitz, T.J. Crowley, 1997: Simulation with an O-AGCM of the influence of variations of the solar constant on the global climate. Clim. Dyn., in press (6.4)

De Cosmo, J., K.B. Katsaros, S.D. Smith, R.J. Anderson, W.A. Oost, K. Bumke and H. Chadwick, 1996: Air-sea exchange of water vapour and sensible heat: The Humidity Exchange Over the Sea (HEXOS) results. J. Geophys. Res. 101, C5: 12,001-12,016 (1.5)

Dennis, A.S., and H.D. Orville, 1997: Comments on "A new look at the Israeli cloud seeding experiments." J. Appl. Met. 36, 277-278 (5.2)

Dickinson, R.E., 1992: Chapter 5, Land Surface. In K.E. Trenberth, ed., Climate System Modeling. Cambridge University Press, 788 pp. (2.4)

Drucker, P.F., 1986: The changed world economy, Foreign Affairs 64:768-791. (8.7)

Dubayah, R., E.F. Wood, M. Zion and K. Czajkowski., 1997: A remote sensing approach to macroscale hydrological modeling, in Schultz, G. and E. Engman, eds., Remote sensing in hydrology and water management, Springer-Verlag. (3.2)

Dubayah, R. and S. Loechel, 1997: Modeling topographic solar radiation using GOES data, J. Applied Meteor., 36:141-154 (3.4)

Dumenil, L., and E. Todini, 1992: A rainfall-runoff scheme for use in the Hamburg climate model, in Advances in theoretical hydrology: A tribute to James Dooge, Eur. Geophys. Soc. Ser. on Hydrol. Sci., vol 1, edited by J. P. O'Kane, pp. 129-157, Elsevier, New York (3.3)

Dunlap, R. E., G. H. Gallup, Jr., and A. M. Gallop, 1993: Health of the Planet. Gallup International Institute, Princeton, New Jersey, USA (9.2)

Eder, K., 1996: The institutionalisation of environmentalism: Ecological discourse and the second transformation of the public sphere, pp.203-223 in: Scott Lash, Bronislaw Szerszynski and Brian Wynne (eds.), Risk, Environment and Modernity. Towards a New Ecology. London: Sage. (8.3)

Enke W., and A. Spekat, 1997: Downscaling climate model outputs into local and regional weather elements by classification and regression. Clim. Res. 8 (in press) (1.4)

Fischer, G., E. Kirk and R. Podzun, 1991: Physikalische Diagnose eines numerischen Experiments zur Entwicklung der grossräumigen atmospärischen Zirkulation auf einem Aquaplaneten. Meteor. Rdsch. 43, 33-42 (1.3)

Flato, G.M. and W.D. Hibler, III, 1992: Modeling pack ice as a cavitating fluid, J. Phys, Oceanogr. 22, 626–651 (2.4)

Flato, G. M., G.J. Boer, W.G. Lee, N.A. McFarlane, D. Ramsden, M.C. Reader, A.J. Weaver, 1998: The Canadian Centre for Climate Modelling and Analysis global coupled model and its climate. In manuscript. (6.4)

Frey-Buness, F., D. Heimann and R. Sausen, 1995: A statistical-dynamcial downscaling procedure for global climate simulations. Theor. Appl. Climatol. 50, 117-131 (1.4)

Fuentes, U. and D. Heimann, 1996: Verification of statistical-dynamical downscaling in the Alpine region. Clim. Res. 7:151-168 (1.4)

Gagin, A., 1971: Studies of factors governing the collidal stability of continental clouds. Preprints, Int. Conf. on Weather Modification, Canberra, Australia, Amer. Meteor. Soc., 5-11 (5.2)

Gagin, A., 1975: The ice phase in winter continental cumulus clouds. J. Atmos. Sci. 32, 1604-1614 (5.2)

Gagin, A., 1986: Evaluation of "static" and "dynamic" seeding concepts

328   *References*

through analyses of Israeli II and FACE-2 experiments. Rainfall Enhancement–
A Scientific Challenge, Meteor. Monogr., No. 43, Amer. Meteor. Soc., 63-70
(5.2)

Gagin, A., and J. Neumann, 1974: Rain stimulation and cloud physics in
Israel. Weather and Climate Modification, W.N. Hess, ed., Wiley-Interscience,
454-494 (5.2)

Gagin, A., and J. Neumann, 1981: The second Israeli randomized cloud
seeding experiment: Evaluation of the results. J. Appl. Meteor. 20, 1301-1311
(5.2)

**Gibson, J.K., Kallberg, P., Uppala, S., Hernandez, A., Nomura,
A., and Serrano, E., 1997:** ECMWF re-analysis project report series, 1.
ERA Description.(4.1)

Giering,R. and T. Kaminski, 1996: Recipes for adjoint code construction. Submitted to ACM Transactions on Mathematical Software (7.2)

Global Environmental Change Report, 1990: GECR climate survey
shows strong agreement on action, less so on warming. Global Environmental Change Report 2, NO. 9, pp. 1-3 (9.4)

Goward, S. N., R. H. Waring, D. G. Dye, and J. Yang, 1994: Ecological remote sensing at OTTER: satellite macroscale observations, Ecological
Appl., 4, 332-343. (3.4)

Grant, L.O., and P.W. Mielke, 1967: A randomized cloud seeding experiment at Climax, CO, 1960-65. Proc. Fifth Berkeley Symp. on Mathematical
Statistics and Probability, Vol, 5, University of California Press, 115-132 (5.2)

**Grant, L.O., C.F. Chappell, L.F. Crow, P.W. Mielke, Jr., J.L. Ras-
mussen, W.E. Shobe, H. Stockwell, and R.A. Wykstra, 1969:** An operational adaptation program of weather modification for the Colorado River
basin. Interim Report to the Bureau of Reclamation, Denver, CO 80225, Colorado State University, 69 pp. (5.2)

Greenpeace International; 1992: Climate scientists fear effects of underestimating global warming, poll shows, Press release, Feb. 9 (9.4)

Grilli, E.R. and M.C. Yang, 1988: Primary commodity prices, manufactured good prices, and the terms of trade of developing countries: What
the long run shows, The World Bank Economic Review 2:1-47. (8.7)

Grove-White, R., 1996: Environmental knowledge and public policy needs:
On humanising the research agenda. Pp. 269-286 in Scott Lash, Bronislaw
Szerszynski and Brian Wynne (eds.), Risk, Environment and Modernity. To-
wards a New Ecology. London: Sage. (8.7)

Grundmann, R. and N. Stehr, 1997: Klima und Gesellschaft, Soziol-
ogische Klassiker und Aussenseiter: Über Weber, Durkheim, Simmel und
Sombart Soziale Welt (forthcoming). (8.4)

**Günther, H., W. Rosenthal, M. Stawarz, Carretero, J.C., M. Gomez,
I. Lozano, O. Serano, Reistad, M., 1997:** The wave climate of the Northeast Atlantic over the period 1955-1994: The WASA wave hindcast. GAOS
(in review) (1.4)

Hallett, J., 1981: Ice crystal evolution in Florida summer cumuli following AgI seeding. Preprints, Eighth Conf. on Inadvertent and Planned Weather Modification, Reno, American Meteorological Society, 114-115 (5.2)

Harding, G., 1968: The tragedy of the commons. Science 162, 1243-1248 (8.6)

Harvey, D., 1993: The Nature of Environment: The Dialectics of Social and Environmental Change", p. 1-49 in Ralph Miliband and Leo Panitch (eds.), The Socialist Register. Real Problems, False Solutions. London: Merlin Press. (8.2)

Hasselmann, K., 1976: Stochastic climate models. Part I. Theory. Tellus 28, 473-485 (1.5, 4.1, 7.7)

Hasselmann, K., 1979: On the signal-to-noise problem in atmospheric response studies. In: Meteorology Over the Tropical Oceans, Shaw BD (ed.), Royal Meteorological Society, Bracknell, Berkshire, England. pp251-259 (6.1)

Hasselmann, K., 1988: PIPs and POPs: The reduction of complex dynamical systems using Principal Interaction and Oscillation Patterns. J. Geophys. Res. 93, 11015-11021 (4.1)

Hasselmann, K., 1990: How well can we predict the climate crisis? In: H. Siebert (ed.) Environmental Scarcity - The International Dimension. J.C.B Mohr, Tübingen, 165-183 (8.6)

Hasselmann, K., 1993: Optimal fingerprints for the Detection of Time dependent Climate Change. J. Clim. 6, 1957-1971 (6.1)

Hasselmann, K., 1997: Multi-pattern fingerprint method for detection and attribution of climate change. Clim. Dyn. 13, 601-611 (6.1)

Hasselmann, K., 1997: Path optimization problems in integrated assessment studies. Proc. Hawaii 'Aha Huliko'a Workshop, Jan.1997 (in press) (7.1)

Hasselmann, K., 1998: Conventional and Bayesian approach to climate change detection and attribution. In manuscript. (6.4)

Hasselmann, K. and S. Hasselmann, 1997: Multi-actor optimization of greenhouse-gas emission paths using coupled integral climate response and economic models, Proc.Potsdam Symp."Earth System Analysis: Integrating Science for Sustainability", 1994 (7.1, 8.6)

Hasselmann, K., R. Sausen, E. Maier-Reimer and R. Voss, 1993: On the cold start problem in transient simulations with coupled atmosphere-ocean models, Climate Dynamics 9, 53-61 (7.2)

Hasselmann, K., L. Bengtsson, U. Cubasch, G.C. Hegerl, H. Rodhe, E. Roeckner, H. von Storch, R. Voss, J. Waszkewitz, 1995: Detection of anthropogenic climate change using a fingerprint method. MPI report 168 and P. Ditlevsen (ed.), Proc 'Modern Dynamical Meteorology', Symposium in Honor of Aksel Wiin-Nielsen, ECMWF Press, pp. 203-221 (6.4)

Hasselmann, K., S. Hasselmann, R. Giering, V. Ocaña and H. von Storch, 1996: Optimization of CO_2 emissions using coupled integral climate response and simplified cost models. A sensitivity study MPI Rep. 192, IIASA workshop series. (7.2, 8.6)

Hasselmann, K., S. Hasselmann, R. Giering, V. Ocaña and H.v. Storch, 1997: Sensitivity study of optimal CO_2 emission paths using a simplified Structural Integrated Assessment Model (SIAM). Climate Change (in press) (7.1)

Hegerl, G.C., H. von Storch, K. Hasselmann, B.D. Santer, U. Cubasch, P.D. Jones, 1996: Detecting Greenhouse-Gas-Induced Climate Change with an Optimal Fingerprint Method. J. Climate 9, 2281-2306 (6.4, 8.3)

Hegerl, G.C., K. Hasselmann, U. Cubasch, J.F.B. Mitchell, E. Roeckner, R. Voss, J. Waszkewitz, 1997: Multi-fingerprint detection and attribution analysis of greenhouse gas, greenhouse gas-plus-aerosol and solar forced climate change. Clim. Dyn. 13, 613-634 (6.1)

Hegerl, G.C. and G.R. North, 1997: Statistically optimal methods for detecting anthropogenic climate change. J. Clim. 10, 1125-1133 (6.7)

Henderson-Sellers, A., A. Pitman, P. Love, P. Irannejad and T. Chen., 1996: The project for the intercomparison of land surface parameters schemes, PILPS: phases 2 and 3, Bull. Amer. Meteor. Soc., 76:489-503 (3.1)

Herrnstein, R.J. and C. Murray, 1994: The Bell Curve. Intelligence and Class Structure in American Life. New York: Free Press. (8.5)

Heyen, H., E. Zorita and H. von Storch, 1996: Statistical downscaling of winter monthly mean North Atlantic sea-level pressure to sea-level variations in the Baltic Sea. Tellus 48 A, 312-323 (1.4)

Hense, A., M. Kerschgens and E. Raschke, 1982: An economical method for computing the radiative energy transfer in circulation models. Quart. J. Roy. Meteor. Soc. 108, 231-252 (1.5)

Hewitson, B.C. and R.G. Crane, 1992: Large-scale atmospheric controls on local precipitation in tropical Mexico. Geophys. Res. Lett. 19, 1835-1838 (1.4)

Hewitson, B.C. and R.G. Crane, 1996: Climate Downscaling: techniques and application Clim. Res. 7, 85-95 (1.4)

Hibler, W.D. III, 1979: A dynamic thermodynamic sea ice model. J. Phys. Oceanogr., 9:815–846 (2.6)

Hindman, E.E., 1978: Water droplet fogs formed from pyrotechnically generated condensation nuclei. J. Wea. Mod. 77-96 (5.4)

Hindman, E.E., II, P.V. Hobbs, and L.F. Radke, 1977a: Cloud condensation nuclei from a paper mill. Part I: Measured effects on clouds. J. Appl. Meteor. 16, 745-752 (5.4)

Hindman, E.E., II, P.M. Tag, B.A. Silverman, and P.V. Hobbs, 1977b: Cloud condensation nuclei from a paper mill. Part II: Calculated effects on rainfall. J. Appl. Meteor. 16, 753-755 (5.4)

Hobbs, P.V., L.F. Radke, and S.E. Shumway, 1970: Cloud condensation nuclei from industrial sources and their apparent influence on precipitation in Washington State. J. Atmos. Sci. 27, 81-89 (5.4)

Hollinger J., J. Pierce, G. Poe., 1990: SSM/I instrument evaluation, IEEE Trans. On Geosc. and Rem. Sens., 28: 781-790, 1990. (3.1)

Huffman, G. J., 1996: GPCP Version 1a combined precipitation data set documentation, Laboratory for Atmospheres, NASA Goddard Space Flight Center, September 9 (3.4)

Hunke, E.C. and J. K. Dukowicz, 1997: An elastic-viscous-plastic model for sea ice dynamics. J. Phys. Oceanogr., in press. (2.6)

Huntington, E., 1925: Civilization and Climate. Yale University Press, 2nd edition. (8.5)

Huntington, E. 1945: Mainsprings of Civilization. New York: John Wiley and Sons. (8.5)

Huntington, E. and S.S. Visher, 1922: Climatic Changes. Yale Univeristy Press, New Heaven, 329 pp.(4.1)

Inaudil, D., X. Collona de Lega, A. Di Tullio, C. Forno, P. Jacquot, M. Lehmann, M. Monti and S. Vurpillot, 1995: Experimental evidence for the Butterfly Effect. Ann. Improb. Res. 1, 2–3 (1.5)

Inglehart, R., 1977: The Silent Revolution. Princeton: Princeton University Press. (8.7)

Inglehart, R., 1987: Value change in industrial society, American Political Science Review 81:1289-1303 (8.7)

International Monetary Fund, 1992: World Economic Outlook. May 1992. Washington, D.C.: The Fund. (8.7)

IPCC, 1990: Climate Change. The IPCC Scientific Assessment (Houghton, J.T. G.J.Jenkins, J.J.Ephraums, eds. Cambridge University Press, Cambridge (7.3)

IPCC, 1991: Climate Change, The IPCC Scientific Assessment Intergovernmental Panel on Climate Change. WMO/UNEP, xi-xxxiii (5.6)

IPCC, 1992: Climate Change 1992. The supplementary report to the IPCC Assessment (J.T. Houghton, B.A. Callander, S.K. Varney., eds. Cambridge University Press, Cambridge (7.3)

IPCC, 1996a: Climate Change 1995. The science of climate change. Contribution of Working Group 1 to the second assessment report of the Intergovernmental Panel on Climate Change. Houghton, J.T., L.G. Meiro Filho, B.A. Callender, N. Harris, A. Kattenberg and K. Maskell, eds., Cambridge University Press, Cambridge, 572 pp. (7.1)

IPCC, 1996b: Climate Change 1995. Economic and social dimensions of climate change. Contribution of Working Group 3 to the second assessment report of the Intergovernmental Panel on Climate Change. J.P. Bruce, H: Lee, E.F.Haites, eds., Cambridge University Press, Cambridge, 448 pp. (7.3)

Jarvis, P. G., 1976: The interpretations of the variations in leaf water potential and stomatal conductance found in canopies in the field. Roy. Soc. London, Phil. Trans. Ser. B 273:593-610 (3.3)

Jenkins, G.M. and D.G. Watts, 1968: Spectral analysis and its application. Holden-Day, 525 pp. (4.1)

Johansen, O., 1975: "Thermal conductivity of soils, Ph.D. Thesis, Trondheim, Norway, Corps of Engineers/CRREL Translation 637, 1977 ADA 044002. (3.3)

Johns, T.C., R.E. Carnell, J.F. Crossley, J.M. Gregory, J.F.B. Mitchell, C.A. Senior, S.F.B. Tett, R.A. Wood, 1997: The second Hadley Centre Coupled Model: description, spinup and validation. Clim. Dyn. 13, 103-134 (6.4)

Jones, P.D., 1995: The Instrumental Data Record: Its Accuracy and Use in Attempts to Identify the "CO2 Signal". In: H. von Storch and A. Navarra (eds.) "Analysis of Climate Variability: Applications of Statistical Techniques", Springer Verlag, 53-76, (ISBN 3-540-58918-X) (1.4)

Jones, P.D. and K.R. Briffa, 1992: Global surface air temperature variations during the twentieth century: Part I, spatial, temporal and seasonal details. The Holocene 2, 165-179 (6.4)

Jones, P.D., T.J. Osborn and K.R. Briffa, 1997: Estimating sampling errors in large-scale temperature averages. J. Clim. 10, 2548-2568 (6.4)

Kaas, E., T.-S. Li and T. Schmith, 1996: Statistical hindcast of wind climatology in the North Atlantic and Northwestern European region. Clim. Res. 7:97-110 (1.4)

Kang I.-K. and Lau K.-M., 1994: Principal modes of atmospheric circulation anomalies associated with global angular momentum fluctuations. J. Atmos. Sci., 51, 1194-1205.(4.1)

Karl, T.R., R.G. Quayle and P.Y. Groisman, 1993: Detecting climate variations and change: New challenges for observing and data management systems. J. Climate 6, 148145-1494 (1.4)

Karoly, D.J., J.A. Cohen, G.A. Meehl, J.F.B. Mitchell, A.H. Oort, R.J. Stouffer, R.T. Wetherald, 1994: An example of fingerprint detection of greenhouse climate change. Clim. Dyn. 10, 97-105 (6.1)

Kates, R.W., 1988: Theories of nature, society and technology. Pp. 7-36 in Erik Baark and Uno Svendin (eds.), Man, Nature and Society. Essays on the Role of Ideological Perceptions. Houndmills, Basingstroke: Macmillan. (8.1)

Katz, R.W. and M. B. Parlange, 1996: Mixtures of stochastic processes: applications to statistical downscaling. Clim. Res. 7:185-193 (1.4)

Kellogg, W.W., 1991: Response to skeptics of global warming. Bull. Amer. Meteor. Soc. 72, 499-511 (5.6)

Kempton, W., J. S. Boster and J. A. Hartley, 1995: Environmental Values in American Culture. The MIT Press, Cambridge, Mass,. USA (9.2)

Kershaw, I., 1973: The great famine and agrarian crisis in England, 1315-1322. Past Present 59, 3-50 (8.6)

Keynes, J.M.,1936: The General Theory of Employment, Interest and Money. London: Macmillan. (8.7)

Koenig, L.R., and F.W. Murray, 1976: Ice-bearing cumulus cloud evolution: Numerical simulations and general comparison against observations.

J. Appl. Meteor. 15, 747-762 (5.3)

Köppen, W., 1923: Die Klimate der Erde. Walter de Gruyter, Berlin (1.1)

La Brecque, M., 1989a: Detecting climate change, I: Taking the world's shifting temperature. MOSAIC 20, No. 4, 2-9 (5.6)

La Brecque, M., 1989b: Detecting climate change, II: The impact of the water budget. MOSAIC 20, 10-16 (5.6)

Lacis, A.A., and J.E. Hansen, 1974: A parameterization for the absorption of solar radiation in the earth's atmosphere, J. Atmos. Sci. 31, 118-133 (2.3)

Lamb, D., J. Hallet, and R.I. Sax, 1981: Mechanistic limitations to the release of latent heat during the natural and artificial glaciation of deep convective clouds. Quart. J. Roy. Meteor. Soc. 107, 935-954 (5.3)

Lasch, C., 1991: The True and Only Heaven: Progress and its Critics. New York: W.W. Norton. (8.7)

Lee, R.B., M.A. Gibson, R.S. Wilson, S. Thomas, 1995: Long-term total solar irradiance variability during sunspot cycle 22. J Geophys Res 100, 1167-1675 (6.5)

Lemke, P.I., G. Flato, M. Harder, and M. Kreyscher, 1997: On the improvement of sea ice models for climate simulations: The sea ice model intercomparision project. Ann. of Glaciol., in press. (2.6)

Leroy, S. 1998: Detecting climate signals: some Bayesian aspects. J. Clim., in press (6.4)

Lettenmaier, D., 1993: Stochastic Modeling of precipitation with applications to climate model downscaling. In: H. von Storch and A. Navarra (eds.) "Analysis of Climate Variability: Applications of Statistical Techniques", Springer Verlag, 197-212 (ISBN 3-540-58918-X) (1.4)

Lettenmaier, D.P., D. Lohmann, E.F. Wood, and X. Liang, 1996: PILPS-2c draft workshop report: Report of a Workshop Held at Princeton University, October 28-31, 1996, Internet http://earth.princeton.edu. (3.2)

Levin, Z., 1994: Aerosol composition and its effect on cloud growth and cloud seeding. Proc. Sixth WMO Scientific Conf. on Weather Modification, Siena, Italy, Worl Meteor. Org., 367-369 (5.2)

Levine, R.A. and M. Berliner, 1998: The statistics of fingerprinting. In manuscript. (6.7)

Liang, X., E.F. Wood, and D.P. Lettenmaier., 1996a: A One-Dimensional Statistical-Dynamic Representation of Subgrid Spatial Variability of Precipitation in the Two-layer VIC Model, J. Geophys. Res., 101(D16):21403-21422, September 27. (3.3)

Liang, X., E.F. Wood and D.P. Lettenmaier, 1996b: "Surface Soil Moisture Parameterization of the VIC-2L Model: Evaluation and Modifications", Global and Planetary Change, 13, 195-206 (3.3)

Liang, X., D.P. Lettenmaier, E.F. Wood and S.J. Burges., 1994: A Simple Hydrologically Based Model of Land Surface Water and Energy Fluxes for General Circulation Models, J. of Geophysical Research, 99, D7

14,415-14,428, July 20. (3.3)

Lohmann, U. and E. Roeckner, 1993: Influence of cirrus cloud radiative forcing on climate and climate variability in a general circulation model. J. Geophys. Res. 100 D, 16305-16323 (1.5)

Lorenz, E.N., 1963: Deterministic nonperiodic flow. J. Atmos. Sci., 20, 130-141 (4.1)

Lorenz, E.N., 1967: The Nature and Theory of the General Circulation of the Atmosphere. World Meteorological Oraginzation, 161 pp. (1.1)

Lunkeit, F., R. Sausen, J.M. Oberhuber, 1995: Climate simulations with the global coupled atmosphere-ocean model ECHAM2/OPYC. Part I: Present-day climate and ENSO events. Clim. Dyn. 12, 195-212 (6.4)

Luoma, J.R., 1991: Gazing into our greenhouse future. Audubon, March 52-59, 124-125 (5.6)

Maier-Reimer, E., 1993: The biological pump in the greenhouse, Global and Planetary Climate Change 8, 13-15 (7.2)

Maier-Reimer, E. and K. Hasselmann, 1987: Transport and storage of CO_2 in the ocean - an inorganic ocean-circulation carbon cycle model, Climate Dynamics 2, 63-90 (7.2)

Maier-Reimer, E., U. Mikolajewicz, and K. Hasselmann, 1993: Mean circulation of the Hamburg LSG OGCM and its sensitivity to the thermohaline surface forcing. J. Phys. Oceanogr., 23, 731-757 (4.1)

Manabe, S., 1969a: Climate and the ocean circulation: I. The atmospheric circulation and the hydrology of the earth's surface. Mon. Wea. Rev., 97:739–774 (2.4)

Manabe, S., 1969b: Climate and the ocean circulation: II. The atmospheric circulation and the effect of heat transfer by ocean currents. Mon. Wea. Rev., 97, 775–805 (2.4)

Manabe, S. and R.J. Stouffer, 1980: Sensitivity of a global climate model to an increase of CO_2 concentration in the atmosphere. J. Geophys. Res. 85, 5529-5554 (6.2)

Manabe, S. and R.J. Stouffer, 1988: Two stable equilibria of a coupled ocean-atmosphere model. J. Climate, 1, 841-866 (4.1)

Manabe, S. and R.J. Stouffer, 1996: Low frequency variability of surface air temperature in a 1000 year integration of a coupled ocean-atmosphere model. J. Clim. 9, 376-393 (6.4)

Marland, G., 1991: CO_2 emissions. In Boden, T.A., P. Kanciruk, and M.P. Farrell, Trends '90: A Compendium of Data on Global Change, Carbon Dioxide Information Analysis Center, Oak Ridge National Laboratory, Oak Ridge, Tennessee, 92-133 (5.7)

Mather, G.K., 1991: Coalescence enhancement in large multicell storms caused by the emissions from a Kraft paper mill. J. Appl. Meteor. 30, 1134-1146 (5.4)

Mather, G.K., M.J. Dixon, J.M. deJager, 1996a: Assessing the potential for rain augmentation—The Nelspruit randomized convective cloud seeding experiment. J. Appl. Meteor. 35, 1465-1482 (5.2)

Mather, G.K., D.E. Terblanche, F.E. Steffens, 1996b: Results of the South African cloud seeding experiments using hygroscopie flares. Preprints, 13th Conf. on Planned and Inadvertent Weather Modification, January 28 - February 2, 1996, Atlanta, GA, AMS, Boston, MA, 121-128 (5.4)

Matsuno, T., 1966: Quasi-geostrophic motions in the equatorial area. J. Met. Soc. Japan, 44, 25-42 (4.1)

Maykut, G.A., and N. Untersteiner, 1971: Some results from a time-dependent thermodynamic model of sea ice. J. Geophys. Res., 76:1550–1575 (2.6)

Mazur, A., 1996: Global environment in the news: 1987-90 vs. 1992-96. Manuscript. (8.3)

Mazur, A. and J. Lee, 1993: Sounding the global alarm: Environmental issues in the US national news. Social Studies of Science 23, 681-720 (8.3)

McCay, B.J. and J.M. Acheson, 1986: Human Ecology of the Commons. Pp. 1-34 in: Allan Schnaiberg et al. (eds.), Distribution Conflict in Environmental Resource Policy. Aldershof: Gower. (8.3)

McCumber, M.C. and R.A. Pielke, 1981: "Simulation of the effect of surface fluxes of hat and moisture in a mesoscale numerical model", J. Geophys. Res., 86(C10), 9929-9938, 1981. (3.3)

McPhee, M.G., 1975: Ice-ocean momentum transfer for the AIDJEX ice model. Aidjex Bull., 29:93–111 (2.6)

McWilliams, J.C., 1996: Modeling the oceanic general circulation. Ann. Rev. Fluid Mech., 28:215–248 (2.5)

Meesen, B.W., F.E. Corprew, J.M.P. McManus, D.M. Myers, J.W. Closs, K.J. Sun, J. Sunday, and P.J. Sellers, 1995: ISLSCP Initiative I-Global Data Sets for Land Atmosphere Models, 1987-1988, Vols. 1-5, NASA, CD-ROM (3.1)

Michaels, P.J., D.E. Sappington, D.E. Stooksbury, and B.P. Hayden, 1990: Regional 500 mb heights and U.S. 1000-500 mb thickness prior to the radiosonde era. Theor. Appl. Climatol. 42, 149-154 (5.6)

Mielke, P.W., Jr., 1995: Comments on the Climax I and II experiments including replies to Rangno and Hobbs. J. Appl. Meteor. 34, 1228-1232 (5.2)

Mielke, P.W., Jr., L.O. Grant, and C.F. Chappell, 1970: Elevation and spatial variation in seeding effects from wintertime orographic cloud seeding. J. Appl. Meteor. 9, 476-488 (5.2)

Mielke, P.W., Jr., L.O. Grant, and C.F. Chappell, 1971: An independent replication of the Climax wintertime orographic seeding experiment. J. Appl. Meteor. 10, 1198-1212 (5.2)

Mielke, P.W., Jr., G.W. Brier, L.O. Grant, G.J. Mulvey, and P.N. Rosensweig, 1981: A statistical re-analysis of the replicated Climax I and II wintertime orographic cloud seeding experiments. J. Appl. Meteor. 20, 643-660 (5.2)

Mikolajewicz, U. and E. Maier-Reimer, 1990: Internal secular variability in an OGCM. Climate Dyn. 4, 145-156 (4.1)

Mulberg, J., 1996: Environmental planning, economic planning and political economy. International Sociology 11, 441-456.

National Academy of Sciences, 1975: Long-Term Worldwide Effects of Multiple Nuclear-Weapon Detonations. Washington, D.C. (5.2)

National Academy of Sciences, 1991: Policy Implications of Greenhouse Warming – Synthesis Panel Report. National Academy Press, Washington, D.C., 127 pp (5.6)

NDU, 1978: Climate Change to the Year 2000. National Defense University, Washington D.C. (9.4)

Nijssen, B., D.P. Lettenmaier, X. Liang, S.W. Wetzel and E. F. Wood, 1998: "Simulation of Runoff from Continental-Scale River Basin using a Grid-Based Land Surface Scheme", accepted Water Resour. Res., 33(4):711-724. (3.3)

Noilhan, J. and S. Planton, 1989: A simple parameterization of land surface processes for meteorological models, Mon. Wea. Rev., 117:536-549 (3.3)

Nordhaus, W.D.; 1991: To slow or not to slow: The economics of the greenhouse effect, Econ. J. 101, 920-937 (7.3, 8.6)

Nordhaus, W.D.; 1993: Rolling the 'DICE': An optimal transition path for controlling greenhouse gases, Resource and Energy Economics 15, 27-50 (7.3)

Nordhaus, W.D., 1994: The ghosts of climate past and the specters of climate future. In: Nakicenovic, Nordhaus, Richels, Toth (ed.): Integrative Assessment of Mitigation, Impact and Adaptation to Climate Change. IIASA, May 1994, 35-62 (8.3)

Nordhaus, W.D., 1997: Climatic Change (in press) (7.2)

Nordhaus, W.D. and Z. Yang, 1996: A Regional Dynamic General-Equilibrium Model of Alternative Climate-Change Strategies. The American Economic Review, 86, pp. 741-765 (7.5)

Norgaard, R., 1985: Environmental economics: An evolutionary critique and a plea for pluralism. Journal of Environmental Economics and Management 12. (8.2)

North, G.R., K-Y. Kim, S.S. Shen, J.W. Hardin, 1995: Detection of Forced Climate Signals, Part I: Filter Theory. J. Clim. 8, 401-408 (6.1)

North, G.R., and K-Y. Kim, 1995: Detection of Forced Climate Signals, Part II: Simulation results. J. Clim. 8, 409-417 (6.4)

North, G.R. and M. Stevens, 1997: Detecting Climate Signals in the Surface Temperature Record. Submitted. (6.1, 6.4)

O'Neill, P. E., Hsu, A. Y., Jackson, T. J., E. F. Wood and M. Zion, 1996: Investigation of the accuracy of soil moisture inversion using microwave data and its impact on watershed modeling. Proc. Int. Geoscience and Remote Sensing Symp.'96, IEEE Publ. 96CH35875:1061-1063. (3.3)

Oort, A.H., 1982: Global circulation statistics. 1958-1973. NOAA Prof

Paper 14. US Government Printing Office, Washington, DC. (6.2)

Orville, H.D., and K. Hubbard, 1973: On the freezing of liquid water in a cloud. J. Appl. Meteor. 22, 137-142 (5.2)

Paehlke, R., 1989: Environmentalism and the Future of Progressive Politics. New Haven: Yale University Press. (8.2)

Parkinson, C.L., and W.M. Washington, 1979: A large-scale numerical model of sea ice. J. Geophys. Res., 84:311–337 (2.1)

Pauwels, V.R.N. and E.F. Wood, 1997: Water and energy balance modeling in BOREAS: Tower and SSA scales, presented at the BOREAS Workshop, Annapolis, MD, March 18-21 (3.3)

Pearce, D., 1991: Evaluating the Socio-Economic Impacts of Climate Change: An Introduction: Pp. 9-20 in OECD, Climate Change. Paris: OECD. (8.2)

Pennebaker, J.W., B. Rime and V.E. Blankenship, 1996: Stereotypes of emotional expressiveness of northerners and southerners: A cross-cultural test of Montesqieu's hypothesis. J. Pers. Soc. Psych. 70, 372-380. (8.5)

Pennell, W.T., T.P. Barnett, K. Hasselmann, W.R. Holland, T.R. Karl, G.R. North, M.C. MacCraken, M.E. Moss, G. Pearman, E.M. Rasmusson, B.D. Santer, H. von Storch, P. Switzer, F. Zwiers, 1993: The detection of anthropogenic climate change, Proc 4th Symp on Global Change Studies, Anaheim (Cal) Jan 17-19, 1993, Am. Meteorol. Soc. 21-28 (6.4)

Peters-Lidard, C.D. and E.F. Wood, 1997: Spatial Variability and Scale in Land-Atmosphere Interactions 2. Model Validation and Results. Submitted to Water Resources Research. (3.3)

Peters-Lidard, C., M. Zion and E.F. Wood, 1997a: A soil-vegetation-atmosphere transfer scheme for modeling spatially variable water and energy balance processes, J. Geophys. Res., Vol. 102(D2), 4303-4324 (3.3)

Peters-Lidard, C., E. Blackburn, X. Liang, E.F. Wood, 1997b: The effect of soil thermal conductivity parameterization on surface energy fluxes and temperatures, J. Atmos. Sci., in press. (3.3)

Pettitt, A.N., 1979: A non-parametric approach to the change-point problem. App.Statist.,126-135 (1.4)

Phillips, N.A., 1957: A coordinate system having some special advantages for numerical forecasting. J. Meteorol., 14:184–185 (2.3)

Pinker, R.T. and I. Laszlo, 1992: Modeling surface solar irradiance for satellite applications on a global scale, J. Appl. Meteor., 31, 194-211 (3.4)

Pitman, A.J., et.al., 1993: Project for intercomparison of land-surface parameterization schemes, PILPS: Results from off-line control simulations, Phase 1a. IGPO 7, World Climate Research Program. (3.3)

Pitman, A.J. et al., 1997: Results from the off-line control simulation phase 1c of the project for intercomparison of land-surface parameterization schemes, PILPS, Clim. Dynamics, in review. (3.3)

Policy Statement of the American Meteorological Society on Global

Climate Change, 1991: Bull. Amer. Meteor. Soc. 72, 57-59 (5.6)

Press, W.H., B.P. Flannery, S.A. Teukolsky and W.T. Vetterling, 1986: Numerical Recipes, Camb.Univ.Press, 818 pp. (7.2)

Prihodko, L., and S.N. Goward, 1997: Estimation of air temperature from remotely sensed observations, Remote Sens. Environ., in review (3.4)

Prince, S. D and S.N. Goward, 1995: Global primary production: A remote sensing appraoch, J. Biogeography, 22:2829-2849 (3.4)

Prince, S. D., Goetz, S. J., Dubayah, R., Czajkowski, K., and Thawley, M., 1997: Inference of surface and air temperature, atmospheric precipitable water and vapor pressure deficit using AVHRR satellite observations: Validation of algorithms. Journal of Hydrology, in press. (3.4)

Rahmstorf, S., 1995: Bifurcations of the Atlantic thermohaline circulation in response to changes in the hydrological cycle. Nature, 378, 145-149 (4.1)

Rahmstorf, S., 1996: On the freshwater forcing and transport of the Atlantic thermohaline circulation. Climate Dym., 12, 799-811.(4.1)

Rangno, A.L., and P.V. Hobbs, 1987: A re-evaluation of the Climax cloud seeding experiments using NOAA published data. J. Clim. Appl. Meteor. 26, 757-762 (5.2)

Rangno, A.L., and P.V. Hobbs, 1993: Further analysis of the Climax cloud-seeding experiments. J. Appl. Meteor. 32, 1837-1847 (5.2)

Rangno, A.L., and P.V. Hobbs, 1995: A new look at the Israeli cloud seeding experiments. J. Clim. Appl. Meteor. 26, 913-926 (5.2)

Rangno, A.L., and P.V. Hobbs, 1997a: Reply. J. Appl. Meteor. 36, 253-254 (5.2)

Rangno, A.L., and P.V. Hobbs, 1997b: Reply. J. Appl. Meteor. 36, 257-259 (5.2)

Rangno, A.L., and P.V. Hobbs, 1997c: Reply. J. Appl. Meteor. 36, 272-276 (5.2)

Rangno, A.L., and P.V. Hobbs, 1997d: Reply. J. Appl. Meteor. 36, 279 (5.2)

Redclift, M., 1988: Economic models and environmental values: A discourse on theory. Pp. 51-66 in R. Kerry Turner, Sustainable Environmental Management: Principles and Practice. London: Belhaven Press. (8.2)

Reynolds, D.W., 1988: A report on winter snowpack-augmentation. Bull. Amer. Meteor. Soc. 69, 1290-1300 (5.2)

Reynolds, D.W., and A.S. Dennis, 1986: A review of the Sierra Cooperative Pilot Project. Bull. Amer. Meteor. Soc., 67, 513-523 (5.2)

Richels, R.G., J. Edmonds, H. Gruenspecht and T. Wigley, 1996: The Berlin Mandate: The design of cost-effective mitigation strategies, Report (draft) Energy Modeling Forum-14, Stanford University (7.5)

Risbey, J.S. and P.H. Stone, 1996: A case study of the adequacy of GCM simulations for input to regional climate change assessments. J. Climate 9, 1441-1467 (1.3, 1.4)

Robinson, J. and J. Tinker, 1997: Reconciling ecological, economic and social imperatives: Towards an analytical framework. In: Ted Schrecker (ed.), Surviving Globalism: Social and Environmental Dimensions. New York: St. Martin's Press (forthcoming). (8.7)

Roeckner, E. and H. von Storch, 1980: On the efficiency of horizontal diffusion and numerical filtering in an Arakawa-type model. Atmosphere-Ocean 18, 239-253 (1.5)

Rogers, J.C. and van Loon, H., 1982: Spatial variability of sea level pressure and 500 mb height anomalies over the Southern Hemisphere. Mon. Wea. Rev., 110, 1375-1392 (4.1)

Romanova, E.N., 1954: The influence of forest belts on the vertical structure of the wind and on the turbulent exchange. Study of the Central Geophysical Observatory, No. 44 (106), Glavnaia Geofizicheskaia Observatoriia, Trudy, Leningrad, 80–90 (2.4)

Rosenfeld, D., 1997: Comments on "A new look at the Israeli cloud seeding experiments." J. Appl. Meteor. 36, 260-271 (5.2)

Rosenfeld, D. and H. Farbstein, 1992: Possible influence of desert dust on seedability of clouds in Israel. J. Appl. Meteor. 31, 722-731 (5.2)

Rosenfeld, D., and W.L. Woodley, 1989: Effects of cloud seeding in West Texas. J. Appl. Meteor. 28, 1050-1080 (5.3)

Rosenfeld, D., and W.L. Woodley, 1993: Effects of cloud seeding in west Texas: Additional results and new insights: J. Appl. Meteor. 32, 1848-1866 (5.3)

Ryan, B.F., and W.D. King, 1997: A critical review of the Australian experience in cloud seeding. Bull. Amer. Meteor. Soc. 78, 239-254 (5.2)

Santer, B.D., T.M.L. Wigley, P.D. Jones, 1993: Correlation methods in fingerprint detection studies. Clim. Dyn. 8, 265-276 (6.1)

Santer, B.D., K.E. Taylor, J.E. Penner, T.M.L. Wigley, U. Cubasch, P.D. Jones, 1995: Towards the detection and attribution of an anthropogenic effect on climate. Clim. Dyn. 12, 77-100 (6.1)

Santer, B.D., K.E. Taylor, T.M.L. Wigley, P.D. Jones, D.J. Karoly, J.F.B. Mitchell, A.H. Oort, J.E. Penner, V. Ramaswamy, M.D. Schwarzkopf, R.S. Stouffer, S.F.B. Tett, 1996a: A search for human influences on the thermal structure in the atmosphere. Nature 382, 39-46 (6.1)

Santer, B.D., T.M.L. Wigley, T.P. Barnett, E. Anyamba, 1996b: Detection of climate change and attribution of causes. In: Houghton, J.T. et al. (eds.) Climate Change, 1995. The IPCC Second Scientific Assessment, Cambridge University Press, pp. 407-444 (6.1)

Sax, R.I., 1976: Microphysical response of Florida cumuli to AgI seeding. Second WMO Scientific Conf. on Weather Modification, Boulder, WMO, 109-116 (5.3)

Sax, R.I., S.A. Changnon, L.O. Grant, W.F. Hitchefield, P.V. Hobbs, A.M, Kahan, and J. Simpson, 1975: Weather modification: Where are we now and where should we be going? An editorial overview. J. Appl. Meteor. 14, 652-672 (5.2)

Sax, R.I., J. Thomas, and M. Bonebrake, 1979: Ice evolution within seeded and nonseeded Florida cumuli. J. Appl. Meteor. 18, 203-214 (5.3)

Sax, R.I., and V.W. Keller, 1980: Water-ice and water-updraft relationships near -10 °C within populations of Florida cumuli. J. Appl. Meteor. 19, 505-514 (5.3)

Schlesinger, M.E., ed., 1991: Greenhouse-Gas-Induced-Climatic Change: A Critical Appraisal of Simulations and Observations, Elsevier Science Publishers, Amsterdam. 615 pp. (6.1)

Schneider, S.H., 1989: The changing climate. Scientific American, September, 70-79 (5.6)

Schneider, S.H., 1990: The global warming debate heats up: An analysis and perspective. Bull. Amer. Meteor. Soc. 71, 1291-1304 (5.6)

Schnur, R. and D. P. Lettenmaier, 1997: Global gridded data set of daily soil moisture for use in General Circulation Models, presented at the 13 th Conference in Hydrology, AMS Meeting, Long Beach, CA, Feb 1997. (3.2)

Scott, B.C., and P.V. Hobbs, 1977: A theoretical study of the evolution of mixed-phase cumulus clouds. J. Atmos. Sci. 34, 812-826 (5.3)

Sellers, P.J., Tucker, C.J., Collatz, G.J., Los, S.O., Justice, C.O., Dazlich, D.A., and Randall, D.A., 1994: A global 1-degree by 1-degree NDVI data set for climate studies. Part2: The generation of global fields of terrestrial biophysical parameters from the NDVI. Int. J. Remote Sensing 15: 3519-3545 (3.1)

Sellers, P., F. Hall, J. Margolis, B. Kelly, D. Baldocchi, G. den Hartog, J. Cihlar, M. G. Ryan, B. Goodison, P. Crill, K. J. Ranson, D. Lettenmaier, and D. E. Wickland. 1995: The Boreal Ecosystem-Atmosphere Study, BOREAS: An Overview and Early Results from the 1994 Field Year. Bull. Am. Met. Soc. 76(9):1549-1577 (3.3)

Semtner, A.J., Jr., 1974: An oceanic general circulation model with bottom topography. Numerical Simulation of Weather and Climate, Tech. Rept. No. 9, University of California, Los Angeles, 99 pp. (2.5)

Semtner, A.J., Jr., 1976: A model for the thermodynamic growth of sea ice in numerical investigations of climate. J. Phys. Oceanogr., 6:379–389 (2.6)

Semtner, A.J., Jr., 1995: Modeling ocean circulation, Science, 269:1379–1385 (2.5)

Semtner, A.J., Jr., 1997: Kirk Bryan's enduring formulation of a world ocean model, in preparation (2.5)

Semtner, A.J., Jr. and R.M. Chervin, 1988: A simulation of the global ocean circulation with resolved eddies. J. Geophys. Res., 93:15502–15522 (2.5)

Semtner, A.J., Jr. and R.M. Chervin, 1992: Ocean general circulation from a global eddy-resolving simulation. J. Geophys. Res., 97:5493–5550 (2.5)

Shuttleworth, J., 1996: GCIP Coupled Modeling Workshop, IGPO publication Series No. 23, Silver Spring, MD, 20 pp. Plus App. (3.3)

Simpson, J., G.W. Brier, and R.H. Simpson, 1967: Stormfury cumulus seeding experiment 1965: Statistical analysis and main results. J. Atmos. Sci. 24, 508-521 (5.3)

Simpson, J., and W.L. Woodley, 1971: Seeding cumulus in Florida: New 1970 results. Science 172, 117-126 (5.3)

Singer, S.F. and J.S. Winston, 1991: IPCC Report Survey, Science and Environmental Policy Project, 1015 18th St., N.W. Suite 300, Washington D.C. 20036 (9.4)

Slade, D.H.; 1989: A survey of informed opinion regarding the nature and reality of a global "greenhouse" warming. David H. Slade Environmental Sciences, 859 Loxford Terrace, Silver Springs, Maryland 20901 (9.4)

Smith, A.F.M., and G.O. Roberts, 1993: Bayesian computation via the Gibbs Sampler and related Markov Chain Monte Carlo methods. J.R. Statist Soc B 55, 3-23 (6.5)

Smith, R.D., J.K. Dukowicz and R.C. Malone, 1992: Parallel ocean general circulation modeling. Physica D, 60:38-61 (2.5)

Smith, R.D., S. Kortas, and B. Meltz, 1996: Curvilinear coordinates for global ocean models. J. Comput. Phys., submitted (2.5)

Sorokin, P., 1928: Contemporary Sociological Theories. New York: Harper & Row. (8.8)

Starr, V.P., 1942: Basic principles of weather forecasting. Harper & Brothers Publishers New York, London, 299 pp. (1.6)

Stehr, N., 1994: Knowledge Societies. London: Sage. (8.7)

Stehr, N., 1996a: The ubiquity of nature: Climate and culture. Journal for the History of the Behavioral Sciences 32, 151-159 (8.5)

Stehr, N., 1996b: Social inequality and knowledge Pp. 41-51 in; David Sciulli (ed.), Comparative Social Research: Supplement 2: Normative Social Action. Greenwich, CT: JAI Press. (8.7)

Stehr, N., 1997: Trust and climate, Climate research 8, 163-169 (8.5)

Stehr, N. and H. von Storch, 1995: The social construct of climate and climate change. Climate Research 5, 99-105 (8.6)

Stehr, N., H. von Storch and M. Flügel, 1996: The 19th century discussion of climate variability and climate change: Analogies for present day debate? World Resources Review 7, 589-604 (8.5)

Stehr, N. and H.v. Storch, 1997: Planned climate change. Paper prepared for the 1997 Open Meeting of the Human Dimensions of Global Environmental Change Research Community, IIASA, Laxenburg, Austria, June 12-14, 1997 (8.3)

Stephens, G.L., 1984: The parameterization of radiation for numerical weather prediction and climate models. Mon. Wea. Rev. 112, 826-867 (2.3)

Stephens, G.L., G.G. Campbell, and T.H. Von der Haar, 1981: Earth radiation budgets. J. Geophys. Res. 86, 9739-9760 (2.3)

Stevens, M., and G.R. North, 1996: Detection of the climate response to the solar cycle. J. Atmos. Sci. 53, 2594-2608 (6.1)

Stewart, T.R., J.L. Mumpower and P. Reagan-Cirincione, 1992: Scientists' Agreement and Disagreement about Global Climate Change: Evidence from Surveys. Research Report, Center for Policy Research, Graduate School of Public Affairs, Nelson A. Rockerfeller College of Public Affairs and Policy (9.4)

Stommel, H., 1961: Thermohaline convection with two stable regimes of flow. Tellus, 13, 224-230 (4.1)

Super, A.B., and B.A. Boe, 1988: Microphysical effects of wintertime cloud seeding with silver iodide over the Rocky Mountains. Part III: Observations over the Grand Mesa, Colorado. J. Appl. Meteor. 27, 1166-1182 (5.2)

Super, A.B., and J.A. Heimbach, 1988: Microphysical effects of wintertime cloud seeding with silver iodide over the Rocky Mountains. Part II: Observations over the Bridger Range, Montana. J. Appl. Meteor. 27, 1152-1165 (5.2)

Super, A.B., B.A. Boe, and E.W. Holroyd, III, 1988: Microphysical effects of wintertime cloud seeding with silver iodide over the Rocky Mountains. Part I: Experimental design and instrumentation. J. Appl. Meteor. 27, 1145-1151 (5.2)

Tahvonen O., 1993: Carbon dioxide abatement as a differential game, Discussion Papers in Economics and Business Studies, No.4, University of Oalu, Finland (7.5)

Tahvonen O., H. von Storch and J. von Storch, 1994: Economic efficiency of CO2 reduction programs, Clim. Res. 4, 127-141 (7.3, 8.6)

Taylor, K.E. and S.J. Ghan, 1992: An analysis of cloud liquid water feedback and global climate sensitivity in a general circulation model. J. Clim. 5, 907-919 (6.6)

Taylor, K.E., and J.E. Penner, 1994: Response of the climate system to atmospheric aerosols and greenhouse gases. Nature 369, 734-737 (6.6)

Tett, S.F.B., J.F.B. Mitchell, D.E. Parker, M.R. Allen, 1996: Human influence on the atmospheric vertical temperature structure: Detection and observations, Science 247, 1170-1173 (6.7)

Thatcher, M., 1990: On long term climate prediction. J. Air Waste Mgt. 40, 1086-1087 (5.6)

Titus, J.G., 1990a: Strategies for adapting to the greenhouse effect. APA Journal, 331-323 (5.6)

Titus, J.G., 1990b: Greenhouse effect, sea level rise and barrier islands: Case study of Long Beach Island New Jersey. Coastal Management 18, 65-90 (5.6)

Trenberth, K.E., ed., 1992: Climate System Modeling. Cambridge University Press, 788pp. (2.1)

Trenberth, K.E., J.R. Christy, 1985: Global fluctuations in the distribution of atmospheric mass. - J. Geophys. Res., 90, D5, 8042-8052 (4.1)

Trenberth, K.E. and D.A. Paolino, 1980: The Northern Hemisphere sea-level pressure data set: Trends, errors and discontinuities. Mon. Wea.

Rev. 108, 855-872 (1.4)

Tukey, J.W., D.R. Brillinger, and L.V. Jones, 1978: The role of statistics in weather resources management. The Management of Weather Resources, Vol. II. Final Report of Weather Modification Advisory Board, Department of Commerce, Washington, D.C. [U.S. Government Printing Office No. 003-0180-0091-1.] (5.2)

Twomey, W., 1974: Pollution and the planetary albedo. Atmos. Environ. 8, 1251-1256 (5.6)

Twomey, S., M. Piepgrass, and T.L. Wolfe, 1984: An assessment of the impact of pollution on global albedo. Tellus 36B, 356-366 (5.6)

Tyler, D.E., 1982: On the optimality of the simultaneous redundancy transformations. Psychometrika 47, 77-86 (1.4)

UCAR/NOAA, 1991: Reports to the Nation. Winter 1991. Published by the UCAR Office for Interdisciplinary Earth Studies and the NOAA Office of Global Programs, 21 pp. (5.6)

van Loon, H. and H.J. Rogers, 1978: The seesaw in winter temperature between Greenland and northern Europe. Part 1: General description. Mon. Wea. Rev. 106, 296-310 (1.4)

Viterbo, P. and A. Beljaars, 1995: An improved land surface parameterization scheme in the ECMWF model and its validation, J. Clim. 8:2716-2748 (3.3)

von Storch, H., 1997: Conditional statistical models: A discourse about the local scale in climate modelling. In P. Müller and D. Henderson (eds.): "Monte Carlo Simulations in Oceanography" Proceedings 'Aha Huliko'a Hawaiian Winter Workshop, University of Hawaii at Manoa, January 14-17, 1997, 59-58 (1.4)

von Storch, H. and G. Hannoschöck, 1985: Statistical aspects of estimated principal vectors ("EOFs") based on small sample sizes. - J. Climate Appl.Met., 24, 716-724 (4.1)

von Storch, H. T. Bruns, I. Fischer-Bruns and K. Hasselmann, 1988: Principal Oscillation Pattern analysis of the 30-60 day oscillation in a GCM equatorial troposphere. - J. Geoph.Res. 93, 11022-11036 (4.1)

von Storch, H., E. Zorita and U. Cubasch, 1993: Downscaling of climate change estimates to regional scales: application to winter rainfall in the Iberian Peninsula. J. Climate 6, 1161-1171 (1.4)

von Storch, H., and A. Navarra (eds.), 1995: Analysis of Climate Variability: Applications of Statistical Techniques, Springer Verlag, (ISBN 3-540-58918-X)(1.4)

von Storch, H. and F.W. Zwiers, 1998: Statistical Analysis in Climate Research. Cambridge University Press (in press) (1.4, 6.5)

von Storch, H., M. Heimann and S. Guess, 1999: Das Klimasystem und seine Modellierung. Eine Einführung. (in preparation) (1.2)

von Storch, J.-S., 1994: Interdecadal variability in a global coupled model. Tellus 46A, 419-432 (6.4)

von Storch, J.-S., 1995: Multivariate statistical modelling: POP model as a first order approximation. In: H. von Storch and A. Navarra (eds.) "Analysis of Climate Variability: Applications of Statistical Techniques", Springer Verlag, 281-297 (ISBN 3-540-58918-X) (4.1)

von Storch, J.-S., 1997a: On the reddest atmospheric modes and the forcings of the spectra of these modes: A description. Submitted to J.Atmos. Sci. (4.1)

von Storch, J.-S., 1997b:: On the dynamics of the reddest atmospheric modes. submitted to J. Atm. Sci.(4.1)

von Storch, J.-S., V. Kharin, U. Cubasch, G.C. Hegerl, D. Schriever, H. von Storch, and E. Zorita, 1997: A description of a 1260-year control integration with the coupled ECHAM1/LSG general circulation model. J. Climate 10, 1526-1544 (4.1, 6.4)

Wallace, J.M. and D.S. Gutzler, 1981: Teleconnection in the geopotential height field during the northern hemisphere winter. Mon. Wea. Rev. 109, 784-812 (4.1)

Walton, J.J., M.C. MacCracken, S.J. Ghan, 1988: A global-scale Lagrangian trace species model of transport, transformation and removal processes. J. Geophys. Res. 93, 8339-8354 (6.6)

WASA group, 1995: The WASA project: Changing storm and wave climate in the Northeast Atlantic and adjacent seas. Proc. Fourth Intern. Workshop on Wave Hindcasting and Forecasting, Banff, Canada, Oct. 16-20, 1995 (1.4)

Washington, W., 1968: Computer simulation of the Earth's atmosphere. Science J. 37-41 (1.3)

Washington, W.M. and C.L. Parkinson, 1986: An Introduction to Three-Dimensional Climate Modelling. University Science Books, 422 pp. (1.3, 2.1)

Weale, A., 1993: Nature versus the state? Markets, states, and environmental protection, Critical Review 6, 153-170 (8.2)

Weare, B.C., 1979: Temperature statistics of short-term climatic change. Mon. Weath. Rev. 107, 172-180 (6.2)

White, Richard, 1996: The nature of progress: Progress and the environment. Pp. 21-140 in Leo Marx and Bruce Malish (eds.), Progress: Fact or Illusion? Ann Arbor: The University of Michigan Press. (8.7)

Wigley, T.M.L., and T.B. Barnett, 1990: Detection of the greenhouse effect in the observations. In: Houghton, J.T., G.J. Jenkins, J.J. Ephraums (eds.): Climate Change, the IPCC scientific assessment. Cambridge University Press, pp.239-255 (6.1)

Wigley, T.M.L., R. Richels and J.A. Edmonds, 1996: Economic and environmental choices in the stabilization of atmospheric CO_2 emissions, 1996, Nature 379, 240-243 (7.3)

Williamson, H., 1770: An attempt to account for the change of climate, which has been observed in the Middle Colonies in North America. Trans.

Amer. Phil. Soc. 1, 272 (8.5)

Willson, R.C., and H.S. Hudson, 1991: The sun's luminosity over a complete solar cycle. Nature 351, 42-44 (6.5)

Wolterek, H., 1938: Klima-Wetter-Mensch. Leipzig: Quelle & Meyer. (8.5)

Wood, E.F., D.P. Lettenmaier, X. Liang, B. Nijssen and S.W. Wetzel, 1997: Hydrological Modeling of Continental-Scale Basins, Annual Reviews of Earth and Planetary Sciences, 25:279-300 (3.3)

Wooddruff, S.D., R.J. Slutz, R.L. Jenne, and P.M. Steuer, 1987: A comprehensive ocean-atmosphere data set. Bull. Amer. Meteor. Soc., 68, 1239-1250 (4.1)

Woodley, W.L, 1997: Comments on "A new look at the Israeli cloud seeding experiments. J. Appl. Meteor. 36, 250-252 (5.2)

Woodley, W., J. Jordan, J. Simpson, R. Biondini, J. Flueck, and A. Barnston, 1982: Rainfall results of the Florida Area Cumulus Experiment, 1970-1976. J. Appl. Meteor. 21, 139-164 (5.2)

Wood, E. F., D. P. Lettenmaier, X. Liang, B. Nijssen and S. W. Wetzel, 1997: Hydrological Modeling of Continental-Scale Basins, Annual Reviews of Earth and Planetary Sciences, 25:279-300 (3.4)

Wyrtki, K., K. Constantine, B.J. Kilonsky, G. Mitchum, B. Miyamoto, T. Murphy, S. Nakahara, and P. Caldwell, 1988: The Pacific Island sea level network. JIMAR Contrib. 88-0137, Data Rep. 002, Univ. of Hawaii, Honolulu, 71 pp. (4.1)

Xu, J.-S., 1993: The joint modes of the coupled atmosphere-ocean system observed from 1967 to 1986. J. Climate, 6, 816-838 (4.1)

Young, J., J. Hagens, and D. Wade., 1995: GOES pathfinder product generation system, 9th Conference on Applied Climatology, Dallas, TX, American Meteorological Society, Boston, MA (3.1)

Zhang, J. and W.D. Hibler, III, 1997: On an efficient numerical method for moderling sea ice dynamics. J. Geophys. Res., in press. (2.6)

Zhang, Y., W. Maslowski, and A. Semtner, 1977: High resolution arctic ocean and sea ice simulations. Part II: Ice model design and early results. Submitted to Journal of Geophysical Research (2.6)

Zhao, R.J., Y.L. Zhang, L.R. Fang, X.R. Liu, and Q.S. Zhang, 1980: The Xinanjiang model, in Hydrological Forecasting, Proceedings Oxford Symposium, IAHS Publ. 129, pp. 351-356 (3.3)

Zorita, E., J. Hughes, D. Lettenmaier, and H. von Storch, 1995: Stochastic characterisation of regional circulation patterns for climate model diagnosis and estimation of local precipitation. J. Climate 8 1023 - 1042 (1.4)

Zwiers, F.W., and S.S. Shen, 1997: Errors in Estimating Spherical Harmonic Coefficients from Partially Sampled GCM Output. Clim. Dyn. 13, 703-715 (6.5)

Zwiers, F.W. and V.V. Kharin, 1998: Changes in the extremes simulated by CCC GCM2 under CO_2 doubling. J. Clim., in press (6.4)

Subject Index